AN INTRODUCTION TO
Genetic Analysis

AN INTRODUCTION TO
Genetic Analysis

David T. Suzuki / Anthony J. F. Griffiths

THE UNIVERSITY OF BRITISH COLUMBIA

W. H. FREEMAN AND COMPANY
San Francisco

Library of Congress Cataloging in Publication Data

Suzuki, David T   1936–
    An introduction to genetic analysis.

    Bibliography: p.
    Includes index.
    1. Genetics. I. Griffiths, Anthony J. F., joint
author. II. Title. [DNLM: 1. Genetics—Education.
QH440 S968i]
QH430.S94   575.1   75-29480
ISBN 0-7167-0574-5

Printed in the United States of America

9 8 7 6 5 4 3 2

To Tara Cullis and Joan Griffiths

A million million spermatozoa,
All of them alive:
Out of their cataclysm but one poor Noah
Dare hope to survive.
And among that billion minus one
Might have chanced to be
Shakespeare, another Newton, a new Donne—
But the One was Me.

ALDOUS HUXLEY

Contents

Preface

This book attempts to teach the reader how to do genetics. It is not designed merely to introduce genetic concepts, but rather lays out a set of ground rules and exercises that will assist the student in becoming a proficient genetic analyst.

Two basic instructive devices are used throughout. First, genetic concepts are presented for the most part in historical sequence, rather than starting with our current knowledge of molecular genetics. This is because it seems to us that a student begins much as biologists did at the turn of the century, asking general questions about the laws governing the inheritance of traits. Only after the ground rules have been established at one level, can one logically go on to pose questions about the next level of hereditary organization. In the text we have extensively used a question-and-answer format to simulate this step-by-step evolution of genetic understanding. Second, a quantitative approach is emphasized because the abstractions of genetics have been based largely on playing with numerical experimental data. However, the only mathematical knowledge assumed is a good grasp of arithmetic and basic algebra. Another instructive device that deserves mention is the use of animation sequences. These sequences occupy the right-hand margins of odd-numbered pages, starting at Chapter 2. By holding the pages in the right hand and letting them flip over at a constant rate, you will see a continuous dynamic image of four genetically important processes: meiosis, mitosis, DNA replication, and translation.

The principles of genetics have already been established to a large degree; consequently, the temptation to load the text with references to the most recent research papers has been resisted. Only a small fraction of the current literature will in time be adjudged classical: unfortunately, we do not yet know which papers will constitute that fraction, and this does not seem an appropriate place to guess. The astonishing rate at which textbooks become "out of date" results primarily from the inclusion of the latest experiments, many of which are soon found to be incorrect, incomplete, or irrelevant. The beauty and simplicity of genetic analysis are timeless, but too often lost under an avalanche of topical detail.

This book is designed for an introductory course in college genetics and grew out of the teaching of a course in fundamental genetics for third year students at The University of British Columbia. The course runs for twelve weeks with three lectures and one "tutorial" session per week. Tutorials are predominantly concerned with problem-solving assignments, which are liberally sprinkled throughout this text. Some of the problems are borrowed, but most are of our own design. They constitute an integral and absolutely necessary part of learning how to perform genetic analyses, and should be worked meticulously, starting with only a blank sheet of paper and a pencil.

In attempting to keep the book to a size that can be covered in a semester, we have restricted the material to that which best illustrates various features of genetic analysis and is generally applicable to other organisms. Consequently, only certain prokaryotes and typical haploid and diploid eukaryotes have been emphasized. A considerable amount of telescoping of subject material has also been inevitable. For example, cytoplasmic inheritance, sex determination, differentiation, and behavior are traditionally treated in separate chapters in genetics texts. Because each is a part of the basic problem of how an egg becomes a complete functional organism, they have been integrated here into a single chapter on development.

Throughout the text, material is summarized in the form of "messages." Not only does this give the reader periodic stopping points from which to orient himself within the chapter, but it also facilitates rapid review of the entire book.

Thanks are due Clayton Person and Tom Kaufman for many helpful discussions and suggestions, Joan Howley for help in proofreading, Rita Rosbergen for typing, and the students and teaching assistants of Biology 334 for their participation in testing the preliminary edition.

We hope that the book will stimulate the reader to do some first-hand experimental genetics, whether as professional scientist, student, amateur gardener, or animal fancier. Failing this, we hope some lasting impressions will be formed of the precision, elegance, and power of genetic analysis.

October 1975

David T. Suzuki
Anthony J. F. Griffiths

Introduction

Genetics in Biology

There is an overall natural tendency in the universe to move toward disorder or disarray. However, there are some isolated eddies in this river of disorder, and one of the most interesting is called *life*. In contrast with the increasing chaos in the rest of the universe, living systems are highly ordered. In fact, life could be defined as persistent order. The word persistent is used here to emphasize the unique ability of living systems to *hand on* to their descendants the instructions necessary for the maintenance of order; in other words, persistent emphasizes the central role of hereditary phenomena in the uniqueness of life. The study of heredity is a science called Genetics.

   The findings of genetics have been monumentally significant in the unification of biology into a coherent science in two basic ways: First, genetics has shown that all life forms on earth—a staggering spectrum including 286,000 species of flowering plants, 500,000 species of fungi, 750,000 species of insects, and so forth—are probably genetically derived from a common ancestor through a process known as evolution. Genetics has revealed precise mechanisms that enable us to understand how this was accomplished. Second, genetics has shown that all life forms are based on a common system of information storage, duplication, copying, and translation.

Furthermore, the science of genetics has provided the life sciences with a powerful approach for studying biological phenomena. This approach, which we will call *genetic dissection* of biological systems, constitutes an important part of this book.

Genetics and Human Affairs

Since the dawn of civilization, when nomadic man began to domesticate plants and animals, the recognition of hereditary phenomena has played a vital role at many levels of human society. Man selected grains with higher yields and vigor and animals with better fur or meat long before biology existed as a scientific discipline. In spite of this longstanding use of selective breeding, the actual basis for inheritance has only been elucidated in the last hundred years. But now, as it does in most areas of science, new knowledge brings with it not only potential benefit for mankind but an equal share of problems.

The recent, much heralded "Green Revolution," brought about by the sophisticated breeding by geneticists of high-yield varieties of dwarf wheat and rice, is a good example of this balance between good and bad. The production of these crops has been found to depend too heavily on extensive cultivation and costly fertilizers in the impoverished countries for which they were intended. Nevertheless, in the present state of overpopulation on Earth, our dependence on high-yield genetic varieties of plant crops and animals for food and other resources has become increasingly obvious. In a very practical sense the stability of society is dependent on the ability of geneticists to juggle the inherited traits that confer higher yield and keep the crops one jump ahead of destructive parasites and predators. A dramatic example of the potential effect of genetics in this connection is the breeding of Marquis wheat in Canada. This strain of high-quality wheat is resistant to disease and, furthermore, matures two weeks earlier than other commercially used strains. Consequently, millions of square miles of fertile soil in northern countries such as Canada, Sweden, and the USSR have been opened up for the growing of wheat.

Already, special genetic strains of fungi and bacteria have been isolated to greatly increase yields of antibiotics and other drugs. The potential future use of specially bred microorganisms is even wider: they may be used to clean up pollutants, to serve as food, to transfer the hereditary determinants conferring the ability to fix atmospheric nitrogen to important crop plants, to provide functions that are missing in people suffering from various kinds of disease, and a whole range of other uses that are no longer simply the wild dreams of science-fiction writers. In another example of the usefulness of genetics in applied technology, areas of heavy insect infestation are being made tolerable for human habitation by the deliberate genetic tampering with the insects' fertility.

But nowhere is the potential impact of an increased knowledge of genetics more exciting and frightening than in man himself. It is becoming increasingly obvious that the brain of man is subject to the same genetic determination as the rest of the body, resulting in certain inherited predispositions to thought and behavior. No longer can we view the mind as a clean slate at birth written upon only by experience. The extent of such inborn constraints on thought and personality, and their relevance to present sociological impasses, is being suggested in such books as *African Genesis, The Territorial Imperative,* and *On Aggression.* The same reasoning is inherent in the current claims for hereditary differences in intelligence among different racial and social groups. This concept is not new, having been debated by Lycurgus in Sparta and Plato in Athens, and reaching its zenith in Nazi Germany. With the pressure for population control mounting, the spectre of legislated sterilization looms large again. If such sterilization were to be of a selective nature, we would be essentially shaping the genetic destiny, or evolution, of our own species, an onerous responsibility that few could confidently undertake.

The sophisticated technology of molecular genetics now places a wide range of new techniques at our disposal for shaping our genetic makeup, with even more bizarre procedures undoubtedly arriving in the near future. This "genetic engineering" differs from conventional breeding procedures in that the genetic apparatus is modified essentially at a chemical level. A rash of popular books warn us of a *Genetic Fix,* a *Biological Time Bomb, Genetic Revolution,* a *Fabricated Man,* and the *Biocrats.* The extent to which such prophecy is realized will depend ultimately on political decisions based on the informed opinions of responsible citizens.

The advances made in the field of genetics have been especially useful in medicine: hereditary diseases can now be diagnosed at an early stage of life when it is possible to provide secondary cures in some cases. Refined techniques such as amniocentesis and fetoscopy (both prenatal) and a battery of postnatal chemical tests have made such cures possible. Furthermore, genetic disease can often be prevented by counseling prospective parents with the help of family pedigrees.

There is a current fear that our increased exposure to chemical food additives, and a vast array of chemicals in other commercial products, is changing our genetic makeup in a very undesirable haphazard way. Other environmental agents capable of causing this random genetic change are fallout from H-bombs, radioactive contamination from nuclear reactors, and radiation from X-ray machines. These agents may be contributing to genetic disease.

The ability to recognize the prevalence of genetic disease in our societies raises an important moral dilemma. It has been estimated that 5% of our population survives with a severe physical or mental genetic defect, and that this percentage will increase with extended exposure to the above environmental agents, and, paradoxically, with improved medical technology. As geneticist Theodosius Dobzhansky has remarked,

> If we enable the weak and the deformed to live and propagate their kind, we face the prospect of a genetic twilight. But if we let them die or suffer when we can save or help them, we face the certainty of a moral twilight.

The extent to which our society will be prepared to shoulder this genetic load will be measured by the amount of money we are prepared to spend in keeping the genetically handicapped alive. A measure of the size of this financial burden on the resources of society is seen in the estimate that 30% of the patients admitted to pediatric hospitals in North America have diseases that can be traced to genetic causes.

From this brief introduction it can be seen that genetics is relevant not only to the biologist but to any thinking member of today's complex technological society, and a working knowledge of the principles of genetics is essential for making informed decisions on many scientific, political, and personal levels. We believe such a working knowledge can come only from understanding how genetic inference is made: that is, from understanding genetic analysis, the subject of this book.

1 / Mendelism

(Or: How to deduce the existence of genes and make conclusions about their location in the cell without even seeing them.)

A basic observation about living organisms that even a child recognizes is the continuity of type from generation to generation. We know that a cat will certainly give birth to kittens and that carrot seeds will grow into carrot plants. In other words, like begets like. Yet within a species there is amazing diversity. We distinguish Rex from Fido, our brothers from our brothers-in-law, and so on, in a gratifying way that enriches all our lives. Genetics is the discipline within biology that attempts to understand how interspecific variation is maintained and how, at the same time, intraspecific variation is generated and inherited. Genetics, then, is about heredity.

Although genetics is about heredity, this is not a good definition of genetics. No geneticists existed before 1865, despite a millenial persistent interest in inheritance, because it wasn't until this time that a way of analyzing information on heredity was invented by an Augustinian monk, Gregor Mendel. His method of analysis is the one we still use today (albeit in an extended form) and call genetics. Probably the crucial advance he made was the identification or recognition of an entity we now call a *gene*. That event occurred in the brain of Gregor Mendel and marked the birth of genetics as a unique way of looking at living organisms and analyzing biological phenomena.

In this chapter we will trace the birth of the gene as a concept, and in later chapters, as a reality. We will see that genetics is an abstract science: most of its entities have begun as hypothetical ones in the minds of geneticists and later on, depending on the soundness of the reasoning that created them, have been identified in physical form.

In Mendel's time, people thought about heredity in a way that can be traced back to the ancient Greeks. Basically, their concept of procreation can be compared to making freeze-dried instant coffee. Eggs and sperm were thought to consist of many essences: of arm, of head, of hair, and so forth. These dehydrated instant building blocks needed only to come together and, with the addition of water, they would burst forth into a new individual. The main point is that it was the building blocks *themselves* that were thought to be handed on, not blueprints for making them.

Consequently, this now-discarded hypothesis is termed *blending inheritance:* just as a mixture of Maxwell House and Yuban coffee results in a blend of both types, so the union of egg and sperm was thought to produce a blend of the essences in each. Mendel was to show that this was wrong, and that inheritance does not result from the union of a teeming multitude of building blocks, but from a few very important particles (now called genes), which *direct* the synthesis of new individuals. This is called *particulate inheritance*.

Unfortunately, the importance of Mendel's work was not appreciated until thirty-five years later (by then he was dead) when it was discovered by three scientists who had independently come to the same conclusion. In a sense, then, Mendel's work was irrelevant to the development of the study of heredity, but this in no way diminishes his achievement nor the exemplary worthiness of his analysis, which we will now pursue.

Mendel studied the garden pea (*Pisum sativum*) for two main reasons: First, peas were available through a seed merchant in a wide array of different forms and colors that are very easily identified and analyzed. Second, peas left to themselves will self-pollinate (or *self*) because the male and female parts of the flower, which produce pollen and eggs respectively, are enclosed in a petal box or keel (Figure 1-1). To cross-pollinate (or *cross*) them, the anthers can be clipped off and pollen from another plant can be transferred to the receptive area with a paintbrush. Consequently, peas can be easily either selfed or crossed.

The first thing Mendel did was to choose several traits to work on and establish *pure lines*. This was a clever beginning because it amounts to a control experiment: he had to be sure of his material. A pure line is a plant pedigree that breeds true or constant for the particular character being studied. For example, he had a line that bred true for purple flowers; that is, when selfed, all the progeny seeds grew into plants that had purple flowers, and when these were selfed, *their* progeny had purple flowers too. Other lines were pure for white flowers, and others for yellow, green, wrinkled, or smooth seeds as well as many other traits.

Let's consider some specific experiments. In one of his early experiments he used a pure line with purple flowers and a pure line with white flowers. If a purple-flowered plant was pollinated by pollen from a white-flowered plant, all the progeny plants had purple flowers (Figure 1-2).

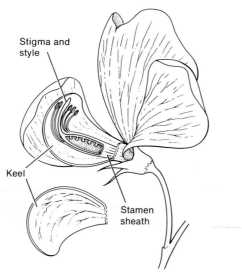

Stigma and
style

Keel

Stamen
sheath

FIGURE **1-1**

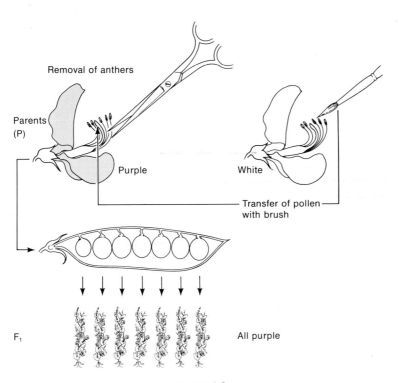

Removal of anthers

Parents
(P)

Purple

White

Transfer of pollen
with brush

F₁

All purple

FIGURE **1-2**

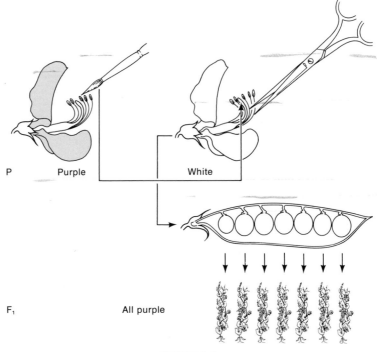

P   Purple   White

F₁   All purple

FIGURE **1-3**

The *reciprocal cross*—that is, a white flower pollinated by a purple-flowered plant—produced the same result (Figure 1-3).

He concluded that it did not make any difference which way the cross was made: when one of the parents (P generation) was purple and the other was white, all the plants in the *first-filial generation* (F₁) were purple. In the F₁ generation, the purple flower color was the same as that of the purple-flowered parent; so in this case, inheritance was obviously *not* like blending purple and white paint to produce a lighter color.

Mendel then selfed the F₁ plants, allowing the pollen of each flower to fall on the stigma within its petal box. From that selfing, he planted 929 peas (the F₂ individuals). Amazingly, although as might have been expected the majority was purple, some grew up to be white-flowered. The white character had reappeared! He then did something that, more than anything else, marks the birth of genetics: he *counted* how many there were of each kind, which may seem trite in today's quantitatively oriented world, but which was the key to the future of genetics. There were 705 purple plants, and 224 white ones. He observed that this was close to a 3:1 ratio (in fact it was 3.1:1).

He repeated this for seven other pairs of pea characteristics and found the same 3:1 ratios in the F₂ generation for all of them. By this time he was undoubtedly beginning to believe in the reality of the 3:1 ratio and felt he had to explain it. Note that, even though the white coloring was completely absent in the F_1 generation, it reappeared with full expression in the F_2. So even though the F_1 flowers were purple, the plants still carried the *potential* to produce progeny with white flowers. Thus, Mendel concluded that the F_1 plants must carry "factors," received from *both* parents, that determine flower color. The factor responsible for purple is *dominant* so that, even in the presence of a factor for white, purple is expressed. The factors for white are said to be *recessive* to the dominant.

Another important observation came from selfing F_2 plants individually. Specifically, he was working in this case with characteristics of the pea seed itself. This allowed him to use much larger numbers because the characteristics can be observed without growing plants from the peas. The pea seed can be thought of as an autonomous progeny individual in its own right in this species. Its appearance is the product of its own constitution and not just that of the mother as in some seed characteristics of other plants. The two pure lines he used had yellow and green seeds, respectively. He made a cross between a plant from each line, and observed that the F_1 peas that developed were all yellow. Symbolically,

P   Yellow × Green
↓
F_1 All yellow

Therefore yellow is dominant and green is recessive.

The F_1 peas were grown into plants and selfed. Of the resulting F_2 peas, 3/4 were yellow and 1/4 were green (the 3:1 ratio again). He then grew 519 F_2 yellow peas into plants and allowed each one to self. When the peas appeared he observed that 166 of the plants had only yellow peas and 353 had both yellow and green peas in a 3:1 ratio.

Therefore, approximately 2/3 of the F_2 yellows were like the F_1 yellows (i.e., produced yellow and green seeds when selfed) and 1/3 were like the pure-breeding yellow parent. Consequently, the 3:1 ratio could be more accurately described as a 1:2:1 ratio.

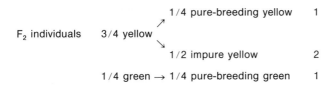

These 1:2:1 ratios were found to be underlying all of the 3:1 ratios he tested. So the problem really was to explain the 1:2:1 ratio.

His explanation was a classical example of a "model" or "hypothesis" derived from observation, which could be subject to testing by further experiments. He deduced the following explanation:

1. There are entities called "hereditary determinants or factors" of a particulate nature. (He saw no blending of characters, so was forced to a "particulate" notion.)

2. Each adult pea plant has two determinants, one from each parent, for each characteristic. (The reasoning here was obvious: the F_1 plants, for example, must have had at least one determinant for the recessive character because it showed up in later generations, and of course they also had the determinant for the dominant character because they showed it.)

3. Each "sex cell" (pollen or egg cell) has only one determinant.

4. During sex-cell formation, either of the pair of determinants of the parent plant passes with equal frequency into the sex cells.

5. The union of sex cells (to form a new individual or *zygote*) is random.

These points can be diagrammatically illustrated, using A to represent the dominant determinant and a the recessive determinant (as Mendel did), much as a mathematician uses symbols to represent entities of various kinds (Figure 1-4).

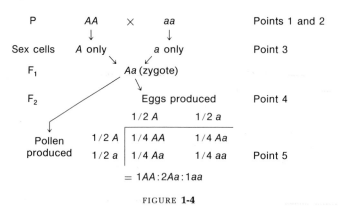

FIGURE **1-4**

The whole thing fitted together very beautifully. However, many beautiful models have been knocked down under test. Mendel's next job was to test it. He did this by taking (for example) an F_1 yellow and crossing it with a green. A $1:1$ ratio of yellow to green seeds could be predicted in the next generation. If we use Y to stand for the yellow determinant and y the recessive to show that they are a pair, we can diagram Mendel's predictions as in Figure 1-5. In this experiment he obtained 58 yellow and 52 green seeds, a very close approximation to the predicted $1:1$ ratio.

Nowadays, the hereditary determinants or factors are called genes, and we will introduce this and some other terms at this point. The various forms of a gene, represented for example by A and a, are called *alleles*. The individuals repre-

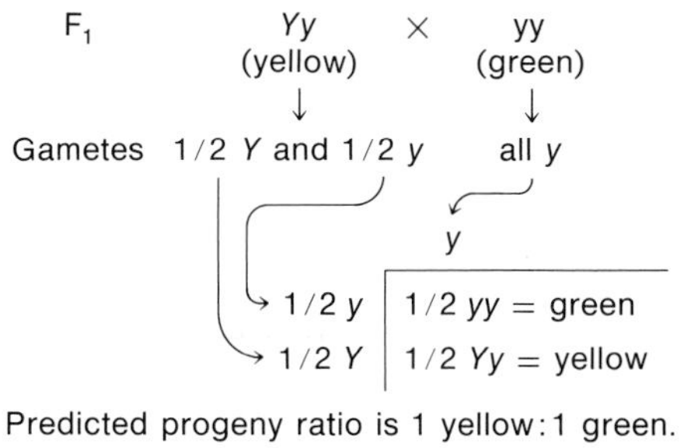

F_1 Yy × yy
(yellow)   (green)
↓   ↓
Gametes   1/2 Y and 1/2 y all y

             y

     1/2 y | 1/2 yy = green
     1/2 Y | 1/2 Yy = yellow

Predicted progeny ratio is 1 yellow : 1 green.

FIGURE **1-5**

sented by *Aa* are called *heterozygotes* (sometimes, hybrids), whereas those in pure lines are called *homozygotes*. Thus, an *AA* plant is said to be homozygous for the dominant allele, and an *aa* plant, homozygous for the recessive. Sex cells are usually called *gametes*. The actual characteristic appearance of an organism is called its *phenotype*. Thus, yellow and green seed colors are different phenotypes. On the other hand, the designated genetic constitution is its *genotype*. Thus, *Yy* and *YY* are different genotypes even though seeds of both types are phenotypically identical (i.e., yellow).

What we have called Point 4 has been given formal recognition as Mendel's first law.

MENDEL'S FIRST LAW

Alleles segregate (i.e., separate) from each other during gamete formation into equal numbers of gametes.

Another point that should be apparent from the discussion thus far is that the starting point in any genetic analysis is the recognition of genes.

MESSAGE

Genes were originally inferred and still are today by observing precise mathematical ratios in the progeny of a cross.

So far we have considered a single gene pair or *monohybrid* system, in which alleles of only one gene affecting one characteristic are considered. The next obvious question is what happens when a *dihybrid* cross is made in which pairs of genes affect two different characteristics? We can use the same symbolism that Mendel used to indicate the genotype of seed color and seed shape. A pure breeding line of plants *yy RR*, on selfing, produces seeds that are green and round. Another pure breeding line is *YY rr*, and upon selfing it produces yellow wrinkled seeds (*r* is a recessive allele of the seed-shape gene and produces a wrinkled seed). When plants from the two lines were crossed, the F_1 seeds were

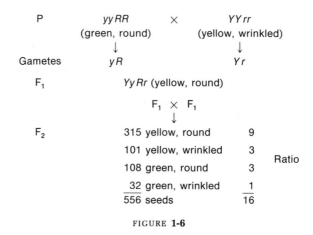

P   *yyRR*   ×   *YYrr*
         (green, round)   (yellow, wrinkled)
                  ↓   ↓
Gametes   *yR*   *Yr*

F₁   *YyRr* (yellow, round)

                  F₁ × F₁
                     ↓

F₂   315 yellow, round   9
           101 yellow, wrinkled   3
                                         Ratio
           108 green, round   3
            32 green, wrinkled   1
           556 seeds   16

FIGURE **1-6**

round and yellow, as expected. The results in the F₂ are complex, as shown in Figure 1-6. Mendel also performed analogous experiments using other pairs of characters in many other dihybrid crosses and always found 9:3:3:1 ratios in each of them. So, he had another phenomenon on his hands: some mathematics to turn into an idea.

He first checked to see whether the ratio for each gene pair was still the same as that obtained in a monohybrid cross. If you look at only the round and wrinkled phenotypes and add up all of the seeds falling into these two classes, the totals are 315 + 108 = 423 round, and 101 + 32 = 133 wrinkled. Hence the monohybrid 3:1 ratio still prevails. Likewise, a 3:1 ratio for yellow:green is also found (315 + 101 = 416:108 + 32 = 140). This gave him the clue that the two systems were independent and he was mathematically astute enough to realize that the 9:3:3:1 ratio was nothing more than two 3:1 ratios combined at random.

This is a convenient point to introduce some elementary rules of probability that will be used a lot throughout this book.

1. *Definition.*

$$\text{Probability} = \frac{\text{the number of times an event happens}}{\text{the number of opportunities for it to happen}}$$
(or the number of trials)

For example, the probability (p) of rolling a "four" on a die in a single trial is written p (one four) = 1/6 because the die has six sides.

2. *The Product Rule.* The probability of two independent events occurring simultaneously is the product of each of their probabilities. For example, with two dice we have independent objects, and

$$p \text{ (two fours)} = 1/6 \times 1/6 = 1/36$$

3. *The Sum Rule.* The probability of either of two mutually exclusive events occurring is the sum of their individual probabilities. For example, with two dice,

$$p \text{ (two fours } or \text{ two fives)} = 1/36 + 1/36 = 1/18$$

In the pea example, the F_2 of a dihybrid cross can be predicted if the mechanism for putting R or r into a gamete is *independent* of the mechanism of putting Y or y into a gamete. The frequency of gamete types can be calculated by determining their probabilities; that is, if you pick up a gamete at random, the *probability* of your picking a certain type of gamete is the same as the frequency of that type in the population.

We know from Mendel's first law that

$$Y \text{ gametes} = y \text{ gametes} = 1/2$$

$$R \text{ gametes} = r \text{ gametes} = 1/2$$

Therefore, in a $Yy\,Rr$ plant:

$$p \text{ (gamete being } R \text{ and } Y) = p\,(R\ Y) = 1/2 \times 1/2 = 1/4 \text{ (product rule)}$$

$$p \text{ (gamete being } R \text{ and } y) = p\,(R\,y) = 1/2 \times 1/2 = 1/4 \text{ (product rule)}$$

$$p \text{ (gamete being } r \text{ and } y) = p\,(r\,y) = 1/2 \times 1/2 = 1/4 \text{ (product rule)}$$

$$p \text{ (gamete being } r \text{ and } Y) = p\,(r\ Y) = 1/2 \times 1/2 = 1/4 \text{ (product rule)}$$

Thus we can represent the F_2 generation by a giant grid known as a Punnett square (Figure 1-7).

♀ \ ♂	RY 1/4	Ry 1/4	ry 1/4	rY 1/4
RY 1/4	$RR\,YY$ 1/16	$RR\,Yy$ 1/16	$Rr\,Yy$ 1/16	$Rr\,YY$ 1/16
Ry 1/4	$RR\,Yy$ 1/16	$RR\,yy$ 1/16	$Rr\,yy$ 1/16	$Rr\,Yy$ 1/16
ry 1/4	$Rr\,Yy$ 1/16	$Rr\,yy$ 1/16	$rr\,yy$ 1/16	$rr\,Yy$ 1/16
rY 1/4	$Rr\,YY$ 1/16	$Rr\,Yy$ 1/16	$rr\,Yy$ 1/16	$rr\,YY$ 1/16

FIGURE 1-7

We have already used the product rule and we use it again to derive the 1/16 values for each box in the square; that is, the probability (or frequency) of $RR\,YY$ will be $1/4 \times 1/4 = 1/16$. Collecting all the types that will look the same (as indicated by the surrounding lines in Figure 1-7), we find our now not-so-mysterious $9:3:3:1$ ratio in all its beauty.

Round, yellow (no outline)	$9/16 = 9$
Round, green (dotted line)	$3/16 = 3$
Wrinkled, yellow (heavy solid line)	$3/16 = 3$
Wrinkled, green (encircled)	$1/16 = 1$

The concept of independence of the round/wrinkled and yellow/green systems is important. From it we state Mendel's Second Law.

MENDEL'S SECOND LAW

Different segregating gene pairs assort independently.

What is beautiful about this is that a few simple assumptions like equal segregation and independent assortment can generate a seemingly baffling complex ratio like $9:3:3:1$. Try to appreciate Mendel's insight.

Of course Mendel insisted on testing his second law. One way in which he did it was to cross an F_1 dihybrid $Yy\,Rr$ with a double-homozygous recessive strain, $yy\,rr$. He predicted that the dihybrid $Yy\,Rr$ should have the gametic types YR, Yr, yR, and yr, in equal frequency, or as shown along one edge of the Punnett square in the frequencies $1/4$, $1/4$, $1/4$, and $1/4$. On the other hand, a $yy\,rr$ plant should produce only one gamete type regardless of equal segregation or independent assortment, because it is homozygous. Thus the progeny phenotypes should be a direct reflection of the gametic types from the $Yy\,Rr$ parent because the yr contribution from the $yy\,rr$ parent does not obscure anything. A $1:1:1:1$ ratio of $Yy\,Rr$, $Yy\,rr$, $yy\,Rr$, and $yy\,rr$ was predicted and was obtained. (A cross with a homozygous recessive is called a *test cross:* in this case it was also a *backcross*, which is a cross of an organism with another that is genotypically identical with one of its parents. We shall meet these kinds of crosses many times.)

The concept of independent assortment was tested on a large number of gene pairs and found to be applicable to every combination in Mendel's published work. It was also verified in a wide variety of other organisms. But, as we shall see later, it is not the whole story.

Of course the deduction of equal segregation and independent assortment as abstract concepts that explain the observed facts leads immediately to the question of what structures or forces are responsible for generating them. The idea of equal segregation seems to indicate that both alleles of a pair are in a paired configuration from which they can separate cleanly during gamete formation (Figure 1-8). If we then imagine another pair behaving independently in the same way, we have independent assortment (Figure 1-9).

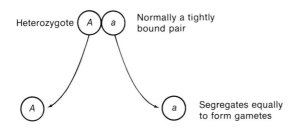

Heterozygote — Normally a tightly bound pair

Segregates equally to form gametes

FIGURE **1-8**

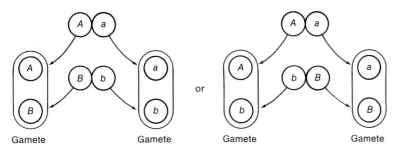

or

Gamete   Gamete   Gamete   Gamete

FIGURE **1-9**

But this is all speculation at this stage. It is mentioned to try to take you back to the era when such thoughts were rife. The actual mechanisms are now known and will be discussed later.

We will finish this chapter with a few words on the working of problems. We have already looked at the Punnett square. Although sure and graphic, it is unwieldy and should be used for illustration only, never for efficient calculation.

Another method favored by some is the "branch" method. For example, the $9:3:3:1$ ratio can be derived by drawing a branch diagram and applying the product rule as follows:

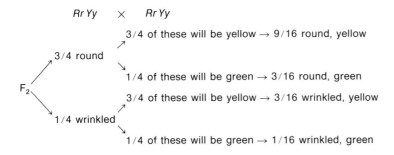

$Rr\,Yy \quad \times \quad Rr\,Yy$

3/4 of these will be yellow → 9/16 round, yellow

3/4 round

1/4 of these will be green → 3/16 round, green

F_2

3/4 of these will be yellow → 3/16 wrinkled, yellow

1/4 wrinkled

1/4 of these will be green → 1/16 wrinkled, green

This can be extended to a trihybrid ratio (e.g., $Aa\,Bb\,Cc \times Aa\,Bb\,Cc$) by drawing another set of branches on the end. However, as the number of pairs of genes increases, the number of identifiable phenotypes rises startlingly, and the number of genotypes climbs even more steeply, as shown in Table 1-1. With

TABLE **1-1**

Number of segregating gene pairs	Number of phenotypic classes	Number of genotypic classes
1	2	3
2	4	9
3	8	27
⋮	⋮	⋮
n	2^n	3^n

such large class numbers, even the branch method flounders. One then resorts to product-rule and/or sum-rule devices. For example, what proportion of progeny from the cross $Aa\,Bb\,Cc\,Dd\,Ee\,Ff \times Aa\,Bb\,Cc\,Dd\,Ee\,Ff$ will be $AA\,bb\,Cc\,DD\,ee\,Ff$? The answer is easily obtained if the gene pairs all assort independently, thereby allowing use of product rule; that is, 1/4 of the progeny will be AA, 1/4 will be bb, 1/2 will be Cc, 1/4 will be DD, 1/4 will be ee, and 1/2 will be Ff, and so multiplying these will give us our answer: $1/4 \times 1/4 \times 1/2 \times 1/4 \times 1/4 \times 1/2 = 1/1024$.

Mendel was the first "genetic surgeon." Using the experimental approach and logic described in this chapter (now called genetic analysis), he was able to identify and distinguish between the various components of the hereditary process, in as convincing and effective a way as if he had microdissected the cell and actually seen those components.

Problems

1. Mum and dad both find the taste of a chemical called phenylthiourea very bitter. But of their four children, two find the chemical tasteless. Assuming the inability to taste to be a monogenic trait, is it dominant or recessive?

2. In humans, the disease galactosemia is inherited as a monogenic recessive trait in a simple Mendelian manner. A woman whose father had galactosemia intends to marry a man whose grandfather was galactosemic. They are worried about having a galactosemic child. What is the probability of this?

3. Suppose that you have two strains of plants—one is *AA BB*, the other *aa bb*. You cross the two and self the F_1 plants. With respect to these two genes, what is the probability that an F_2 plant will have half from one grandparent and half from the other? All from one grandparent?

4. Huntington's chorea is a rare, fatal disease which usually develops in middle age. It is caused by a dominant allele. A phenotypically normal man in his early twenties learns that his father has developed Huntington's chorea:

   a. What is the probability that he will develop the symptoms himself later on?
   b. What is the probability that his son will develop the symptoms in later life?

5. Achondroplasia is a form of dwarfism inherited as a simple monogenic trait. Two achondroplastic dwarfs working in a circus marry and have a dwarf child and later have a second child who is normal.

   a. Is achondroplasia produced by a recessive or a dominant allele?
   b. What are the genotypes of the two parents in this mating?
   c. What is the probability that their next child will be normal? Dwarf?

6. Suppose that on a hike into the mountains you notice a beautiful harebell plant that has white flowers instead of the usual blue. Assuming this to be caused by a single gene, outline *precisely* what you would do to find out if the allele causing white flowers is dominant or recessive to that causing blue.

7. Suppose that a husband and wife are both heterozygous for a recessive gene for albinism. If they have dizygotic (two-egg) twins, what is the probability that both of the twins will be of the same phenotype with respect to pigmentation?

8. In dogs, dark coat color is dominant over albino, and short hair is dominant over long hair. If these effects are caused by two independently assorting genes, write the genotypes of the parents in each of the following crosses, using the symbols *C* and *c* for the dark and albino coat-color alleles, and *S* and *s* for the short- and long-hair alleles, respectively. Assume homozygosity unless there is evidence otherwise.[1]

	Phenotypes of offspring			
Parental phenotypes	Dark short	Dark long	Albino short	Albino long
a. Dark short × dark short	89	31	29	11
b. Dark short × dark long	18	19	0	0
c. Dark short × albino short	20	0	21	0
d. Albino short × albino short	0	0	28	9
e. Dark long × dark long	0	32	0	10
f. Dark short × dark short	46	16	0	0
g. Dark short × dark long	21	31	9	11

[1] Reprinted with the permission of Macmillan Publishing Co., Inc., from *Genetics* by M. Strickberger. Copyright © Monroe W. Strickberger 1968.

9. In the mountains of British Columbia a small group of Sasquatches was discovered. A study of four matings occurring in the group in the course of several years produced the following results:

Mating	Parent 1	Parent 2	Progeny
1	Bowlegs, hairy knees	Bowlegs, hairy knees	3/4 bowlegs, hairy knees, 1/4 knock-knees, hairy knees
2	Bowlegs, smooth legs	Knock-knees, smooth legs	1/2 bowlegs, smooth legs, 1/2 knock-knees, smooth legs
3	Bowlegs, hairy knees	Knock-knees, smooth legs	1/4 bow smooth, 1/4 bow hairy, 1/4 knock hairy, 1/4 knock smooth
4	Bowlegs, hairy knees	Bowlegs, hairy knees	3/4 bowlegs, hairy knees, 1/4 bowlegs, smooth knees

a. How many genes are involved in these phenotypes?

b. Which character differences are controlled by which alleles of these genes?

c. Which alleles are dominant or recessive?

d. There were in fact only five parents participating in these matings. Give the genotypes of these five individuals.

e. Draw a Sasquatch.

10. We have dealt with only two pairs of genes but the same principles hold for more than two at a time; so try this: In the cross $Aa\ Bb\ Cc\ Dd\ Ee \times aa\ Bb\ CC\ Dd\ ee$, what proportion of progeny will phenotypically resemble

a. the first parent?

b. the second parent?

c. either parent?

d. neither parent?

What proportion will *genotypically* resemble (a) the first parent, (b) the second, and so forth?

11. In Canada there are people who collect "radar bills." These are dollar bills whose seven-digit serial number forms a numerical palindrome (e.g., 4618164). How rare are radar bills (that is, what is the probability of finding one)? You are probably wondering what the genetic content of this question is; there is none, although palindromes do occur in genetic material. It is simply meant to illustrate the usefulness of a knowledge of probability.

12. Peach trees have fuzzy fruits. Nectarine trees have smooth fruits (no fuzz). Most commercial varieties of peaches and nectarines have yellow-fleshed fruits but some have white flesh. Suppose that you have some peach and some nectarine trees and plan to live a long time; so you decide to do some genetic experiments with them. You make the following crosses and get the following progenies:

Cross	Parents	Progeny
1	White peach 1 × yellow nectarine 1	12 yellow peach trees and 10 white peach trees
2	Yellow nectarine 1 × yellow nectarine 2	15 yellow nectarine trees
3	White peach 1 × yellow nectarine 2	14 yellow peach trees

Represent the allele for fuzzy fruits by f^+, the allele for smooth fruits by f^0; the allele for yellow flesh by y^+, the allele for white flesh by y^0.

a. What are the genotypes of white peach 1? of yellow nectarine 1? of yellow nectarine 2?

b. What *proportion* of phenotypes would you expect in the progeny if you selfed yellow nectarine 1?

_____ yellow peaches   _____ yellow nectarines

_____ white peaches   _____ white nectarines

c. What *proportion* of phenotypes would you expect in the progeny if you selfed one of the yellow peaches in the progeny from cross 3?

_____ yellow peaches   _____ yellow nectarines

_____ white peaches   _____ white nectarines

d. What can you conclude from these experiments about the genetic difference between peaches and nectarines?[2]

---

[2] Problem 12 courtesy of Fred Ganders.

Key to Animated Sequences

Meiosis

—○— = chromosome

○ = centromere

——— = cell membrane

Mitosis

—○— = chromosome

○ = centromere

——— = cell membrane

DNA replication

——— = single strand of DNA

▼ = progress of polymerization process

⟩⟨ = action of ligase

Translation

● = amino acid

——— = messenger RNA

= transfer RNA

= ribosome with two binding sites

---

2 / Chromosome Theory of Inheritance

(Or: How the location of the hypothetical Mendelian determinants was inferred.)

The beauty of Mendel's analysis is that data derived from genetic crosses can be interpreted by means of the laws of segregation and independent assortment. Furthermore, it then becomes possible to make predictions on the outcome of further crosses. All of this is possible by simply representing abstract hypothetical factors of inheritance, or genes, with symbols without any concern about their physical basis or location in a cell. Nevertheless, although the validity of Mendelian principles is verified in many different organisms and genotypes, the obvious next question is: What structure(s) within cells might correspond to genes?

Cytologists in studying cells had carefully recorded the sequence of events that could be detected microscopically when cells divide. Thus they noted the apparent distinction within the cell of the *nucleus* and the *cytoplasm* and the appearance of organelles called *chromosomes* within the nucleus at the time of division.

Meiosis

Mitosis followed by DNA Replication

Translation

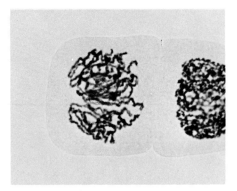

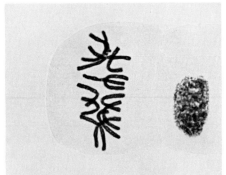

A. *Prophase.* Chromosomes are contracting and becoming visible. Each chromosome has already duplicated into two daughter chromatids, but these will not be clearly visible until metaphase.

B. *Metaphase.* All chromosomes assemble on the equatorial plane of the cell. Chromatids are now visible and centromeres have duplicated, too. Note that the chromosomes do *not* pair during mitosis.

C. *Anaphase.* Centromeres segregate (or disjoin), pulling one daughter chromatid to each cell pole.

D. *Anaphase.* Disjunction is complete. Compare the number of chromosomes in each cell with that in the original cell.

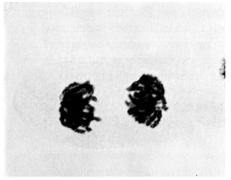

E. *Telophase.* The nucleus reorganizes and cell division is almost complete.

F. Metaphase that has been disrupted by chemical treatment to show the individual chromosomes more clearly. Can you see any homologous ones?

FIGURE 2-1
Mitosis in *Hepatica acutiloba* (2n = 14).

Division in somatic cells proceeds by a process called *mitosis,* as distinct from cell division in reproductive cells called *meiosis.* A single fertilized human egg will ultimately produce, by mitosis, an adult consisting of 60 trillion (plus or minus a few billion) cells! Within the gonads of a postpubertal male, millions of sperm are produced daily as the result of meiosis. These two processes are illustrated in Figures 2-1 and 2-2. You have undoubtedly been bludgeoned with explanations of mitosis and meiosis and, at this point, the fine details of the changes within the cell are not important. Just recognize that the processes are dynamic and that it is only for the convenience of description that arbitrary stages are selected and defined. For the present discussion, the basic feature of mitosis is that each chromosome duplicates into *sister chromatids,* which are initially attached to one another at a point referred to as the *centromere* or *kinetochore.* Fibers attach to each of the centromeres and extend away to anchor points at opposite "poles." These are the "spindle" fibers (so called because the fibers form a shape like a spindle). The fibers exert force on the centromeres of the sister chromatids, pulling one chromatid to each pole with the chromatid arms dragging behind them. This results in the formation of two nuclei—one at each pole—each consisting of the same number and type of chromosomes as the other and as the original cell before division.

In meiosis, on the other hand, two successive divisions characteristically occur in rapid succession. During the first meiotic division, duplicated chromosomes pair and each member of the pair is called a *homolog* of the other. The paired homologs are almost always very similar in size and shape so there appears to be an important element of relatedness between the paired chromosomes.

In meiosis, spindle fibers from opposite poles attach to the centromeres of each homolog. In this case, contraction of the fibers pulls the duplicated homologs apart but does not split sister chromatids. This division is often referred to as *reductional division* because the number of separate centromeres is reduced by one-half in the two daughter cells. In the second meiotic division, spindle fibers from each pole attach to the centromeres and the sister chromatids are separated in a manner resembling that in mitosis. This division is called *equational* because the number of centromeres in the daughter cells remains the same as in the cell generating them. The important end result of meiosis is the production of cells having half of the original chromosome number; each cell now has one member of each of the originally paired homologs. Obviously, fertilization of one meiotic product or gamete with another restores the original number and types of chromosomes present in premeiotic cells.

The regularity with which chromosomes duplicate, separate, and are restored in number after fertilization suggested to cytologists before the twentieth century that chromosomes must be biologically important organelles. Indeed, if we look at large numbers of individuals *within* a given species, with very rare exceptions the chromosome number is constant from individual to individual (sometimes the number may vary between sexes but in a very regular way). On

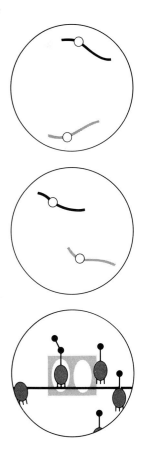

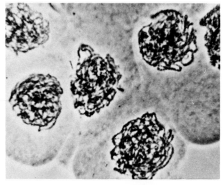

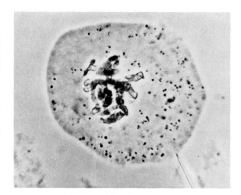

A. *Early prophase I.* The chromosomes in several cells are becoming visible as long threads. Each chromosome is already duplicated into two daughter chromatids, but in this set of photographs the chromatids are not visible until anaphase I.

B. *Late prophase I.* The chromosomes have shortened (in this species the chromosomes become very squat) and the pairing of homologs can be discerned.

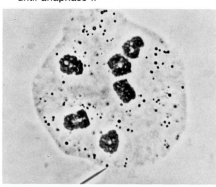

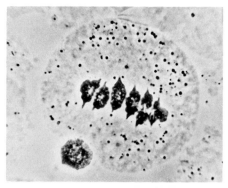

C. *Metaphase I.* The pairing of homologous units is now clearly visible. Note that each of the six groups consists of four homologous chromatids and two undivided centromeres.

D. *Anaphase I.* The groups of homologs are lined up on the equatorial plane of the cell and segregation, or disjunction, of the centromeres is beginning.

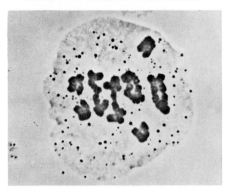

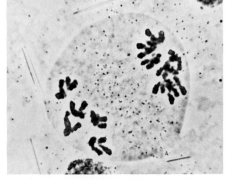

E. *Anaphase I.* Different stages of disjunction are visible in different groups. Note that daughter chromatids are now visible, still attached at their centromere.

F. *Late anaphase I.* The division of the nuclear contents is now almost complete. How many centromeres does each new nucleus have?

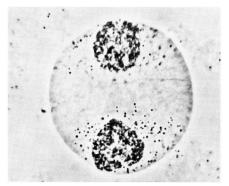

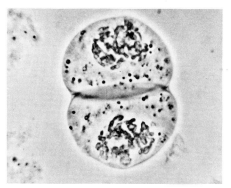

G. *Telophase I.* Reorganization of the nuclei has occurred and cell division is almost complete. The chromosomes elongate again.

H. *Prophase II.* Both cells begin the second meiotic division. The chromosomes contract.

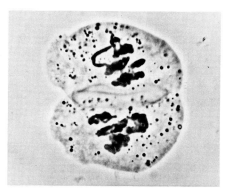

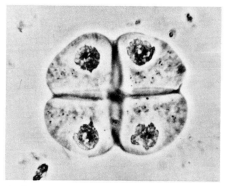

I. *Metaphase II.* The centromeres have divided and the daughter chromatids are now disjoining.

J. *The four products of meiosis.* Each nucleus is haploid and has six chromatids, which are now the new chromosomes. Each cell will develop into a pollen grain, or male gamete.

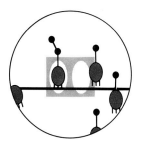

FIGURE 2-2
Meiosis in the anthers of *Tradescantia paludosa* (2n = 12).

TABLE 2-1

Organism	Haploid chromosome number
Man	23
Dog	39
Horse	32
Cat	19
Mallard	40
Chicken	39
Alligator	16
Cobra	19
Bullfrog	13
Goldfish	47
Starfish	18
Fruitfly	4
Housefly	6
Neurospora	7
Sphagnum moss	23
Field horsetail	108
Giant sequoia	11
Tobacco	24

the other hand, chromosome number *between* different species varies remarkably (Table 2-1). One chromosome set is called a *genome*.

With the rediscovery of Mendel's laws in 1900, it was soon recognized by an American graduate student, Walter Sutton, and the great German biologist Theodor Boveri that the behavior of chromosomes during meiosis paralleled the behavior of Mendel's hypothetical units. To account for this, Sutton and Boveri in 1902 postulated that Mendel's factors of inheritance "are on chromosomes" (Figure 2-3).

MESSAGE

The parallel behavior of Mendelian factors and chromosomes with respect to segregation and independent assortment led to the suggestion that what are now called genes are on chromosomes.

To students of today who have had any biology this may not seem very earthshaking, but early in the twentieth century this suggestion, which potentially united cytology and the infant field of genetics, was a bombshell. Of course the immediate response to any hypothesis is to try to pick holes in it and for years after there was a raging controversy over the validity of what became known as the Sutton-Boveri Chromosome Theory of Heredity.

Mendel's factors Chromosomes

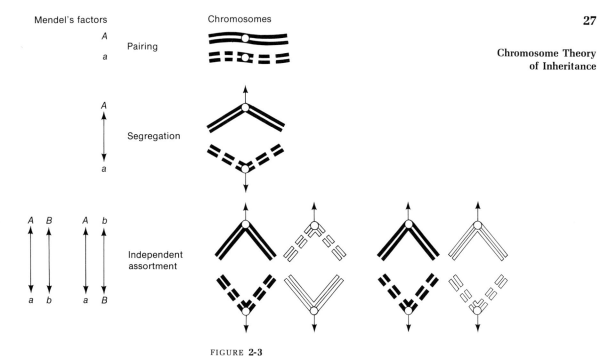

FIGURE **2-3**

It is worth considering what some of the objections were. For example, during interphase chromosomes are not detectable, and it was only through some very diligent studies of chromosome position before and after interphase that Boveri was able to argue that, although cytologically invisible, chromosomes retain their continuity through interphase. Another objection was that, in many organisms, chromosomes do not look all that different and so perhaps they were pairing willy-nilly, whereas Mendel's laws absolutely required segregation of alleles. However, where chromosomes differ in size and shape, it was verified that similar chromosomes do occur in pairs and the homologs pair and segregate during meiosis. It was also argued that chromosomes appeared as stringy structures and one could not tell whether they differed in any qualitative way from each other. Perhaps they were all just more or less of the same stuff. That nonhomologous chromosomes are different from each other was proved by Alfred Blakeslee who studied the Jimsonweed (*Datura*), which has 12 pairs of chromosomes. He constructed twelve different strains, each of which had 12 chromosome pairs plus one more of one of each of the chromosomes, and showed that each strain was phenotypically distinct from the others. This would not be expected if the chromosomes were all alike.

Finally, whether chromosomes do indeed assort independently was tested when Elinor Carothers found an unpaired chromosome in the testes of a species of grasshopper, along with a chromosome pair in which the two chromosomes were not identical (called *heteromorphic*). Thus, by looking at anaphase nuclei,

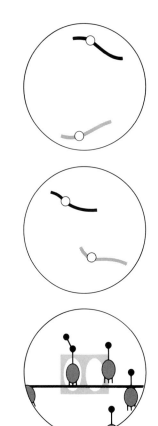

the number of times each of the dissimilar chromosomes of the heteromorphic pair migrated to the same pole as the unpaired chromosome could be counted (Figure 2-4). She found that the two patterns of chromosome separation (which

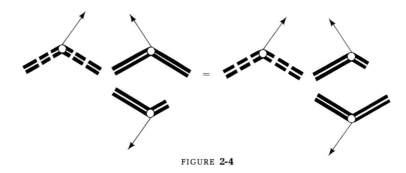

FIGURE **2-4**

result in four different gametic types) occurred with equal frequency; that is nonhomologous chromosomes assort independently.

So far, then, the behavior of chromosomes and that of genes parallel each other very closely. This of course makes the Sutton-Boveri proposal attractive but does not constitute *proof* that genes are on chromosomes. Proof of the hypothesis came from further observations.

Sex Linkage

All crosses discussed up to this point yield similar progeny, regardless of whether the parents from each strain in a cross are male or female; that is, reciprocal crosses (e.g., A female × B male and A male × B female) yield similar progeny. The first exception to this came from experiments by L. Doncaster and G. H. Raynor in 1906. They were studying wing color in the magpie moth (*Abraxas*), using two different strains—one with light wings, the other with dark. If light-winged females are crossed with dark-winged males, all the progeny have dark wings, thereby showing that the allele for light wings is recessive. However, in the reciprocal cross (dark female × light male) all the female progeny have light wings and the males have dark. So here we encounter reciprocal crosses that do not give similar results and, in the second cross, the wing phenotypes are associated with the sex of the moths. Note that the female progeny are phenotypically similar to their fathers and the males to their mothers. This is often referred to as *criss-cross inheritance*. How can we explain these results? Let's consider another example before attempting this.

William Bateson had been studying the inheritance of feather pattern in chickens. One strain had feathers with alternating stripes of dark and light coloring, a phenotype called barred. Another strain, nonbarred, has feathers that were uniformly colored. In the cross barred male × nonbarred female, all the progeny were barred, thus showing that nonbarred is recessive. However, the reciprocal cross (barred female × nonbarred male) gave barred males and nonbarred females. Again, criss-cross inheritance is observed. How can we explain the similar results obtained by Doncaster and Raynor and by Bateson?

An explanation came from the laboratory of Thomas Hunt Morgan, who began studying inheritance in a fruitfly (*Drosophila melanogaster*), in 1909. Because this organism has played a key role in the study of inheritance, it is worthwhile digressing to briefly discuss the beast. The life cycle of *Drosophila* typifies the life cycles of many insects (Figure 2-5).

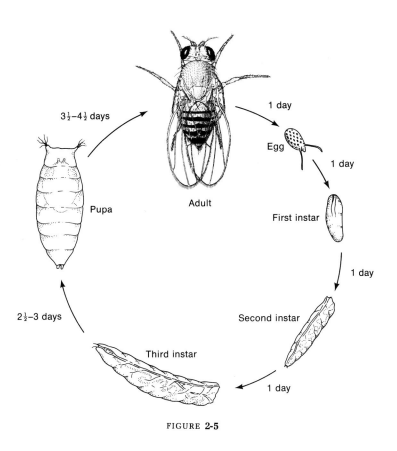

FIGURE **2-5**

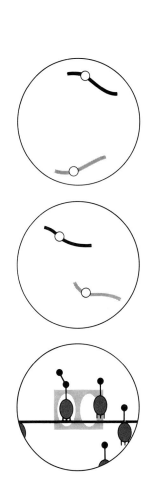

The flies grow vigorously in the laboratory. In the egg, the early embryonic events lead to the production of a larva called the first "instar." Growing rapidly, the larva molts twice and the third instar larva then pupates. In the pupa, the larval carcass is replaced by adult structures and an "imago" or adult emerges from the pupal case, ready to mate within twelve to fourteen hours. The adult fly is about two millimeters in length so it takes up very little space. The life cycle is very short (twelve days at room temperature) in comparison with that of a human, a mouse, or a corn plant; so many generations can be reared in a year. Moreover, the flies are extremely prolific—a single female is capable of laying several hundred eggs. Perhaps the beauty of the animal when observed through a microscope added to its early allure. In any case, as we shall see later, the choice of *Drosophila* was a very fortunate one for geneticists and especially for Morgan, who won a Nobel Prize for his work in 1934.

The normal eye color of *Drosophila* is bright red. Early in his studies, Morgan discovered a male with completely white eyes. When this male was crossed with red-eyed females, all the F_1 progeny were found to have red eyes, thereby showing that the allele for white is recessive. On crossing the red-eyed F_1 males and females, he obtained a $3:1$ ratio of red- to white-eyed flies, but all the white-eyed flies were males. The ratio of red-eyed females to red-eyed males was $2:1$. What is going on?

When Morgan crossed white-eyed males with red-eyed female progeny of the cross white males $\times$ red females, he obtained red- and white-eyed males and females in equal numbers. Finally, in a cross of white females and red males (which is the reciprocal of the cross of the original white-eyed male), all the females were red-eyed and all the males, white-eyed. This is criss-cross inheritance again. However, note that criss-cross inheritance was observed in the experiments on chickens and moths when the parental males carried the recessive genes, whereas in the *Drosophila* cross it is seen when the female parent carries the recessives.

Now before turning to Morgan's explanation of the *Drosophila* results, we should look at some of the cytological information that he had to draw on for his interpretations. In males of a species of Hemiptera (the true bugs), H. Henking had observed in 1891 that meiotic nuclei had 11 pairs of chromosomes and an element that segregated to one of the poles from no other element. Henking called this an "X" body, which he interpreted as a nucleolus but which was later found to be a chromosome. Clarence McClung saw that, in testes of grasshoppers, the consequence of meiosis was the production of equal numbers of sperm with and without an X element and suggested that it was concerned with sex determination. In 1905, Edmund Wilson noted that in *Protenor*, another Hemipteran, females had 7 pairs of chromosomes whereas males had 6 pairs and a single X element. Obviously, the female had a pair of X's. On the other hand, in the same year, Nettie Stevens found that the males and females of the beetle *Tenebrio* had the same number of chromosomes but one of the pairs in males was heteromorphic. One of the heteromorphic chromosomes appeared to be the X chromosome of which the females had two and the other heteromorphic element was called the Y chromosome by Wilson. Stevens showed that *Drosophila melanogaster* has 4 pairs of chromosomes with one of the pairs being heteromorphic in males.

With that as background, Morgan interpreted his genetic data in the following way. He knew that *Drosophila* females had 4 chromosome pairs, whereas males had 3 pairs and a heteromorphic X-Y pair. Thus, in meiosis, all females produce eggs bearing one X chromosome whereas males produce two types of sperm, one bearing an X, and the other a Y. Union of an egg with an X-bearing sperm will produce an XX female and with a Y will yield an XY male. He recognized that the data could be explained by assuming that the alleles for red and white eye color are on the X chromosome, with none on the Y chromosome. The red eyes of all F_1 progeny in the original cross of the white-eyed male with red-eyed females shows that the gene for red eyes is dominant; so we can represent the two alleles as W (red) and w (white). Thus, if we designate the X chromosomes as X^W and X^w to indicate the alleles supposedly carried by them, we can diagram the two reciprocal crosses as shown in Figure 2-6.

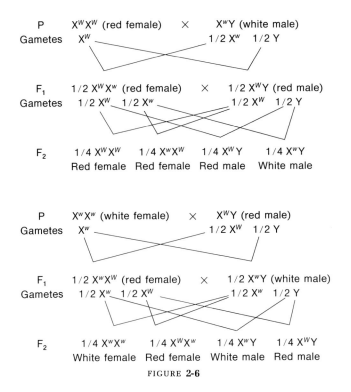

FIGURE 2-6

As you can see, the genetic results of the two reciprocal crosses are completely consistent with the known meiotic behavior of the X and Y chromosomes. This experiment strongly supports the notion of chromosomal location of genes but still is only a correlation and does not constitute definitive proof.

If you try to use the same XX and XY chromosome theory for the earlier crosses made using chickens and moths, you will find it does not work. Richard Goldschmidt recognized immediately that an explanation similar to Morgan's could be made by assuming that *males* had pairs of identical chromosomes whereas females had a heteromorphic pair. To distinguish this from the X-Y

situation in *Drosophila,* Morgan suggested that the heteromorphic chromosomes be called W-Z, with males being ZZ and females WZ. Thus, if the genes in the chicken and moth crosses were on the Z chromosome, they can be diagrammed as shown in Figure 2-7.

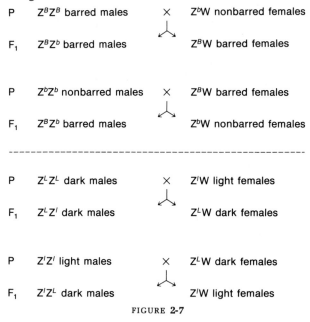

P $Z^B Z^B$ barred males   × Z^bW nonbarred females

F₁ $Z^B Z^b$ barred males Z^BW barred females

P $Z^b Z^b$ nonbarred males   × Z^BW barred females

F₁ $Z^B Z^b$ barred males Z^bW nonbarred females

P $Z^L Z^L$ dark males   × Z^lW light females

F₁ $Z^L Z^l$ dark males Z^LW dark females

P $Z^l Z^l$ light males   × Z^LW dark females

F₁ $Z^l Z^L$ dark males Z^lW light females

FIGURE 2-7

Again this interpretation was consistent with the genetic data. In this case, cytological evidence came after the suggestion that, in moths and birds, females had a heteromorphic chromosome pair. In 1914, J. Seiler verified that, in moths, both chromosomes in all pairs were identical in males, whereas females have a heteromorphic pair.

MESSAGE

The special inheritance pattern of some genes makes it extremely likely that they are borne on the chromosomes associated with sex which show a parallel inheritance pattern.

These correlations were consistent with the supposition that genes do indeed reside on chromosomes but this was still heatedly debated for many years after. The critical proof of this hypothesis came from experiments performed in Morgan's laboratory in connection with the Ph.D. thesis of one of his most brilliant students, Calvin Bridges. But before going into his work, we should introduce some symbols. Rather than writing out the words for each sex, ♀ is used for female (the astrological symbol for Venus) and ♂ is used for male (the astrological symbol for Mars). Often the plural is indicated by ♀♀ or ♂♂. In addition, symbols for alleles were introduced for *Drosophila* and this system is now used by most geneticists except those studying plants. (Plant geneticists continue to show dominance and recessiveness with capital and small letters, respectively.) However, for a given *Drosophila* character the allele that is found most frequently in natural populations is designated as the standard or *wild type* and all other alleles then are "nonwild type." As we'll see much later in the book, this definition will come under some fire. However, at the phenotypic level we're considering now, this is a workable definition of wild type. The symbol for the wild-type allele is indicated by the presence of a $+$. The name of a gene comes from the first nonwild-type allele found. Thus, for example, the wild-type eye color is red and a nonwild-type is white. We can call the nonwild-type allele white and symbolize it as w. The wild-type allele then is w^+. But the wild-type allele is not always dominant over a nonwild-type allele. We show the dominance relationship between wild and nonwild alleles by using the capital- or small-letter symbol of the nonwild type. Thus, for the two alleles w and w^+, we know the wild type is dominant over white because the symbol for white is a small letter. In another case, the wild-type condition of a fly's wing is straight and flat. A nonwild-type gene causes the wing to be curled and is therefore called Curly and designated as Cy because it is dominant to its wild-type allele, Cy^+. Had Curly been recessive to the wild-type allele, it would be designated as cy. (*Note: Cy* is one gene, not two.)

Now let's return to Bridges' work. As we now know, in the cross (using our new symbolism) X^wX^w (white)♀ $\times X^{w^+}Y$ (red)♂, the progeny are $X^{w^+}X^w$ (red)♀♀ and X^wY (white)♂♂. But Bridges discovered that, if he made this cross on a large scale, rare exceptions were found. About 1 of every 2,000 F$_1$ progeny were white-eyed females or red-eyed males. Because these exceptional progeny resemble the parents of their sex, the phenotype of females is said to be *matroclinous* and of the males, *patroclinous*. They are called *primary exceptional progeny*. The patroclinous males were always found to be sterile. However, when Bridges crossed the primary exceptional white-eyed females with normal red-eyed males, 4% of the progeny were matroclinous white-eyed females and patroclinous red-eyed males that were fertile. Thus, exceptional offspring were again recovered, but at a higher frequency and the males were fertile. These exceptional progeny of primary exceptional mothers are called *secondary exceptional offspring* (Figure 2-8). How do we explain the exceptional progeny?

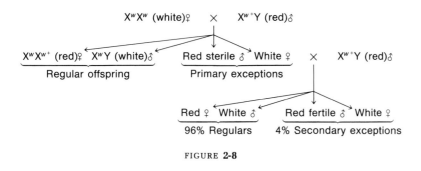

FIGURE 2-8

It is obvious that the matroclinous females—which, like all females, have two X chromosomes—must get both of them from their mothers since they are homozygous for w. Similarly, patroclinous males must derive their X chromosomes from their fathers since they carry w^+. Bridges hypothesized that, during meiosis in the females, rare mishaps occur whereby the paired X chromosomes fail to separate during either the first or second divisions. This would result in meiotic nuclei containing either two X's or no X at all. Such a failure to separate is called nondisjunction and produces an XX and a nullo-X (containing no X) nucleus. Fertilization of these two types of nuclei will produce four zygotic classes as shown in Figure 2-9.

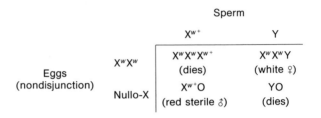

FIGURE 2-9

If we assume that the XXX and YO classes die, then the two types of exceptional progeny can be expected to be $X^w X^w Y$ and $X^{w^+} O$ chromosomally. The sterility of primary exceptional males is explicable if we can assume that the Y chromosome must be present in order to have male fertility. If, during meiosis in the XXY females, the two X's pair and disjoin most of the time whereas the Y remains unpaired, then equal numbers of X- and XY-bearing eggs will result from their formation. However, we know that in males the X and Y chromosomes can pair; so, if in approximately 16% of the pairings in $X^w X^w Y$ females, the Y successfully pairs with an X^w, then the other X^w will be free to separate to either pole. Half of these pairings will result in X^w and $X^w Y$ eggs, but the other half (8%) will result in $X^w X^w$ and Y eggs (Figure 2-10).

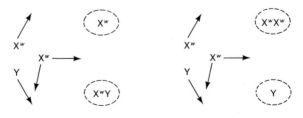

FIGURE 2-10

Half of the time, fertilization of the latter two gametic types will produce $X^wX^wX^{w+}$ and YY zygotes, which we presume die, and the other half will be secondary exceptions, which are X^wX^wY and $X^{w+}Y$. Now we can see why secondary exceptional males are fertile: they receive a Y chromosome from their XXY mothers (Figure 2-11).

$$X^wX^wY \qquad \times \qquad X^{w+}Y$$

(primary
exceptional ♀)

Sperm

		X^{w+}	Y	
X-Y pairing (infrequent)	X^wX^w	$X^wX^wX^{w+}$ (dies)	X^wX^wY (white ♀)	Secondary exceptional progeny
	Y	$X^{w+}Y$ (red fertile ♂)	YY (dies)	
	X^w	$X^{w+}X^w$ (red ♀)	X^wY (white ♂)	
	X^wY	$X^wX^{w+}Y$ (red ♀)	X^wYY (white ♂)	"Regular" (expected) progeny
X-X pairing	X^wY	$X^wX^{w+}Y$ (red ♀)	X^wYY (white ♂)	
	X^w	X^wX^{w+} (red ♀)	X^wY (white ♂)	

Eggs

FIGURE **2-11**

So far we have made assumptions about the chromosome location of w and w^+ and hypothesized nondisjunction to explain the exceptional progeny. If w and w^+ are indeed on the X chromosome, we can now make testable predictions.

1. Primary exceptional females and males detected genetically should be XXY and XO when studied cytologically. Bridges found that they are.

2. Secondary exceptional females and males detected genetically should be XXY and XY cytologically. Bridges found that they are.

3. Half of the red-eyed daughters of exceptional white-eyed females should be XXY and half should be XX. Bridges found that they are.

4. Half of the white-eyed sons of exceptional white-eyed females should themselves give exceptional progeny and all of those that do should be XYY. Bridges found that this is the case.

Thus, by hypothesizing that w and w^+ are indeed on the X chromosome, all the testable predictions arising from the unexplained but inferable process of nondisjunction are verified. This provides unequivocal evidence that genes are associated with chromosomes.

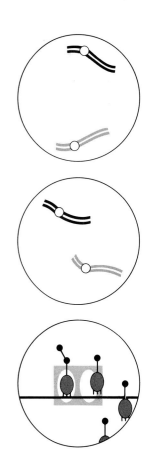

MESSAGE

When Bridges used the chromosome theory to successfully predict the outcome of certain genetic analyses, the chromosome location of genes was established beyond reasonable doubt.

At this point, you might ask whether the Y chromosome carries any genes that can be detected. In the fish *Lebistes,* the Y chromosome carries a gene called *maculatus,* which determines a pigmented spot at the base of the dorsal fin. Because the Y is passed directly from father to son, only males express the maculatus phenotype. In *Drosophila,* Bridges' work showed that the Y chromosome is necessary for male fertility but it in no way affects females. Curt Stern found that the Y does carry a gene that is also located on the X chromosome. The nonwild-type allele causes a phenotype of shorter, more slender bristles called *bobbed* (*bb*). If an $X^{bb}X^{bb}$ female is crossed with a wild-type $X^{bb^+}Y^{bb^+}$ male, then all the F_1 progeny are wild type: $X^{bb^+}X^{bb}$ females and $X^{bb}Y^{bb^+}$ males. All the males that result from crossing the F_1 generation are wild type, whereas half the females are bobbed and half are wild type. In humans, as we shall see later, the presence of the Y chromosome determines maleness. Furthermore, there is apparently a gene on the Y that is not present on the X, an allele of which results in a phenotype called hairy ears (Figure 2-12). In this case, like the maculatus

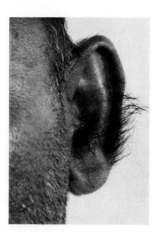

FIGURE **2-12**

phenotype in the fish, hairy ears are passed on by and expressed only in males.

Once again, our discussion can be greatly simplified by the introduction of a few terms. Those chromosomes that appear to be different in number or type in each sex are called the *sex chromosomes.* Thus, and X and Y (and in other organisms the W and Z) chromosomes are sex chromosomes. All of the other chromosomes are called *autosomes.* Genes such as w and w^+, *bb* and bb^+ that are located on sex chromosomes are said to be *sex-linked*. The sex that produces two types of gametes with respect to the sex chromosomes is called the *heterogametic sex*. *Drosophila* males, therefore, are heterogametic because they produce X- and Y-bearing sperm, whereas, in birds and moths, females are the heterogametic sex. The sex that produces only one kind of gamete with respect to sex chromosomes is called *homogametic*. So *Drosophila* and human females are homogametic in that all normal ova carry X chromosomes. Finally, if a gene has no homologous allele because the chromosome has no pairing partner (as in XO male grasshoppers) or pairs with a heteromorphic element such as the Y chromosome, the organism is said to be *hemizygous* for the gene. Hemizygous genes can show either *X-linkage* (as in *w* in *Drosophila*) or *Y-linkage* (as in hairy ears in humans). Henceforth, these terms will be used in the text.

Did Mendel Cook His Results?

Mendel reported on 7 gene pairs (allele pairs) controlling 7 different characteristics. He indicated that he got independent assortment in every combination he tried and this supported his second law. However, we now know that independent assortment can be seen optimally only if the gene pairs are on separate chromosomes. It happens that in *Pisum* there are only 7 chromosome pairs and we also know now that Mendel did in fact choose one gene pair from each chromosome pair! The probability of doing this can be compared to having seven buttons to push randomly in the dark, and by chance pushing each one once in seven trials. The probability of that happening is illustrated in the following schematic:

The first push is free; any of the seven will be acceptable	$= 7/7$
The second push is restricted because only six remain to be pushed	$= 6/7$
The third push is restricted because only five remain to be pushed	$= 5/7$
The fourth push is restricted because only four remain to be pushed	$= 4/7$
The fifth push is restricted because only three remain to be pushed	$= 3/7$
The sixth push is restricted because only two remain to be pushed	$= 2/7$
The seventh push is restricted because only one remains to be pushed	$= 1/7$

Using the product rule, we can say that the probability of pushing them all in seven tries is $7!/7^7 = 0.0061$ or 0.61%. (The symbol ! stands for factorial.)

Another "suspicious" feature of Mendel's work is that his ratios are too good to be true. Just by random chance one expects that ratios will be off the mark quite often, but Mendel's rarely were. Did he select only the best data? Did he have a smart gardener? Did God help him? Or was he just lucky? We believe it was a didactic device: he just did not want his publication cluttered with less unambiguous material that might detract from his main message.

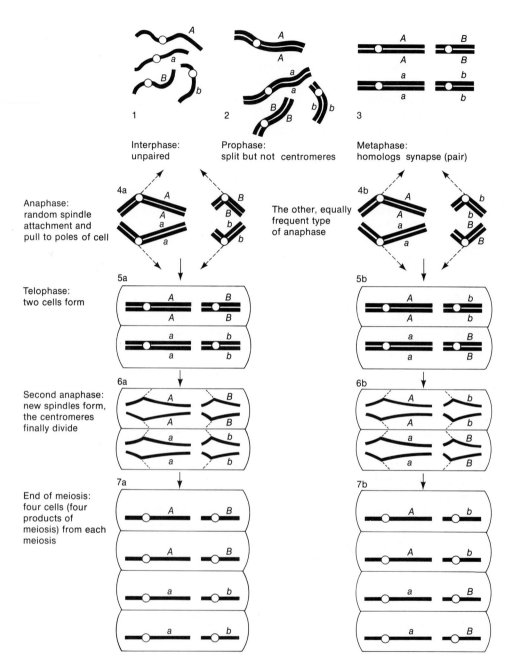

FIGURE 2-13

Having (we hope) convinced you of the great truth of the chromosome theory of inheritance, we'll end the chapter with a meiotic diagram (Figure 2-13), including some hypothetical genes, to summarize the conclusions of the chapter.

We have been considering diploid organisms, those containing two whole sets of chromosomes. The diploid (or "2n") number can range from four to hundreds, depending on the organism; for the purpose of this demonstration we'll assume that 2n = 4. (This means that the haploid [n] gametes will contain 2 chromosomes.) Into this organism we'll put some genes, in fact, we'll make it *Aa Bb*, where the two different gene pairs are on separate homologous chromosome pairs.

Each product of meiosis becomes a gamete: if in a male, sperms; if in a female, eggs.

Because the 4a pattern equals 4b in frequency, it is easy to see that the gametic frequencies of $AB = ab = Ab = aB = 1/4$, which is the distribution we obtained from inference about the abstract models.

Notice that the clean 1:1 segregation of genes is based on metaphase pairing and anaphase segregation of chromosomes and that the independent behavior of gene pairs is based on the independent movement of different chromosome pairs by random spindle attachment.

LET'S NOT FORGET HAPLOIDS

Haploids have genes and chromosomes too, but only one set of each rather than two as in diploids. Also, many diploid plants (and some animals) have haploid stages in their life cycles. For example, in ferns, meiosis does not produce haploid gametes, but haploid gametophytes. The n product of meiosis grows into a small inconspicuous haploid plant, which later produces haploid gametes by mitosis.

Three important life cycles based on chromosome number can be distinguished: (1) diploids (e.g., peas, *Drosophila*, and humans) as shown in Figure 2-14; (2) haploids (e.g., yeast, single-celled algae, molds, and mushrooms) as

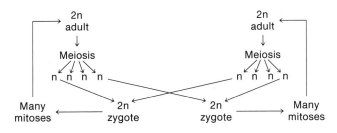

FIGURE **2-14**

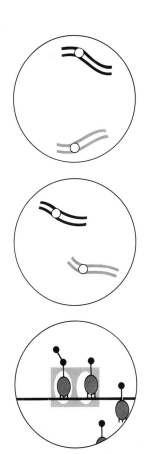

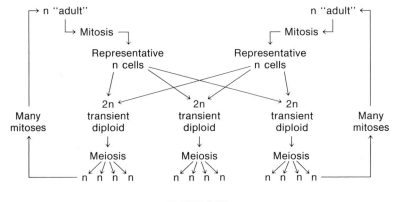

FIGURE **2-15**

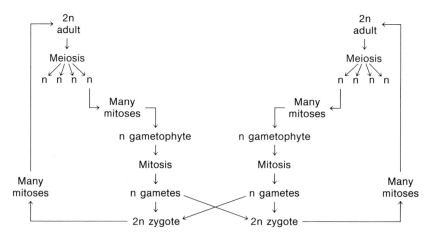

FIGURE **2-16**

shown in Figure 2-15; and (3) diploid/gametophyte alternation (e.g., ferns, liverworts, and mosses) as shown in Figure 2-16.

It is interesting to wonder if Mendel would have deduced his laws had he been working with a haploid organism instead of diploid peas. In later chapters, we'll be considering the genetics of the bread mold *Neurospora*. Let's assume that Mendel was working with *Neurospora*, which has a typical haploid cycle as shown in Figure 2-15).

The normal wild-type phenotype of a *Neurospora* culture is orange and fluffy. Let's assume that Mendel obtained an albino (white) form and crossed it (by mixing) with a normal orange. The progeny from this cross would have been 50% orange and 50% albino. From this he would probably have deduced the particulate nature of determinants (since the offspring are not "blends" of orange and white). Furthermore, the 1:1 ratio would have given him the clue about the *pairing* of the determinants and their *equal segregation* (his first law).

If he then obtained a nonfluffy variant of *Neurospora* (let's call it colonial) and crossed that with a fluffy, he would also have obtained a 1:1 ratio in the progeny. A dihybrid cross of an albino fluffy with an orange colonial would have given:

1/4 albino fluffy   1/4 albino colonial

1/4 orange colonial   1/4 orange fluffy

He would have noted that the 1:1 ratios of both gene pairs still prevail and that the 1/4:1/4:1/4:1/4 or 1:1:1:1 ratio really reflects *independent assortment* (or his second law). It is derived as follows:

1/2 albino
↗ 1/2 fluffy   = 1/4 albino fluffy
↘ 1/2 colonial = 1/4 albino colonial

1/2 orange
↗ 1/2 fluffy   = 1/4 orange fluffy
↘ 1/2 colonial = 1/4 orange colonial

Using symbols, $al\ col^+ \times al^+\ col$ (it is haploid), where $col^+ =$ fluffy and $al^+ =$ orange, gives:

1/4 $al\ col^+$ 1/4 $al\ col$

1/4 $al^+\ col$ 1/4 $al^+\ col^+$

If you had been Mendel working with *Neurospora,* how would you have tested your hypothesis?

Problems

1. Duchenne's muscular dystrophy is sex-linked and usually affects only boys. Victims of the disease become progressively weaker, starting early in life.
   a. What is the probability that a woman whose brother has Duchenne's disease will have an afflicted child?
   b. If your mother's brother (your uncle) had Duchenne's disease, what is the probability that you carry the gene?
   c. If your father's brother had the disease, what is the probability that you carry the gene?

2. Suppose that in the karyotype of a human male, two interesting *rare* cytological abnormalities are discovered: on *one* of the chromosome pair 4 there is an extra piece or satellite, and one *one* of the chromosome pair 7 there is an abnormal pattern

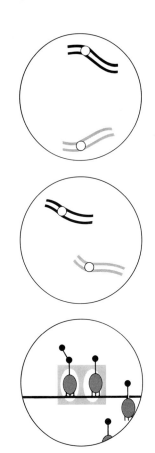

of staining. Assuming that all his gametes are equally viable, what proportion of his children will have the same visible karyotype as he has? (A karyotype is a chromosome complement.)

3. The following pedigree is concerned with an inherited dental abnormality, amelogenesis imperfecta. (*Note:* In pedigree analysis, a square indicates a male, a circle a female, and a shaded unit an affected individual.)

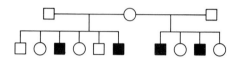

   a. What mode of inheritance *best* accounts for the transmission of this trait?

   b. Draw the genotypes of the individual members according to your hypothesis.

4. A sex-linked recessive gene *c* produces red-green color blindness in humans. A normal woman whose father was color-blind marries a color-blind man.

   a. What genotypes are possible for the mother of the color-blind man?

   b. What are the chances that the first child from this marriage will be a color-blind boy?

   c. Of the girls produced by these parents, what percentage is expected to be color-blind?

   d. Of all the children (sex unspecified) from the parents, what proportion can be expected to be normal?

5. Male house cats are either black or yellow and females are black, tortoise-shell pattern, or yellow.

   a. If these colors are governed by a sex-linked gene, how can these results be explained?

   b. Using appropriate symbols, determine the phenotypes expected in the offspring from crossing a yellow female with a black male.

   c. Do the same for the reciprocal cross of part b.

   d. Half of the females produced by a certain kind of mating are tortoise-shell and half are black; half the males are yellow and half are black. What colors are the parental males and females in such crosses?

   e. Another kind of mating produces offspring, one-quarter of which are yellow males, one-quarter yellow females, one-quarter black males, and one-quarter tortoise-shell females. What colors are the parental males and females in such crosses?

6. A condition known as icthyosis hystrix gravior appeared in a boy in the early eighteenth century. His skin became very thick and formed loose spines that were sloughed off at intervals. When he grew up, this "porcupine man" married and had six sons, all of whom had this condition, and several daughters who were normal. For four generations, this condition was passed from father to son. What can you postulate about the location of this gene?

7. Assume the pedigree presented at the top of the next page to be straightforward, with no complications such as illegitimacy. Trait W, found in individuals represented by the shaded symbols, is rare in the population at large.

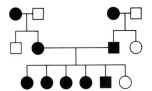

Tell which of the following patterns of transmission for W are consistent with or excluded by this pedigree:

    a. Autosomal recessive.

    b. Autosomal dominant.

    c. X-linked recessive.

    d. X-linked dominant.

    e. Y-linkage.[1]

8. Suppose that meiosis occurs in a haploid organism of chromosome number n. What is the probability that an individual haploid resulting from the meiotic division will have a complete parental set of centromeres (i.e., all from one parent or all from the other parent). Derive the formula.

9. A wild-type female schmoo who is graceful (G) is mated to a nonwild-type male who is gruesome (g). Their progeny are made up solely of graceful males and gruesome females. Interpret these results and give genotypes.[2]

10. What do you think are the advantages and disadvantages of having genes organized into chromosomes? Why don't genes float free in the nucleus or cell? (Try to remember to return to this question when you have finished the book and see if your answers have changed.)

11. *Answer this problem in sequence.* Suppose that you have two homyozygous strains of *Drosophila*, one found in Vancouver (strain A), the other in Los Angeles (strain B). Both strains have bright scarlet eyes, a phenotype quite distinct from the dull red eyes of the wild type.

    a. When you cross strain-A males with strain-B females, you obtain 200 wild-type males and 198 wild-type females in the F_1 generation. From this result, what can you say about the inheritance of eye color in the two strains?

    b. When you cross strain-B males with strain-A females, you obtain 197 scarlet-eyed males and 201 wild-type females in the F_1 generation. What do you learn about the inheritance of eye color from this result?

    c. When you intercross the F_1 generation of part a, you obtain in the F_2:

151 wild females	126 scarlet males
49 scarlet females	74 wild males

Diagram the genotypes of the parents and of the F_1 offspring. Indicate the expected ratios of F_2 genotypes and phenotypes.

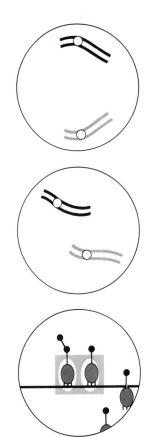

---

[1] Problem 7 is from *General Genetics*, 2d ed., by A. M. Srb, R. D. Owen, and R. S. Edgar. W. H. Freeman and Company. Copyright © 1965.

[2] From E. H. Simon and J. Grossfield, *The Challenge of Genetics*, 1971, Addison-Wesley, Reading, Mass.

12. In mice there is a mutant allele that causes a bent tail. From the cross results given in the accompanying table, deduce the mode of inheritance of this trait.
    a. Is it recessive or dominant?
    b. Is it autosomal or sex-linked?
    c. What are the genotypes of parents and progeny in all crosses?

	Parents		Progeny	
Cross	Female	Male	Female	Male
1	Normal	Bent	All bent	All normal
2	Bent	Normal	1/2 bent, 1/2 normal	1/2 bent, 1/2 normal
3	Bent	Normal	All bent	All bent
4	Normal	Normal	All normal	All normal
5	Bent	Bent	All bent	All bent
6	Bent	Bent	All bent	1/2 bent, 1/2 normal

13. Suppose you have the following human pedigree in which a dot represents the occurrence of an extra finger and a shaded area represents the occurrence of an eye disease.

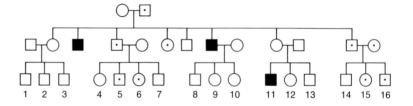

    a. What can you tell about the inheritance of an extra finger?
    b. What can you tell about the inheritance of the eye disease?
    c. What were the genotypes of the original parents?
    d. What is the probability that a child of child number 4 will have an extra digit? The eye disease?
    e. What is the probability that a child of child number 12 will have an extra digit? The eye disease?

3 / Extensions of Mendelian Analysis

(Or: Exceptions can prove the rules.)

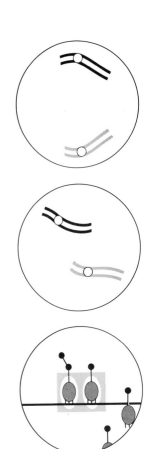

It has already been shown how Mendel's laws form a very accurate and useful base for predicting the outcome of simple crosses. However, it is only a base and there are many complications and apparent exceptions to the original rules. We have already considered one of these: sex linkage. Although not at odds with the Mendelian principles, sex linkage is an obvious special case, or extension of them. Of course, rather than creating a more complex and hopeless situation, these exceptions are often the source of new insights into the workings of any given system and the ones to be covered in this chapter are good illustrations of that principle.

Multiple Alleles

Early in the study of genetics it became clear that it is possible to have more than two allelic forms of a gene. Although only two can exist in any diploid cell, the number of different alleles that may be detected can often be quite large. The ABO human blood group alleles afford a modest example.

There are four blood "groups" or phenotypes in the ABO system. These are shown in the following chart:

Blood phenotype	Genotype
O	ii
A	$I^A I^A$ or $I^A i$
B	$I^B I^B$ or $I^B i$
AB	$I^A I^B$

It is obvious from the genotype column that there are three alleles of this gene, I^A, I^B, and i, but in any one person only two can be present. So, if Fred is $I^A i$ and Mary is $I^B i$, their children can have any one of the four blood groups.

MESSAGE

A gene can exist in several different states, an observation referred to as multiple allelism, and the alleles are said to form an allelic series.

Another example is the coat color of rabbits, which is affected by a series of multiple alleles symbolized as: C (full color), c^{ch} (chinchilla), c^h (Himalayan) and c (albino). Note that the superscript indicates an allele of the basic gene, in this case C. Once again only two are present in a normal diploid animal. You have probably been wondering about dominance in this situation; as written, alleles on the left are dominant over those on the right:

$$c^{ch}c^h \text{ is chinchilla}$$

$$Cc^h \text{ is full color}$$

$$c^h c \text{ is Himalayan}$$

and so forth.

In any cross of heterozygotes for multiple alleles, the progeny will always be recovered in a $3:1$ or a $1:2:1$ ratio. Try it and see.

Partial Dominance

Sometimes a heterozygote exhibits a phenotype that is intermediate between those of the pure-breeding parents. For example, there are two strains of four-o'clock plants, one (AA) has red flowers, whereas the other (aa) has white. In the cross red $\times$ white, all the F_1 plants have pink flowers. No examples of this partial or incomplete dominance were reported by Mendel. Of course, the

F₂ four-o'clocks will give a modified Mendelian ratio of 1 red : 2 pink : 1 white. It is probable that the observation of this type of incomplete dominance was responsible for the kind of logic behind the idea of blending inheritance.

Codominance

Codominance is really no dominance in that both alleles express their phenotypes in a heterozygote. We have already had an example of this, although it was sidestepped at the time: people who are blood type AB are heterozygotes exhibiting the phenotypes of both the I^A and the I^B alleles. There is another gene that controls a different blood-group classification, the MN system. Individuals whose genotype is $L^M L^M$ are blood type M, those who are $L^N L^N$ are type N, and, together, the alleles are codominant, $L^M L^N$ individuals having blood type MN. In other words, heterozygotes for codominant alleles are phenotypically similar to both parental types. The distinction between codominance and incomplete dominance may seem artificial to you at this point: isn't pinkness a joint manifestation of redness and whiteness? True—the real distinction lies in the way in which genes act. Briefly (for now), codominance can be observed if both alleles are active, and incomplete dominance can generally be observed in cases in which one allele (the recessive) is completely inactive. Molecular reasons for this will be given in chapter 11. For the time being, it is interesting to ponder the fact that

---

MESSAGE

*Complete dominance and recessiveness are not fundamental aspects
of Mendel's laws; those laws were more concerned with the
inheritance patterns of genes than with their nature and/or function.*

Lethal Genes

Lucien Cuénot studied a strain of mice having yellow fur in contrast to the grey wild-type mice. On mating yellow mice with grey mice from a pure line, he observed a 1 : 1 ratio of yellow to grey mice. This suggests that the allele for yellow coat color is dominant over the allele for grey. When he crossed yellow mice together, he recovered yellow and grey offspring in a 2 : 1 ratio, a drastic deviation from a Mendelian 3 : 1 ratio. In numerous crosses of the yellow offspring of a yellow × yellow cross, not one pair of mice produced only yellow progeny. In other words, yellow mice always appear to be heterozygous. This provides us with a clue to explain the 2 : 1 ratio. Because yellow homozygotes are never born, the 2 : 1 ratio could be part of a 1 : 2 : 1 ratio with one-quarter of the progeny somehow being lost if homozygous for the allele for yellow.

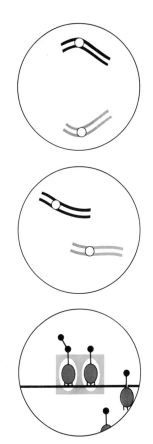

In fact, the $1:2:1$ ratio is converted into a $2:1$ ratio by the death of the homozygous yellow class in utero. This was verified when spontaneously aborted fetuses were observed in the uteri of females in the above cross. Notice that Y is *dominant* in its effects on the coat-color phenotype (the normal allele, y, must be present in yellow individuals), yet it is *recessive* in its lethality, since the allele must be homozygous before the fetus dies. We say that Y is a dominant visible and recessive lethal gene. Its effects are *pleiotropic;* that is, it has the ability to affect different aspects of the phenotype of the mouse, both its color and its survival. That does not mean that the effect on color and survival is not the result of the same basic cause.

Gene Interaction

Genes do not express their phenotypes in a cellular vacuum. They can be greatly affected by other genes present in the cell, and the expression of an allele of one gene pair is often dependent on the presence of a specific allele of another pair. For example, different pure-breeding strains of peas may be phenotypically identical in having white-petalled flowers. When the two strains are crossed, however, all the F_1 plants have red flowers; when these are crossed, the F_2 plants are in a ratio of 9 red:7 white. This is a striking deviation from a $9:3:3:1$ dihybrid expectation, but the ratio suggests a variation of that theme. The basis of this inheritance pattern is as follows (arbitrarily using a and b to indicate two different genes):

White strain 1   ×   White strain 2
AAbb   ↓   *aaBB*

F_1 *AaBb* (red)
     ↓

F_2 9   *A-B-*   Red = 9
      3   *A-bb*
      3   *aaB-*   White = 7
      1   *aabb*

Overall, a $9:7$ ratio is seen. Note that in complete dominance, AA and Aa are phenotypically indistinguishable. Therefore, we lump these two genotypes together by classifying them as A-, the hyphen indicating that the second allele could be either A or a. In other words, we can explain the $9:7$ ratio simply by assuming that both A and B have to be present to give a red pigment to the petal, an effect called *complementary gene interaction.* (What would you predict if you ground up petals from the two white strains and mixed them together in a test tube? Think about it until we get to a later chapter.) The intention here is not to figure out how genes could interact to give a modified Mendelian ratio (although modified ratios can often give important clues about the functions of genes), but simply to indicate that the ratios can in fact be made consistent with Mendelian expectations.

A different kind of dependence is exemplified by coat color in some strains of dogs. If a homozygous albino dog is crossed with a homozygous brown animal, all the F_1 progeny have black fur. Crosses of the F_1 progeny with each other produce F_2 in a ratio of 9 black:4 white:3 brown. Again, this appears to be a variation of a 9:3:3:1 dihybrid ratio and can be diagrammed as:

$BBcc$ (albino) $\times$ $bbCC$ (brown)
↓
F_1 $BbCc$ (black)
↓
F_2 9 $B\text{-}C\text{-}$ Black = 9
        3 $B\text{-}cc$ Albino
        3 $bbC\text{-}$ Brown = 3   4
        1 $bbcc$ Albino

Overall, a 9:4:3 F_2 ratio is seen. In this case, the alleles B and b give black and brown coats, respectively, but only in the presence of the dominant allele (C) of a separate, independently assorting locus. To distinguish this interaction from a relationship of dominance and recessiveness, we say that cc is *epistatic* to either B or b; that is, homozygosity for c prevents any pigment formation irrespective of other color alleles. The effect that cc has on B is called *recessive epistasis*. There is also dominant epistasis (What ratio would be produced?), as well as other epistases with their attendant complicated terminology, which was invented before the nature of genes was understood and so we do not wish to stress it at all. The main point is that by determining F_2 phenotypic ratios, the Mendelian inheritance of the interacting genes can still be derived by assigning arbitrary genotypes to each phenotypic class.

Because practice is the best way to understanding, let's take another example of interacting genes, in this case, affecting the production of a compound called malvidin in certain species of plant of the genus *Primula*. If two pure strains, one lacking malvidin and the other having it, are crossed, the F_1 plants are found to have no malvidin. Crossing the F_1 plants produces plants that lack malvidin and plants that have it in a ratio of 13:3. Again, this appears to be a variation of the dihybrid ratio and can be explained as:

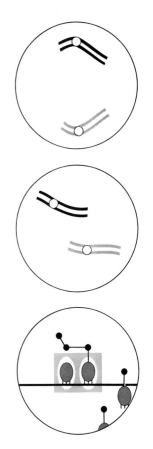

$KKdd$ (malvidin) $\times$ $kkDD$ (none)
↓
F_1 $KkDd$ None
↓
F_2 9 $K\text{-}D\text{-}$ None
        3 $K\text{-}dd$ Malvidin = 3   13
        3 $kkD\text{-}$ None
        1 $kkdd$ None

Overall, a 13:3 F_2 ratio is seen. The allele designated K is necessary for the production of malvidin, but its action can be suppressed by an allele D of another gene. *Suppressor* alleles (Which can also be recessive) may or may not have their own phenotypic effects other than that of acting as suppressor.

The last example of interacting genes introduces a concept that will be developed later in the chapter. It concerns the genes that control fruit shape in the plant shepherd's purse. Two different pure strains have fruits that differ in shape: one "round" and one "narrow." On crossing the two strains, all the F_1 plants have round fruits. But crossing F_1 plants produces F_2 plants of which 15/16 develop round fruit and 1/16 produces narrow fruit. A dihybrid basis for these results is shown in Figure 3-1. The presence of at least one dominant

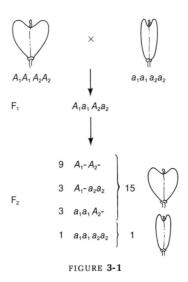

$A_1A_1\,A_2A_2$ $\times$ $a_1a_1\,a_2a_2$

F_1 $A_1a_1\,A_2a_2$

F_2

9   $A_1\text{-}A_2\text{-}$

3   $A_1\text{-}a_2a_2$ } 15

3   $a_1a_1\,A_2\text{-}$

1   $a_1a_1\,a_2a_2$ } 1

FIGURE **3-1**

allele, either A_1 of one gene or A_2 of the other independent gene will produce a round fruit. The genes A_1 and A_2, although different, appear to be identical in function and are therefore called *duplicates*.

MESSAGE

Genes can interact to give modified Mendelian ratios.

The Environment

Genes can interact not only with other genes but also with the environment to produce the final phenotype. *The phenotype, then, equals "genotype plus environment."* This is an important catchphrase and it leads up directly to the concepts of *penetrance* and *expressivity*. A genotype that does not always produce a specific phenotype is said to be *incompletely penetrant*. This lack of penetrance can be due to other genes in the genome (suppressors, for example), or to the environment. The twins in Figure 3-2 are identical genotypically, both having the genotype for harelip. (Is this an environmental or a genetic modification of the harelip genotype?)

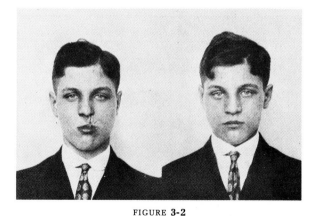

FIGURE **3-2**

Even if a gene is penetrant, the degree of expression, or expressivity, may vary. For example, the harelip may be mild or severe. Again genetic or environmental factors may be involved.

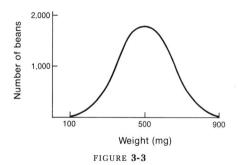

FIGURE **3-3**

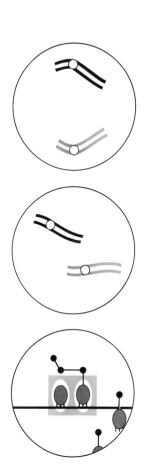

The words genotype and phenotype were coined by Wilhelm Johannsen, who first realized the importance of the environment. His analysis sprang from an initial experiment in which he weighed a large number of individual bean seeds and obtained the distribution shown in Figure 3-3. He was impressed by the very large number of bean weights possible and could not imagine how the genes postulated by Mendel could be responsible for what was essentially a continuous distribution of these weights. The weight of a bean seed could be *anywhere* from 100 to 900 mg, whereas Mendel had stressed *discontinuous* traits that fell into discrete classes (e.g., dwarf versus tall, or round versus wrinkled) determined by a few particulate hereditary factors. On the face of things it looked as though the continuous bell-shaped distribution, which is very often found for phenotypic characteristics in biological populations (e.g., height, skin color, diameter of skull, and I.Q.) and is called a *normal distribution,* could only be explained by some kind of blending (paint-mixing) theory of inheritance. Johannsen knew that beans are self-fertilizing; so any individual bean seed was probably homozygous for most of its gene pairs. How did he know this? The theory behind it had in fact been worked out by Mendel and it goes as follows:

HOW PURE-BREEDING PLANTS ARE GENERATED BY SELFING

Assume: all plants are heterozygous for one locus; that is, *Aa.*

Assume: each plant has four progeny.

Assume: only selfing is allowed in each generation.

Each *Aa* plant will have, on the average, 1 *AA*, 2 *Aa*, and 1 *aa* progeny, and of course half of these are homozygous (either *AA* or *aa*) and will breed true. The 2 *Aa* heterozygotes will have, on the average, 2 *AA*, 4 *Aa,* and 2 *aa* progeny.

$$
\begin{aligned}
&1\ AA \rightarrow 4\ AA \\
&2\ Aa \rightarrow 2\ AA + 4\ Aa + 2\ aa \\
&\underline{1\ aa \rightarrow \qquad\qquad\qquad 4\ aa} \\
&\qquad 6\ AA + 4\ Aa + 6\ aa = \text{a 3:2:3 ratio}
\end{aligned}
$$

Thus at every generation of selfing, the heterozygous class whittles away at itself and produces some homozygous offspring. The average expected distribution of each genotypic class in successive generations is given in Table 3-1. After only

TABLE **3-1**

Generation	AA	Aa	aa
0	0	2	0
1	1	2	1
2	3	2	3
3	7	2	7
4	15	2	15
5	31	2	31
n	$(2^n - 1)$	2	$(2^n - 1)$

five generations, for example, 62/64 of the plants are expected to be pure-breeding (either *AA* or *aa*).

This conclusion can be looked at another way. If, after five generations, 62/64 of the plants are pure for the *A/a* gene pair (i.e., either *AA* or *aa*), then any *one* plant will be pure-breeding at 62/64 of all gene pairs, which were heterozygous in the original, or zero, generation.

Johannsen selected nineteen original bean seeds of quite different weights and from them grew nineteen different lines that he could be reasonably sure were genotypically pure because of the above logic. He then selfed plants in each pure line and looked at the weights of the seeds produced by each. His results were something like those given in Figure 3-4 (the actual numbers have been simplified). Take lines 3, 11, 12, and 17 as illustrative examples. The diagram

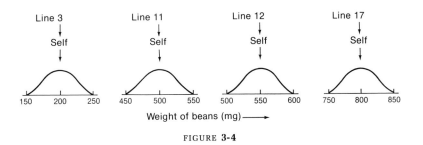

FIGURE 3-4

shows the distributions of seeds in the various weight classes and the mean (average) weight in each of the four lines.

He then selected individual bean seeds from the progeny in each line, grew them into plants, selfed them, and derived distribution curves for *their* progeny seeds. Surprisingly, the weight distributions were exactly the same as those produced by their parents. For example, a plant grown from a 150-mg bean seed from the line-3 progeny, on selfing, gave seeds with a mean weight of 200 mg and a distribution ranging from 150 mg to 250 mg. Similarly, a 250-mg bean from the line-3 distribution, on growing up and selfing, gave seeds for which the distribution ranged from 150 mg to 250 mg with a mean of 200 mg. In other words, *each line bred true for the mean weight.* He therefore attributed the mean weight to the genotype of the line, whereas *the variations within each line* he attributed to the environment.

You can see how such experiments led easily to the concept of genotype and phenotype. Notice that some of the progeny in lines 11 and 12 have the same phenotype (say, 540 mg) but different genotypes. In summary, then, Johannsen could visualize how the normal curve could be the result of the superimposition of hundreds of small genotype curves, rendered indistinct by environmental factors, as shown in Figure 3-5. (Why are most of the genotype curves near the center of the normal curve? Wait until the next section to see.)

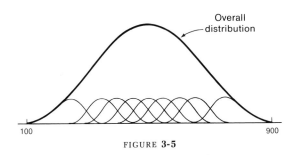

FIGURE 3-5

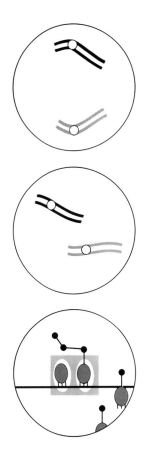

Polygenes

Although Johannsen's experiments pointed to the fact that a bean's place on the normal curve (i.e., its weight) depends on the interaction of the environment and its genotype, his analysis offers no clue to what the genotypes were that could produce nineteen different pure-breeding lines (and presumably more) in his sample alone. The work of Herman Nilsson-Ehle on wheat was to throw light on this question.

Nilsson-Ehle crossed pure-breeding red-seeded wheat with pure-breeding white-seeded wheat. The color of the seeds is dependent in this case on the genotype of the plant, not the seed, because the pigment is derived from maternal tissue. He found that all the F_1 plants had red seeds. He selfed these F_1 plants and (was he expecting a 3:1 ratio?) found that all seventy-eight F_2 plants also had red seeds (too bad!). He then set about selfing the F_2 plants individually and found that the seventy-eight F_2 plants could be sorted into four groups according to the ratios produced in the selfings.

Group one produced plants having red seeds and plants having white ones in a 3:1 ratio; so he reasoned that they were heterozygous for one gene pair (there were eight plants in this group). Group two produced a 15:1 ratio; so he reasoned that these were heterozygous for 2 gene pairs (rather like the shepherd's purse example) and there were fifteen of this kind. Group three produced a 63:1 ratio and therefore were heterozygous for 3 gene pairs [p (three homozygous recessive) $= (1/4)^3 = 1/64$] and there were five of these. The remaining fifty gave progeny having red seeds so he reasoned that they were homozygous for the dominant allele for redness of at least one gene pair. Hence he was basically dealing with a triplicate gene situation (analogous to the duplicate genes of the shepherd's purse) in which a dominant allele of any one of three separate genes (A, B, or C) could cause redness. The original cross must have been $AA\,BB\,CC$ (red) $\times$ $aa\,bb\,cc$ (white) and the F_1 $Aa\,Bb\,Cc$. The formal analysis is presented in Table 3-2, which shows the precise F_2 genotypes and their theoretical probabilities and observed frequencies.

As you can see, he was unlucky not to have picked up the $aa\,bb\,cc$, but the results fit the triplicate-gene model very well. In general, this kind of situation in which several different genes act on a given characteristic is referred to as *polygenic inheritance*.

In the seeds we have been calling "red," Nilsson-Ehle actually observed a lot of variation in the amount of redness, some being very dark red, others a light pink. There were many intermediate shades; in fact, so many as to make accurate groupings impossible. However, in a dihybrid selfing (say, $Aa\,Bb\,cc \times Aa\,Bb\,cc$), he *was* able to classify the progeny plants into five groups with the

TABLE 3-2

55

Extensions of
Mendelian Analysis

Ratio on selfing	Genotype of F$_2$ plants	p(genotype)		Observed ratio
3:1	Aa bb cc	$1/2 \times 1/4 \times 1/4 = 1/32$		
	or aa Bb cc	$1/4 \times 1/2 \times 1/4 = 1/32$		
	or aa bb Cc	$1/4 \times 1/4 \times 1/2 = 1/32$		
		$3/32 = 6/64$		8/78
15:1	Aa Bb cc	$1/2 \times 1/2 \times 1/4 = 1/16$		
	or Aa bb Cc	$1/2 \times 1/4 \times 1/2 = 1/16$		
	or aa Bb Cc	$1/4 \times 1/2 \times 1/2 = 1/16$		
		$3/16 = 12/64$		15/78
63:1	Aa Bb Cc	$1/2 \times 1/2 \times 1/2 = 1/8 = 8/64$		5/78
1:0	AA bb cc	$1 - 6/64 - 12/64$		
	or AA BB cc	$-8/64 - 1/64 = 37/64$		50/78
	or aa BB cc, etc.			
0:1	aa bb cc	$1/4 \times 1/4 \times 1/4 = 1/64$		0/78

following relative frequencies:

Dark red	1	
Medium dark red	4	
Medium red	6	15
Light red	4	
White	1	

He hypothesized that the *total number* of dominant alleles of both gene pairs was determining the *degree* of pigmentation. In other words, each dominant gene could be considered to be contributing one dose of redness, and the effect of each dose was *equal* and *additive*.

This idea fit the observed ratio well: forgetting about the third gene pair (it is irrelevant here), we can calculate the probability of getting 0, 1, 2, 3, and 4 doses of dominant alleles from the cross $Aa\,Bb \times Aa\,Bb$ as shown in Figure 3-6.

For A	For B	For A and B	Total number of dominant "doses"	Total probability
	BB = 1/4	AA BB = 1/16		
AA = 1/4	Bb = 1/2	AA Bb = 1/8	4	1/16
	bb = 1/4	AA bb = 1/16		
			3	4/16
	BB = 1/4	Aa BB = 1/8		
Aa = 1/2	Bb = 1/2	Aa Bb = 1/4	2	6/16
	bb = 1/4	Aa bb = 1/8		
			1	4/16
	BB = 1/4	aa BB = 1/16		
aa = 1/4	Bb = 1/2	aa Bb = 1/8	0	1/16
	bb = 1/4	aa bb = 1/16		

FIGURE 3-6

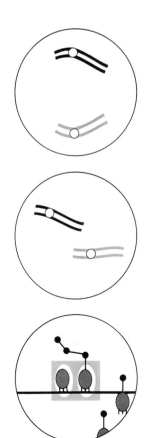

It is thus easy to see where the $1:4:6:4:1$ ratio comes from. This quantitative additivity of gene effect on phenotype was a revolutionary idea (it is sometimes called *quantitative inheritance*); most important, it provides a genotypic basis for the normal curve often seen in natural quantitatively varying characters like height, weight, and color. This can be seen by plotting the duplicate-gene situation as a histogram (Figure 3-7). The dotted line shows that the distribution of plants in each pigment class begins to look like a bell-shaped normal curve, even in plotting only two pairs of genes: the curve is even closer to a bell shape for triplicate genes (Figure 3-8). In fact, one can visualize Johannsen's environmental factors "knocking the edges off" the histogram to form a smooth bell.

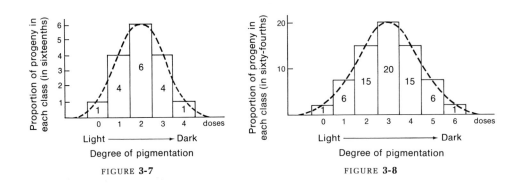

FIGURE **3-7**   FIGURE **3-8**

MESSAGE

Continuous distributions can be produced by environmental modifications of polygenic ratios if the effect of different genes is similar and additive.

(Note that, if enough polygenes participate, the bell becomes smooth even without the environmental effects.) Many polygenic situations of this kind are known: perhaps the most familiar is that for skin pigmentation in humans in which at least four separate gene pairs participate (quadruplicate genes). Assuming dominance of the pigment-producing genes and simple additivity of effect of each dominant, there are nine different pigment classes in which the number of dominant genes ranges from 0 to 8. So we have seen that basic Mendelian genetics can explain what is apparently a continuous variation of characters.

The Binomial Distribution

We derived the $1:4:6:4:1$ ratio for wheat seed color the hard way but declined to derive the $1:6:15:20:15:6:1$ ratio for triplicate genes. The fact is that there is an easy way to figure out such ratios, using *binomial* mathematics; it is a very useful set of mathematics for geneticists to know about because of its applicability to continuous distributions and because it will lead conveniently into a discussion of the χ^2 (chi-square) test, a statistical test widely used by geneticists.

Let's examine the triplicate-gene case. An *Aa Bb Cc* plant will produce $2^3 = 8$ gamete types equally frequently. By now you ought to be able to figure out (the long way) what these classes are (the first is *A B C*, . . . , the last is *a b c*). However, it is instructive to write them in a form that indicates whether a gamete carries the dominant (D) or recessive (r) alleles of each gene pair shown in Table 3-3. The frequency of gametes with 3D is 1/8, with 2D is 3/8, with 1D is also 3/8, and with 0D is 1/8.

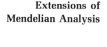

TABLE **3-3**

Frequency	Locus		
	A/a	*B/b*	*C/c*
1/8	D	D	D
1/8	D	D	r
1/8	D	r	D
1/8	r	D	D
1/8	D	r	r
1/8	r	D	r
1/8	r	r	D
1/8	r	r	r

Without knowing it, what we have done here is to expand (or multiply out) a binomial expression. A binomial expression is a mathematical description of the process of sampling if there are only *two* alternative choices to be made. The expansion tells us the frequency (probability) of each kind of sample of a specific size. In general terms, we have expanded $(a + b)^n$ where

$$a = p \text{ (dominant allele at one locus)} = 1/2$$
$$b = p \text{ (recessive allele at one locus)} = 1/2$$
$$n = \text{number of gene pairs} = 3$$

$$(a + b)^3 = 1a^3 + 3a^2b + 3ab^2 + 1b^3$$

The first term in this expansion gives us the probability of a gamete having $3D = 1 \times 1/2 \times 1/2 \times 1/2 = 1/8$. The second term gives us the probability of $2D = 3 \times 1/2 \times 1/2 \times 1/2 = 3/8$, and so on. This obviously will give us the 1/8:3/8:3/8:1/8 ratio that we derived previously. Because $(a + b)$ always equals 1, all binomial expressions regardless of the value of *n*, also add up to 1. (It should be emphasized here that the name of each allelic class—1D, 2D, or 3D—indicates the number of dominants from any combination of the three genes.)

There are two convenient ways of expanding binomial expressions that bypass

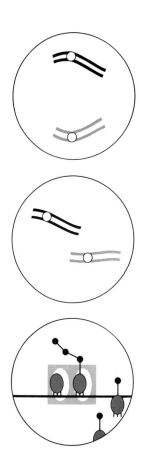

$\frac{n}{1}$					1		1				
2				1		2		1			
3			1		3		3		1		
4		1		4		6		4		1	
5	1		5		10		10		5		1
6	1	6		15		20		15		6	1

FIGURE 3-9

the laborious algebra involved when the number of genes becomes very large. The first of these is the Pascal Triangle (Figure 3-9), which gives us the number preceding each term in the expansion. (You can derive the rest yourself by observing that any number in the triangle is the sum of the two immediately above it on the right and the left.) If $n = 3$, we obtain the familiar $1:3:3:1$ ratio, which we could have read off directly, and we would have known that these numbers go in front of a^3, a^2b, ab^2 and b^3, respectively, because this is always the form of every binomial expression.

The second way is to use the binomial formula that states that in n binomial "trials," the probability of one event occurring s times and the other event occurring t times is:

$$\frac{n!}{s!t!} \times a^s b^t$$

where $s + t = n$, and

$$a = p(\text{one event})$$
$$b = p(\text{the other event})$$
$$(a + b) = 1$$

The expression $n!$ means $n \times (n - 1) \times (n - 2) \cdots \times 1$; for example, $3! = 3 \times 2 \times 1 = 6$. In our example, $a = 1/2$, $b = 1/2$, $n = 3$. Let's give s a value of 2 and t a value of 1, and use the formula to give us the term in the expansion corresponding to 2D:

$$\frac{3 \times 2 \times 1}{(2 \times 1) \times 1} \times \left(\frac{1}{2}\right)^2 \times \left(\frac{1}{2}\right)^1 = 3 \times \left(\frac{1}{2}\right)^3 = \frac{3}{8}$$

Now that we have expanded the triplicate-gene gametic situation in three different ways, we can return to the problem at hand, which is to use this expansion to allow us to calculate the distribution of allele doses among the *progeny* (not the gametes) in a triplicate-gene cross. It is done quite simply: the frequency distribution for either the male or the female gamete is given by $(a + b)^3$; therefore we match frequency of each type of female gamete with each type of male gamete (Figure 3-10).

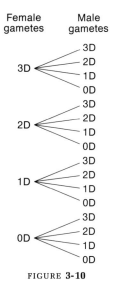

Female gametes	Male gametes

3D ⟨ 3D, 2D, 1D, 0D

2D ⟨ 3D, 2D, 1D, 0D

1D ⟨ 3D, 2D, 1D, 0D

0D ⟨ 3D, 2D, 1D, 0D

FIGURE 3-10

Because both 3D:2D:1D:0D gametic distributions are described by $(a + b)^3$, the zygotic distribution is $(a + b)^3 \times (a + b)^3 = (a + b)^6$.

By simply looking at the row for which $n = 6$ in Pascal's Triangle, we find the ratio 1:6:15:20:15:6:1, the numbers that we have been trying to justify. Alternatively, we could have used the formula to calculate each term in the expansion; in this case, however, the Triangle is simpler. Notice that $1 + 6 + 15 + 20 + 15 + 6 + 1 = 64$; so if, as in this case, $a = b = 1/2$, the terms in the expansion are all in sixty-fourths. Of course, a is not always equal to b, and then the calculations are harder.

Another example of the use of the binomial distribution is seen in the following calculation. What proportion of families of five children will have three boys and two girls (3B + 2G)? The order of birth doesn't matter: for example, BBBGG and BGBGB are equivalent in terms of the 3B:2G sex ratio. The binomial situation is where

$$p(B) = 1/2 = a$$
$$p(G) = 1/2 = b$$
$$n = 5$$

We want the value that goes in front of the term a^3b^2, and the row for which $n = 5$ in Pascal's Triangle tells us that this is 10; $10a^3b^2 = 10 \times (1/2)^5 = 10/32$. So 10/32 of such families will have the 3:2 ratio of boys to girls.

A RELEVANT PARADOX

A philosopher asks, "I have two children, and one of them is a boy: what is the probability that the other is a boy?" The answer is 1/3, not 1/2. Think about it.

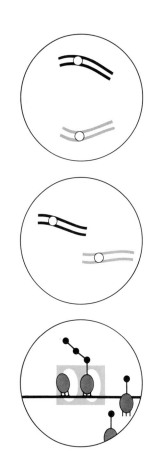

Coin tossing (with an unloaded coin) also gives us a binomial situation where

$$a = p(\text{heads}) = 1/2$$
$$b = p(\text{tails}) = 1/2$$
$$n = \text{number of tosses}$$

If $n = 100$, we would get a relatively smooth approximation of the normal curve, as shown in Figure 3-11, in which the hump of the curve occurs at 50 heads and 50 tails. This is called the *mean* of the curve and it is the *single most*

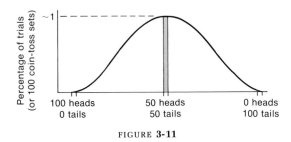

FIGURE **3-11**

frequent outcome in sets of 100 tosses. However, the area represented by $100!/50!50! \times (1/2)^{50}(1/2)^{50}$ is very small compared with that lying under the rest of the curve. In fact, it amounts to about 1%; in other words, most of the time we *do not* get our "expected" ratio, or mean.

MESSAGE

Expected ratios are rarely actually obtained. We get a clustering of values around the expected ratio, but seldom on the button.

Representing probabilities as areas seems reasonable in a histogram, but many people find this hard to grasp when the areas lie below a smooth curve. It is helpful to think of such curves as imperfect, the result of a series of slightly stepped adjacent columns.

Realizing this point about probability allows us to divide the curve into two areas that we will call *probable* (quite likely deviations from the mean) and *improbable* (unlikely deviations from the mean). But what *quantitative* measure can be given to "likely" and "unlikely"? The answer is that there is no good theoretical ground at all; so what is done is to use our intuition and say that our dividing line shall be $1/20 / 19/20$, or 5% / 95% as shown in Figure 3-12. In other words, an event that we would expect to occur less than once in twenty times is unlikely, and once or more in twenty times is likely. The area comprising 5%, or 1/20 of the total area, is divided between the two tails because unlikely outcomes occur at both ends of the range. Let's return to the coin-

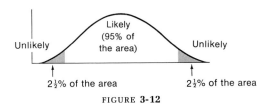

FIGURE **3-12**

tossing example and focus attention on a particular coin. We toss it 100 times and get 80 heads and 20 tails. This creates a dilemma: either (1) the coin is not loaded and we have simply come up with a chance deviation from the expected 1:1 ratio; or (2) the coin is loaded (perhaps lead has been imbedded in the side on which the "tail" appears).

These two possibilities could be considered hypotheses. The first is called a *null* hypothesis and is easier to test (as are most null hypotheses), because it provides us with a precise expectation (a 1:1 ratio, or 50 heads:50 tails), whereas the "loaded" hypothesis does not tell us *to what degree* it is loaded. Of course, disproving the null hypothesis does not mean that the second hypothesis is correct—other hypotheses are possible (the wind? sticky thumbs?). *Is* the null hypothesis believable? To test this, it is necessary to see if our result (80H:20T) falls in the likely area (an acceptable chance deviation from the expected 1:1 mean) or the unlikely (an unacceptable deviation—in other words, too much to believe). We could test this by using the binomial formula for calculating the sum of the areas,

$$\frac{100!}{80!20!}\left(\frac{1}{2}\right)^{80}\left(\frac{1}{2}\right)^{20} + \frac{100!}{81!19!}\left(\frac{1}{2}\right)^{81}\left(\frac{1}{2}\right)^{19} \cdots \frac{100!}{100!0!}\left(\frac{1}{2}\right)^{100}\left(\frac{1}{2}\right)^{0}$$

(a grisly piece of arithmetic), and then multiplying the whole thing by 2. What this would give us is the area of the binomial curve to the right of the observed deviation plus an equal deviation on the other side of the mean; that is, the total shaded area of the curve as shown in Figure 3-13. In other words, to be realistic, we have decided not to deal with the probability of getting this specific deviation, but to deal with the probability of getting a deviation this big or bigger. To avoid this horrendous arithmetic, we employ a neat device called the χ^2 test, and all it really is is a fast way to calculate those areas. If the area turns out to be less than 5%, we're in the "unlikely area"; if greater than 5%, we're in the "likely" area.

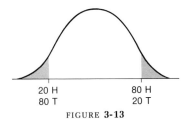

FIGURE **3-13**

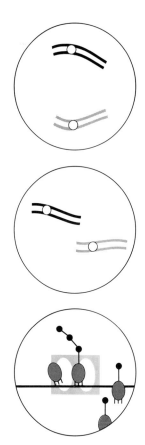

There is a magic formula for calculating χ^2: $\chi^2 = $ total of $(O - E)^2/E$ where $O = $ the number observed in a class and $E = $ the number expected in a class. In our coin-tossing example:

Class	O	E	$(O - E)^2$	$\dfrac{(O - E)^2}{E}$
H	80	50	900	18
T	20	50	900	18
	100	100		$36 = \chi^2$

Another magic item needed is the number of degrees of freedom, without which χ^2 cannot be used. The number of degrees of freedom (d.f.) can be defined simply in the present context as *the number of classes − 1*.

In our example, we have $2 - 1 = 1$ d.f. The next step is to look at the table of χ^2 values and see where a χ^2 of 36 comes out on the 1 d.f. line (Table 3-4 on page 64). As you can see, the corresponding probability value (area below the curve) of getting a deviation as big as this or bigger must be less than 0.005, or 0.5%, which is less than 5% of course. (Note that the χ^2 value or 36 does not appear in the 1 d.f. row in Table 3-4; however, if p values of much less than 0.005 had been included, the χ^2 of 36, or even higher values, would appear in that row.) So, we have calculated the shaded areas in Figure 3-13 and find that they total less than 0.5% and therefore know we were very much into the unlikely reaches of the binomial/normal distribution. We therefore reject the null hypothesis. Maybe the coin is loaded—who knows? We certainly cannot tell.

What is the relevance of all of this to genetics? The penny example could easily have been a testcrossed monohybrid F_1. As another genetic example, let's assume that in an F_2 we obtain seventy of one phenotype and thirty of another. Is this a chance deviation from a $3:1$ ratio? We can test this hypothesis (it, too, is null: we test the notion that there is no deviation):

Class	O	E	$(O - E)^2$	$\dfrac{(O - E)^2}{E}$
Phenotype 1	70	75	25	0.33
Phenotype 2	30	25	25	1.00
	100	100		$1.33 = \chi^2$,
				with $(2 - 1) = 1$ d.f.

Consulting the χ^2 table, we see at once that this is in the likely area of the curve; so the hypothesis that we are observing a $3:1$ ratio is accepted. But what curve

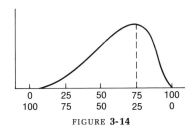

| 0 | 25 | 50 | 75 | 100 |
| 100 | 75 | 50 | 25 | 0 |

FIGURE **3-14**

are we looking at here? It is obviously not a normal curve because the mean will be about 75 : 25 (Figure 3-14). However, as long as this curve is not too skewed, the χ^2 still works well.

Note that the above example is still concerned with a binomial situation where $a = 3/4$ and $b = 1/4$. What about testing deviations from a 9 : 3 : 3 : 1 ratio or other situations for which a simple binomial situation is not applicable? These are basically multinomials and, to cut a long story short, let's simply say that, although it is hard to graph a multinomial, the χ^2 can still be used to test observed ratios against expected results, with higher degrees of freedom.

As a penultimate cautionary note, remember always to use real numbers, such as the number of yellows, wrinkled, heads, tails, goals by a hockey team, and so forth, when applying the χ^2 test. Never use percentages in a χ^2 test. The reason is that the χ^2 value depends on the *sample size* (try testing 6 : 4 against a 1 : 1 ratio and then try 60 : 40 against a 1 : 1 ratio and convince yourself of the truth of this—very different χ^2 values result but the percentages are the same in these data.) Finally, remember that 1/20 of the time you will reject a valid hypothesis on the basis of the χ^2 test! Good luck.

Heritability

One of the things we have seen in this chapter is how Mendelian particulate genetics can be applied to traits showing continuous variation. This is an appropriate point at which to discuss another concept that is useful in the analysis of complex continuous traits: heritability.

Although heritability is quite an easy concept to define, it can be very difficult to handle on a practical level. It is, quite simply, that proportion of phenotypic variation in a population that is due to genetic factors. The normal curve describes the total phenotypic variation: we have seen that Johannsen's experiments tell us that part of this is attributable to polygenic genotypic variation among individuals, and part to variation in the environments to which different individuals are exposed. It is often necessary to estimate the size of the genetic component: for example, in the breeding of heavier, or bigger, or more proteinaceous, or faster-growing varieties of commercial crops and livestock, it is imperative to establish whether the characteristics you intend to select for are in fact going to be sensitive to selection. If all the variation in a population of Douglas fir trees, for example, were environmental, it would be of little use to

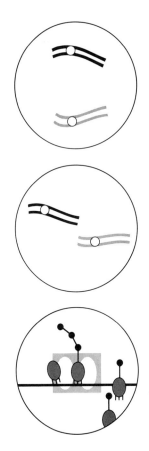

TABLE **3-4**

Critical values of the chi-square distribution

d.f.	p 0.995	0.975	0.9	0.5	0.1	0.05	0.025	0.01	0.005	d.f.
1	.000	.000	0.016	0.455	2.706	3.841	5.024	6.635	7.879	1
2	0.010	0.051	0.211	1.386	4.605	5.991	7.378	9.210	10.597	2
3	0.072	0.216	0.584	2.366	6.251	7.815	9.348	11.345	12.838	3
4	0.207	0.484	1.064	3.357	7.779	9.488	11.143	13.277	14.860	4
5	0.412	0.831	1.610	4.351	9.236	11.070	12.832	15.086	16.750	5
6	0.676	1.237	2.204	5.348	10.645	12.592	14.449	16.812	18.548	6
7	0.989	1.690	2.833	6.346	12.017	14.067	16.013	18.475	20.278	7
8	1.344	2.180	3.490	7.344	13.362	15.507	17.535	20.090	21.955	8
9	1.735	2.700	4.168	8.343	14.684	16.919	19.023	21.666	23.589	9
10	2.156	3.247	4.865	9.342	15.987	18.307	20.483	23.209	25.188	10
11	2.603	3.816	5.578	10.341	17.275	19.675	21.920	24.725	26.757	11
12	3.074	4.404	6.304	11.340	18.549	21.026	23.337	26.217	28.300	12
13	3.565	5.009	7.042	12.340	19.812	22.362	24.736	27.688	29.819	13
14	4.075	5.629	7.790	13.339	21.064	23.685	26.119	29.141	31.319	14
15	4.601	6.262	8.547	14.339	22.307	24.996	27.488	30.578	32.801	15
16	5.142	6.908	9.312	15.338	23.542	26.296	28.845	32.000	34.267	16
17	5.697	7.564	10.085	16.338	24.769	27.587	30.191	33.409	35.718	17
18	6.265	8.231	10.865	17.338	25.989	28.869	31.526	34.805	37.156	18
19	6.844	8.907	11.651	18.338	27.204	30.144	32.852	36.191	38.582	19
20	7.434	9.591	12.443	19.337	28.412	31.410	34.170	37.566	39.997	20
21	8.034	10.283	13.240	20.337	29.615	32.670	35.479	38.932	41.401	21
22	8.643	10.982	14.042	21.337	30.813	33.924	36.781	40.289	42.796	22
23	9.260	11.688	14.848	22.337	32.007	35.172	38.076	41.638	44.181	23
24	9.886	12.401	15.659	23.337	33.196	36.415	39.364	42.980	45.558	24
25	10.520	13.120	16.473	24.337	34.382	37.652	40.646	44.314	46.928	25
26	11.160	13.844	17.292	25.336	35.563	38.885	41.923	45.642	48.290	26
27	11.808	14.573	18.114	26.336	36.741	40.113	43.194	46.963	49.645	27
28	12.461	15.308	18.939	27.336	37.916	41.337	44.461	48.278	50.993	28
29	13.121	16.047	19.768	28.336	39.088	42.557	45.722	49.588	52.336	29
30	13.787	16.791	20.599	29.336	40.256	43.773	46.979	50.892	53.672	30
31	14.458	17.539	21.434	30.336	41.422	44.985	48.232	52.192	55.003	31
32	15.135	18.291	22.271	31.336	42.585	46.194	49.481	53.486	56.329	32
33	15.816	19.047	23.110	32.336	43.745	47.400	50.725	54.776	57.649	33
34	16.502	19.806	23.952	33.336	44.903	48.602	51.966	56.061	58.964	34
35	17.192	20.570	24.797	34.336	46.059	49.802	53.203	57.342	60.275	35
36	17.887	21.336	25.643	35.336	47.212	50.998	54.437	58.619	61.582	36
37	18.586	22.106	26.492	36.335	48.363	52.192	55.668	59.893	62.884	37
38	19.289	22.879	27.343	37.335	49.513	53.384	56.896	61.162	64.182	38
39	19.996	23.654	28.196	38.335	50.660	54.572	58.120	62.428	65.476	39
40	20.707	24.433	29.051	39.335	51.805	55.758	59.342	63.691	66.766	40
41	21.421	25.215	29.907	40.335	52.949	56.942	60.561	64.950	68.053	41
42	22.139	25.999	30.765	41.335	54.090	58.124	61.777	66.206	69.336	42
43	22.860	26.786	31.625	42.335	55.230	59.304	62.990	67.460	70.616	43
44	23.584	27.575	32.487	43.335	56.369	60.481	64.202	68.710	71.893	44
45	24.311	28.366	33.350	44.335	57.505	61.656	65.410	69.957	73.166	45
46	25.042	29.160	34.215	45.335	58.641	62.830	66.617	71.202	74.437	46
47	25.775	29.956	35.081	46.335	59.774	64.001	67.821	72.443	75.704	47
48	26.511	30.755	35.949	47.335	60.907	65.171	69.023	73.683	76.969	48
49	27.250	31.555	36.818	48.335	62.038	66.339	70.222	74.920	78.231	49
50	27.991	32.357	37.689	49.335	63.167	67.505	71.420	76.154	79.490	50

Source: Values from 1 to 30 degrees of freedom from C. M. Thompson, *Biometrika* 32(1941):188–189, with permission of the publisher. Values from 31 to 50 degrees of freedom from *Statistical Tables* by F. James Rohlf and Robert R. Sokal, W. H. Freeman and Company, Copyright © 1969.

select seed from the tallest trees in the hope of producing bigger trees that will yield more lumber in the next generation. How is heritability measured? A rigorous treatment of this subject requires statistical techniques that we won't consider in this book, but it is instructive to consider some simple approaches to answering the question, since these adequately illustrate the type of problems encountered.

In a controlled experimental situation, an estimate of heritability can be obtained by selecting individuals from a given position in the normal distribution, and then examining the distribution produced in the progeny population (basically the approach used by Johannsen). If heritability is 100% (or 1), then all variation is genetic and the mean for the progeny will be equal to the position on the curve from which the parents were selected. If all variation is environmental, the heritability is 0% (or 0) and the mean for the progeny will equal the mean of the original curve. Thus we can measure heritability as the "gain" divided by the "selection differential" as shown in Figure 3-15.

It is often necessary to try to estimate heritability in human populations, but because humans are not experimental organisms a controlled situation is never possible, and judicious guesses have to be made that are often the subject of controversy. The most intriguing characters are usually the most complex, and the most difficult to handle, especially such behavioral characteristics as schizophrenia and intelligence.

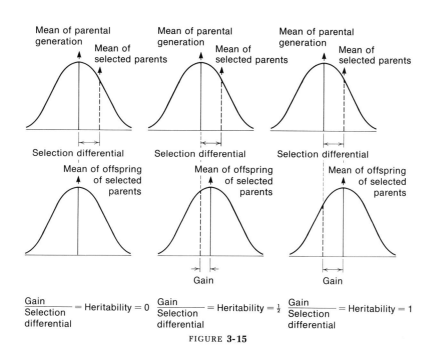

$$\frac{\text{Gain}}{\text{Selection differential}} = \text{Heritability} = 0 \qquad \frac{\text{Gain}}{\text{Selection differential}} = \text{Heritability} = \tfrac{1}{2} \qquad \frac{\text{Gain}}{\text{Selection differential}} = \text{Heritability} = 1$$

FIGURE 3-15

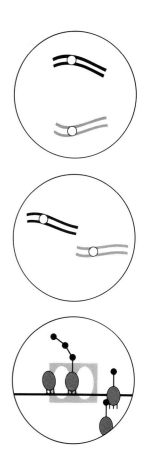

What is the heritability of intelligence? The simple answer is that it is not known, in large part because there isn't even a widely accepted definition of intelligence. Why is it not possible to do an analysis, such as the preceding one, in a human population? A study that looks like one has in fact been performed. Forty thousand English families were grouped according to the occupational class of the parents. The I.Q. (intelligence quotient, which is a predictive test of certain types of performance whose relationship with intelligence is not known) scores of the parents and their children were compared, as shown in Figure 3-16.

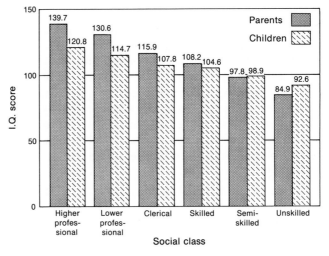

FIGURE 3-16

You will appreciate that this chart can be looked at as simply a way of sampling from various points on the I.Q. distribution curve. It can be seen that there is some regression to the mean (100) in the offspring of every class; thus heritability appears to be neither 100% nor 0%, but the amount of gain is not accurate for the following reason. Whereas in a controlled experiment the progeny are thrown back into the entire gamut of environmental exposures, in the class experiment this is obviously not true because there is an inevitable tendency for children to be reared in the same environment as their parents. The children of parents whose I.Q. scores are high, for example, tend to be exposed to a more intellectual environment than those of parents whose scores are low. This would act as a "brake" on the regression to the mean, causing an exaggeratedly high heritability value.

Because heritability values cannot be adequately determined *within* populations, it would seem to be sheer folly to make comparisons *between* populations. Yet this has been attempted in comparing black and white North American populations. The two populations show different I.Q. distributions, with the distribution mean of the blacks being from 10 to 20 I.Q. points below that of the whites (Figure 3-17).

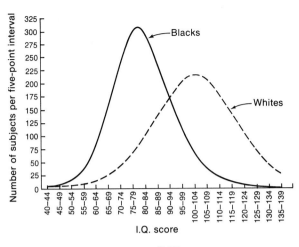

FIGURE **3-17**

Note that, even if high heritability of I.Q. performance could be determined in both black and white populations, this would in no way imply a genetic basis for the difference in the means of the two populations. Because the environmental living conditions and educational opportunities are, on the average, inferior in black populations, the differences in the distributions are more readily explained by these factors than by some vague notion of inferior black genes. This is particularly true since there are good grounds for believing that the interaction of genotype and environment on I.Q. is *not* additive, but *multiplicative.* This, of course, increases phenotypic variation and confounds attempts to separate its genetic and environmental components.

Another difficulty in this issue lies in the unique nature of the intelligence quotient. Whereas it is easy for anybody to accurately measure the weight of a bean or the height of an individual human, to within several decimal places, no two people can agree on the right way to measure intelligence. In fact, few people can even agree on what intelligence is. Thus I.Q. is truly an elastic yardstick measuring an elastic object, a fact that adds tremendously to the variability of I.Q. scores in populations, and further obscures attempts to portion the variability between genetic and environmental components.

MESSAGE

Because we are uncertain of how much I.Q. variation within populations is genetic, we cannot make valid statements about genetic variation between populations.

It should be stressed that heritability is a *population* concept: what it *does* do is measure genetic variation within a population; what it does *not* do is measure the contributions of genotype and of environment to the phenotype of the

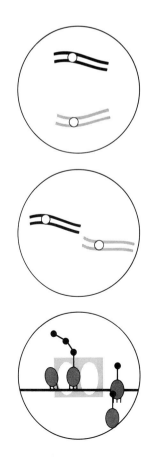

LOW HERITABILITY, HIGH GENETIC COMPONENT

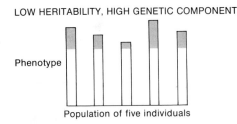

Phenotype

Population of five individuals

HIGH HERITABILITY, LOW GENETIC COMPONENT

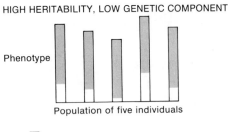

Phenotype

Population of five individuals

☐ Environmental component
☐ Genetic component

FIGURE **3-18**

individual. (Perhaps the choice of the word heritability was unfortunate.) This distinction can be seen if two small hypothetical populations are compared, one that has low heritability, yet a high genetic determination in individuals, and a second that has high heritability, yet a low genetic determination (Figure 3-18).

Is it possible to measure the amount of genetic determination in a complex polygenic trait such as intelligence? The question can be approached through the study of the I.Q. scores of pairs of one-egg twins. Such twins are produced from a single zygote that divides into two cells, each cell itself acting as a zygote. Of course these cells are genetically identical and the resulting twins are often called identical twins. As we all know, identical twins tend to be reared in very similar environments; they tend to be treated the same, dressed the same, fed the same, and so on. There are cases, however, in which identical twins have been raised apart, and these cases provide an opportunity for evaluating the degree of genetic determination of various traits, including intelligence. Because a pair of twins is genetically identical, any difference in their intelligence *must* be due to environmental differences. We have only to measure *phenotypic* similarity and compare it with their 100% *genotypic* similarity to obtain a good idea of the genetic determination of intelligence. Phenotypic similarity in a continuously varying trait is measured by the *correlation coefficient.* Briefly, the correlation coefficient is equal to 1 if the compared twins are always equal or if one value is always perfectly proportional to the other, and it is 0 if the compared values are completely independent. Figure 3-19 shows correlation coefficients in intelligence derived from many studies of different genetic relationships, including one-egg twins raised apart. Each dot is derived from one study, and the vertical lines indicate the mean values. We see that the mean for one-egg twins raised apart is from about 0.7 to 0.8. This suggests a high degree of genetic determination of this trait, which is also shown by the other relationships in the chart. If genetic determination were small, any correlation might be expected to be obliterated by environmental variation. One would have to conclude that the environment does affect intelligence, but not as strongly as heredity does. The method of comparing phenotypic similarity in related pairs of known genetic similarity is a general one that can be used to estimate genetic determination in other complex traits.

Correlation studies of twins sometimes involve complex characteristics that

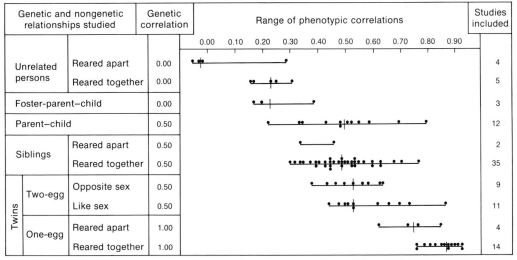

Genetic and nongenetic relationships studied		Genetic correlation	Range of phenotypic correlations	Studies included
Unrelated persons	Reared apart	0.00		4
	Reared together	0.00		5
Foster-parent–child		0.00		3
Parent–child		0.50		12
Siblings	Reared apart	0.50		2
	Reared together	0.50		35
Twins — Two-egg	Opposite sex	0.50		9
	Like sex	0.50		11
Twins — One-egg	Reared apart	1.00		4
	Reared together	1.00		14

Horizontal lines show range of correlation coefficients.
Vertical lines show averages.

FIGURE 3-19

are "all-or-none," in which case the twins can be classified as *concordant* (similar) or *discordant* (dissimilar). The percentage of concordance is also, of course, indicative of the degree of genetic determination of a phenotype. Some values from one-egg and two-egg twins are shown in Table 3-5.

TABLE 3-5

Pathology	Percentage of concordance	
	Monozygotic twins	Dizygotic twins
Tuberculosis	54	16
Cancer at same site	7	3
Clubfoot	32	3
Measles	95	87
Scarlet fever	64	47
Rickets	88	22
Arterial hypertension	25	7
Manic-depressive syndrome	67	5
Death from infection	8	9
Rheumatoid arthritis	34	7
Schizophrenia (1930s)	68	11
Criminality (1930s)	72	34
Feeble-mindedness (1930s)	94	50

Source: *Heredity, Evolution, and Society*, 2d ed., by I. M. Lerner and W. J. Libby. W. H. Freeman and Company. Copyright © 1976.

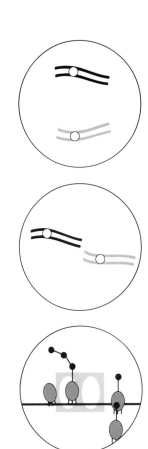

MESSAGE

*Studies of similarity between relatives can reveal genetic
determination of a complex, presumably polygenic, phenotypic trait.*

Problems

1. Suppose that you have three homozygous strains of *Drosophila* having black, yellow, and brown body color, respectively. You suspect that the genes controlling body color in these strains are allelic. What crosses would you perform to test this and, assuming that the genes are allelic, what results would you expect?

2. In the pedigree shown here the shaded symbols represent an inherited disease. The gene for the disease *must* be sex-linked. True or false?

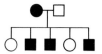

3. Normal *Drosophila melanogaster* have deep red eyes. Suppose that you establish two homozygous strains, one having bright scarlet eyes, the other dark brown eyes. When "scarlet" flies are crossed with "brown" ones, all the F_1 offspring have deep red eyes. An $F_1 \times F_1$ cross produces the following offspring:

432 red eyes

158 scarlet eyes

139 brown eyes

52 white eyes

Explain these results using symbols.

4. Consider the following pedigree for a dominant, autosomal gene.

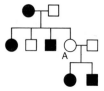

How is this possible and what can you deduce about A's genotype?

5. To solve this problem you need to know that the M, N, and MN blood groups are determined by two alleles L^M and L^N, and that Rh$^+$ (Rhesus positive blood group) is caused by a dominant allele R. In a court case concerning a paternity dispute,

each of two men claimed three children to be his own. The blood groups of the men, the children, and their mother were:

Husband	O	M	Rh$^+$
The wife's lover	AB	MN	Rh$^-$
Wife	A	N	Rh$^+$
Child 1	O	MN	Rh$^+$
Child 2	A	N	Rh$^+$
Child 3	A	MN	Rh$^-$

From this evidence, can the paternity of the children be established?

6. When purebred brown dogs are mated with purebred white dogs, all the F_1 pups are white. When $F_1 \times F_1$ crosses are made, the results are 118 white, 32 black, and 10 brown F_2 pups. What is the genetic basis for these results?

7. Suppose that you have been given a single virgin *Drosophila* female and you notice that the bristles on her thorax are much shorter than normal. You mate her with a normal male, which has long bristles, and obtain in the F_1: 1/3 short-bristle females, 1/3 long-bristle females, and 1/3 long-bristle males. A cross of the F_1 long-bristle females with their brothers gives only long-bristle F_2. A cross of short-bristle females with their brothers gives 1/3 short-bristle females, 1/3 long-bristle females, and 1/3 long-bristle males. Explain these data.

8. *Answer this problem in sequence.* In *Drosophila melanogaster,* wild-type eyes are deep red in color. Suppose that you have been given two strains of *Drosophila,* strains A and B, in which the eyes are white.

   a. When strain-A males are crossed with strain-B females, you obtain: 435 red-eyed males and 428 red-eyed females. What can you conclude about strains A and B?

   b. When strain-B males are crossed with strain-A females, you obtain: 420 white-eyed males and 405 red-eyed females. What can you conclude about strains A and B?

   c. When the F_1 flies of part a are crossed, you obtain the following F_2 (bright scarlet and brownish are new eye colors):

Male	Female
86 white	88 wild
44 wild	34 bright scarlet
14 bright scarlet	29 brownish
16 brownish	11 white

When the F_1 flies of part b are crossed, you obtain the following F_2:

Male	Female
180 white	175 white
88 wild	90 wild
28 bright scarlet	29 bright scarlet
24 brownish	26 brownish

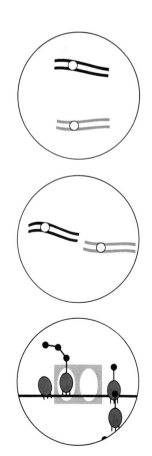

Explain these results. Diagram the crosses in parts a, b, and c. Using χ^2 analysis, determine whether the F_2 males of part b fit the predictions based on your explanation.

9. In *Drosophila* a dominant allele H reduces the number of body bristles, giving rise to a "hairless" condition. In the homozygous condition H is lethal. At an independently assorting locus, the dominant allele S has no effect on bristle number except in the presence of H, in which case a single dose of S will suppress the hairless phenotype, thus restoring the hairy condition. However, S is also lethal in the homozygous (SS) condition.

   a. What ratio of hairy to hairless individuals would you find in the live progeny of a cross between two hairy flies both carrying H in the suppressed condition?

   b. If the hairless progeny are backcrossed with a parental fly, what phenotypic ratio is expected in the live progeny?

10. Certain varieties of flax have been found to show different resistances to specific races of fungus called flax rust. For example, flax variety "770B" is resistant to rust race 24 but susceptible to rust race 22, whereas flax variety "Bombay" is resistant to rust race 22 and susceptible to rust race 24. When 770B and Bombay were crossed, the F_1 hybrid was found to be resistant to both rust races. When selfed, it produced an F_2 containing the following phenotypic proportions:

		Rust race 22		
		Resistant	Susceptible	
Rust race 24	Resistant	184	63	Observed F_2 numbers
	Susceptible	58	15	

   a. Propose a hypothesis to account for the genetic basis of resistance in flax for these particular rust races. In doing so,
      i. Make a concise statement of the hypothesis.
      ii. Define any gene symbols you use.
      iii. Show your proposed genotypes of the 770B, Bombay, F_1 and F_2 flax plants.
   b. Test your hypothesis using the χ^2 test.
      i. Give expected values.
      ii. Give χ^2 value to 2 decimal places.
      iii. Give appropriate probability value.
      iv. Exactly what is iii the probability of?
      v. Do you accept or reject your hypothesis on the basis of the χ^2 test?

11. Suppose that a heterozygous kandiyohi frog (Kk) is crossed with a wild one (kk). What are the probabilities of the following combinations in a progeny population of eight?

   a. All kandiyohi.
   b. The sequence kandiyohi, kandiyohi, wild, wild, kandiyohi, kandiyohi, wild, wild.
   c. The sequence wild, kandiyohi, kandiyohi, kandiyohi, wild, wild, kandiyohi, wild.
   c. Four kandiyohi and four wild.
   e. One or more kandiyohi.
   f. Two or more kandiyohi.
   g. No fewer than two kandiyohi *or* two wild.

12. From repeated matings between Erma and Harvey, the barnyard pigs, who are really rather cute, each having a tail length of 25 cm, the following little piglets resulted that, although cute, were found to have tails that differed from those of their parents as follows:

Length of tail	15 cm	20 cm	25 cm	30 cm	35 cm
Number of piglets	9	37	57	34	12

Note, however, that most piglets had intermediate tail lengths like Harvey's and Erma's.
   a. How many pairs of genes would regulate this character?
   b. What kinds of genetic mechanism would account for this result?
   c. Give the expected ratio of each type of tail length and the proportion of genes in each genotype that would give you the above results.
   d. What offspring phenotypes would be expected from a mating between 15-cm and 30-cm progeny?

13. Ralph Emerson and Edward East crossed two strains of corn, one having long ears, and the other short ears. They obtained F_1 and F_2 populations that looked like this: How would you explain these data?

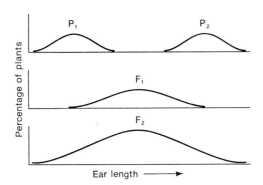

14. Identical twins are the products of the first mitosis of the zygote and are therefore genotypically identical. If they are similar in phenotype they are concordant, if not they are discordant. Identical twins show 91% concordance for cigarette smoking, whereas two zygote twins only 65% concordance. Does this prove cigarette smoking has a high genetic determination?

15. In *Neurospora* (a haploid), a mutant gene *td* results in an inability of the fungus to make its own tryptophan, and consequently tryptophan must be supplied in the medium for growth to occur; *su* is an allele of a separate gene, unlinked to *td*, whose only known effect is to suppress the td phenotype. Consequently, strains carrying *td* and *su* do not require tryptophan. If a *td su* strain is crossed with a genotypically wild-type strain,
   a. What genotypes are expected in the progeny and in what proportions?
   b. What will be the ratio of tryptophan-dependent to tryptophan-independent progeny?

16. Many polygenic systems do not show continuous phenotype distribution, but show sudden changes in phenotype at specific "threshold" gene doses. For example, in

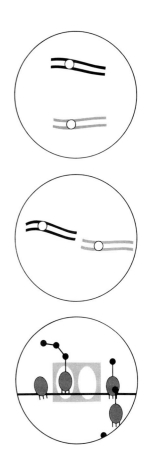

one species of *Penstemon,* the number of flowers per plant is controlled by four independently assorting gene loci as follows: when two or less doses of any dominant alleles are present, the plant has one flower; when three to five doses are present, the plant has two flowers; and when six or more doses are present, the plant has three flowers. If a plant that is heterozygous at all four loci (call it $A_1a_1 A_2a_2 A_3a_3 A_4a_4$) is selfed,

a. What proportion of the progeny plants will have one flower?
b. What proportion of the progeny plants will have two flowers?
c. What proportion of the progeny plants will have three flowers?

17. On a fox ranch in Wisconsin, a mutation arose that gave a "platinum" coat color. The platinum color turned out to be greatly desired by the fox-coat-buying public, but the breeders could not develop a pure-breeding platinum strain. Every time two platinums were crossed, some normal foxes appeared in the progeny. For example, in repeated matings of the same pair of platinums (which was typical of all other such matings), a total of 82 platinums and 38 normals was produced.

a. State a *concise* genetic hypothesis to account for these findings.
b. Test your hypothesis with the χ^2 test, using the numerical data from the cross described. (Be sure to state the predicted ratio from your hypothesis, the χ^2 value, the number of degrees of freedom, the probability value, what the probability value means, and your conclusion.)

18. In corn all three dominant factors, *A*, *C*, and *R*, are necessary for colored seeds. Genotypes *A- C- R-* are colored; all others are colorless. A colored plant is crossed with three tester plants with the following results:

With *aa cc RR*, it produced 50% colored seeds.

With *aa CC rr*, it produced 25% colored seeds.

With *AA cc rr*, it produced 50% colored seeds.

What is the plant's genotype?

19. A panther fancier has developed two strains of panther: one is pink and spotted (spotted pink), the other is blue and without spots (solid blue). Reciprocal crosses between these strains always give the same result:

1/2 solid pink

1/2 solid blue

The solid pink panthers produced in this way were then intercrossed, and a large number of progeny were reared. They were of four kinds, in the following proportions:

1/2 solid pink

1/6 spotted pink

1/4 solid blue

1/12 spotted blue

Explain this peculiar phenotypic ratio and assign genotypes, using symbols of your own choosing.[1]

20. A fungus disease of beans called bean anthracnose shows the following interactions between two different bean varieties, A and B, and two different strains of parasitic fungus, α and β.

	A	B
α	+	−
β	−	+

where a plus indicates disease and a minus indicates no disease. The cross A $\times$ B is made and an F_1 and F_2 obtained. Both generations are treated with a mixed inoculum containing both α *and* β. The F_1 shows no disease reaction and the F_2 shows a 9:7 ratio of no disease to disease.

a. Interpret these results genetically.
b. If the two parasite strains were crossed, how would you guess their progeny would react with A? with B? Make a testable prediction. (Assume for simplicity that the fungus is haploid.)

---

[1]Problem 19 courtesy of Clayton Person.

4 / Linkage

(Or: Genes are hooked together like beads on a necklace. Aren't they?)

We have already established the basic principles of segregation and assortment and correlated them with chromosomal behavior during meiosis. Thus, in a dihybrid F_2 from the cross $Aa\,Bb \times Aa\,Bb$, we expect a $9:3:3:1$ ratio of genotypes. As we saw in Bridges' study of nondisjunction, exceptions to simple Mendelian expectations can direct the experimenter's attention to new discoveries, and an exception involving a dihybrid cross provided the clue to the processes described in this chapter.

MESSAGE

Exceptions to predicted behavior may often be the source of important new insights.

William Bateson and R. C. Punnett were studying inheritance in the sweet pea. They had two different pairs of genes, one affecting flower color (P, purple, and p, red), the other the shape of pollen grains (L, long, and l, round). They made a dihybrid cross, $PP\,LL$ (purple, long) $\times pp\,ll$ (red, round), and selfed the F_1 $Pp\,Ll$ heterozygotes to get an F_2. The proportions of each class of F_2 plants recovered was:

$$284 \text{ purple, long} \quad (P\text{-}L\text{-})$$
$$21 \text{ purple, round} \quad (P\text{-}ll)$$
$$21 \text{ red, long} \quad (ppL\text{-})$$
$$55 \text{ red, round} \quad (ppll)$$

This clearly is a striking deviation from the expected $9:3:3:1$ ratio. What is going on? Note that the class carrying at least one of each dominant allele makes up about 75% of the F_2 plants and the homozygous recessive class, expected to be the least frequent by Mendelian predictions, is more than 14%. Bateson and Punnett called this tendency of both dominants or both recessives in a dihybrid cross to be recovered more often *coupling*. This does not look like something that can be explained as a special type of Mendelian ratio.

The explanation for these results had to await the development of *Drosophila* as a genetic tool. Morgan found a deviation from Mendel's second law in studying two different pairs of autosomal genes. One pair affects eye color (pr, purple, and pr^+, red) and the other, wing length (vg, vestial, and vg^+, normal). He crossed $prpr\,vgvg$ flies with $pr^+pr^+\,vg^+vg^+$ and testcrossed the heterozygous F_1 females: $prpr^+\,vgvg^+\,♀ \times prpr\,vgvg\,♂$. Note again that in a testcross, because one parent contributes gametes carrying only recessive alleles, the phenotype of the offspring represents the gametic genotype contributed by the other parent. The results obtained were:

$$
\begin{array}{ll}
pr^+\,vg^+ & 1{,}339 \\
pr\ vg & 1{,}195 \\
pr^+\,vg & 151 \\
pr\ vg^+ & \underline{154} \\
& 2{,}839
\end{array}
$$

Obviously, this is a drastic deviation from the Mendelian prediction of a $1:1:1:1$ ratio and again indicates a coupling of genes. The most common classes are the two gene combinations, $pr^+\,vg^+$ and $pr\,vg$, originally introduced by the P_1 flies.

If, in a cross, each parent is homozygous for one of the recessive genes, a different result is obtained:

$$P_1 \quad pr^+pr^+\,vgvg \;\times\; prpr\,vg^+vg^+$$
$$\downarrow$$
$$F_1 \qquad pr^+pr\,vg^+vg\,♀ \;\times\; prpr\,vgvg\,♂$$

The following progeny of the testcross are obtained:

$$
\begin{array}{ll}
pr^+\,vg^+ & 157 \\
pr\ vg & 146 \\
pr^+\,vg & 965 \\
pr\ vg^+ & \underline{1{,}067} \\
& 2{,}335
\end{array}
$$

Again, this is not anywhere near a $1:1:1:1$ ratio. But now the most common classes have one or the other dominant gene. This was called *repulsion* by Bateson and Punnett. In this case, the most frequent classes are also the same as the original P_1 genotypes. How can we explain these results?

Morgan suggested that both pairs of genes being studied were located *on the same homologous chromosomes*. Thus, when *pr* and *vg* were introduced from one parent, they were physically located on the same chromosome, as were *pr*+ and *vg*+ on the homolog from the other parent (Figure 4-1). This would apply to

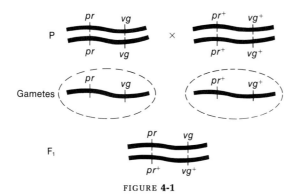

FIGURE **4-1**

repulsion as well, where one P_1 chromosome carries *pr* and *vg*+ and the other has *pr*+ and *vg*. This explains why P_1 gene combinations remain together, but how do we explain the nonparental combinations?

Morgan postulated that, during meiosis when homologous chromosomes pair, there may occasionally be a physical exchange of chromosome parts by a process called *crossing over*. What he pictured then is shown in Figure 4-2, in which the black box simply indicates our ignorance about how crossing over occurs, but the end result is a physical exchange of chromosome segments. The original P_1 gene arrangement on the two chromosomes is called the parental combination and the two new combinations are called *crossover types, exchange products,* or *recombinants.*

Is there any cytologically observable process that could account for something

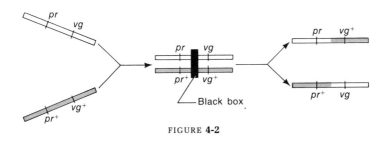

FIGURE **4-2**

like crossing over? F. Janssens reported observations on meiotic chromosomes in amphibians and then in Orthoptera that suggested a cytological basis for a crossover event. He often saw that during meiosis, when duplicated homologous chromosomes are paired with each other, two nonsister chromatids cross each other while the other two do not. He called such a cross-shaped structure a *chiasma* (plural *chiasmata*). These chiasmata are just what one might expect a crossover event to look like (Figure 4-3). (*Note:* This would indicate that

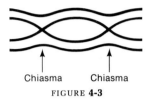

Chiasma   Chiasma

FIGURE **4-3**

crossing over occurs between chromatids, not unduplicated chromosomes. We'll return to this point later.) For the present we will accept that chiasmata *are* the cytological counterparts of crossovers. It turns out that it does not affect the art of genetics very much whether they are or are not, but it does provide a pictorial image for the interpretation of genetic data.

MESSAGE

Chiasmata are probably the visible manifestations of crossovers.

Gene pairs that are located on the same pair of homologous chromosomes, then, are said to be *linked.* In a cross between *AA BB* and *aa bb*

$$P_1 \quad AA\,BB \quad \times \quad aa\,bb$$
$$Aa\,Bb \longleftarrow \times \quad aa\,bb$$

the two genotypes *A B* and *a b* are parental combinations and *A b* and *a B* are *recombinants*. Note that recombination of the original P_1 genes can be due either to Mendelian assortment if the two gene pairs are on different chromosomes or to crossing over if the genes are linked. Thus, crossing over generates recombinant chromosomes, but recombinants are not necessarily derived from crossovers.

DEFINITION

Recombination is any meiotic process that generates a haploid product of meiosis whose genotype is different from the two haploid genotypes that constituted the meiotic diploid.

This definition has been stated this way so that it holds for haploid life cycles too. (In a diploid such as *Drosophila* or peas the haploid product of meiosis is of course the gamete.) So you can see that, by stating the definition in this way, in a study of recombination we look at the genotypes of the *output* (or products) of meiosis and compare them with the *input* genotypes (or the two haploid types that made the organism in which the meiosis is occurring).

THE TWO PROCESSES OF RECOMBINATION

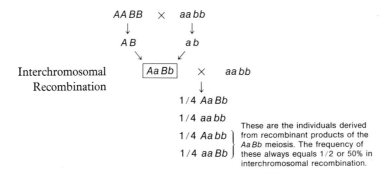

Note how useful a testcross is. If we had made the cross *Aa Bb* × *Aa Bb*, it would have been doubly difficult to analyze the results in that we would have had two meioses to juggle and a progeny type, say *AA Bb* (a recombinant), would be impossible to distinguish from *AA BB* (a parental) phenotypically.

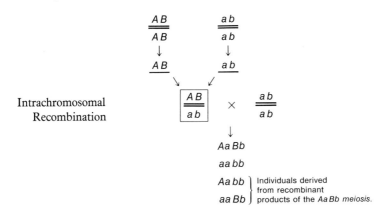

We have seen that the recombinant frequency can be less than 50% [$(151 + 154)/2839 = 10.7\%$] in the data given on page 78.

Our symbolism for describing crosses now becomes cumbersome with the discovery of linkage. We can show the genetic constitution of each chromosome in the *Drosophila* cross

$$\frac{pr \quad vg}{pr^+ \quad vg^+}$$

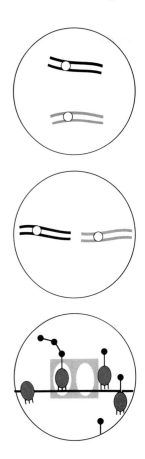

where each line represents a chromosome, the genes above being on one chromosome and those below on the other. A crossover is represented by $\times$ between the two chromosomes so that

$$\frac{pr \qquad vg}{\underset{pr^+ \qquad vg^+}{\times}}$$

is read

We can simplify the genotypic designation of linked genes by simply drawing a single line, those on each side being on the same chromosome; so it becomes

$$\frac{pr \quad vg}{pr^+ \quad vg^+}$$

But this gives us problems in typing and writing, so let's tip the line to $pr\,vg/pr^+\,vg^+$, still keeping the genes on one chromosome on one side of the line, and those of the other homolog on the other side. Finally, we always designate linked genes on each side in the same order, so it is always $a\,b/a\,b$, never $a\,b/b\,a$. That being the case, we can indicate the wild-type allele with a plus sign ($+$), and $pr\,vg/pr^+\,vg^+$ becomes $pr\,vg/+\,+$. From now on, we'll use this kind of designation unless its use creates an ambiguity.

If we go back and think about the results obtained by Bateson and Punnett, we can now explain the coupling phenomenon by means of the concept of linkage. Their results are more complex because they did not do a testcross. Try to see if you can derive estimated numbers for the recombinant and parental types in the gametes. If, instead of autosomal genes, we were to look at sex-linked genes, a testcross would not be necessary. Think about it. A female, heterozygous for two different sex-linked genes, will produce males hemizygous for those genes so that the gametic genotype will be the phenotype. Let's take an example using the genes y (yellow body) and y^+ (brown body) and w (white eye) and w^+ (red eye). In the following cross, $Y = $ the Y chromosome; do not confuse it with the y/y^+ genes.

$$P_1 \quad y+/y+♀ \times +w/Y♂$$
$$F_1 \quad y+/+w♀ \times y+/Y♂$$

The number of F_2 males in each phenotypic class is:

$y\ w$	43
$+\ w$	2,146
$y\ +$	2,302
$+\ +$	22
	4,513

These classes reflect the products of meiosis in the F_1 female. The system amounts to a testcross because the F_2 males obtain only a Y chromosome from the F_1 $y+/Y$ males. The recombinant frequency is then $(43 + 22)/4513 = 1.4\%$, whereas the frequency of recombinants for the autosomal genes we studied (i.e., pr and vg) was $[(151 + 154)/2839] \times 100 = 10.7\%$ of the progeny. It would seem then that the amount of crossing over between the linked gene pairs is not constant; what's more, there is no reason to expect crossing over between different linked gene pairs to occur with the same frequency. In fact, as Morgan studied more linked genes, he saw that the number of progeny in recombinant classes varied considerably relative to those in the parental classes. Morgan felt that these variations in crossover frequency might somehow reflect the actual distances separating genes. He had another student, Alfred Sturtevant, who, like Bridges, became one of the great geneticists. Sturtevant was still an undergraduate at the time and was asked by Morgan to make some sense of the data on crossing over between different linked genes. In one night, Sturtevant developed a method for describing linkage relationships between genes that is still used today. Take, for example, a testcross from which we obtain the following results:

$$pr\,vg/++\,♀ \qquad \times \qquad pr\,vg/pr\,vg\,♂$$

Parental or noncrossover classes	$\begin{cases} pr\,vg/pr\,vg \\ ++/pr\,vg \end{cases}$	165 191
Crossover or recombinant classes	$\begin{cases} pr+/pr\,vg \\ +vg/pr\,vg \end{cases}$	23 $\underline{21}$ 400

Of the progeny in this example, which represent 400 female gametes, 44 or 11% $[(44/400) \times 100]$ are recombinant. Sturtevant suggested that we can use this *percentage of recombinants* as a quantitative index of the "distance" between two gene pairs on a *genetic map*.

MESSAGE

One genetic map unit (m.u.) is the distance between gene pairs on a genetic map such that one product of meiosis out of one hundred is recombinant. A recombinant frequency (RF) of 0.01 or 1% ≡ 1 m.u.

The place on this map that represents a gene pair is called the *gene locus* (plural *loci*). The locus of the eye-color gene pair and the locus of the wing-shape gene pair, for example, are 11 map units apart and are usually diagrammed as follows:

pr   11.0   vg

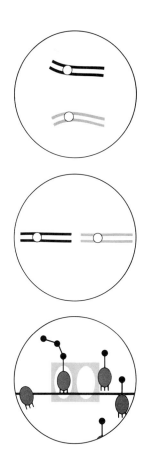

although they could be diagrammed equally well like this:

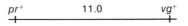

or like this:

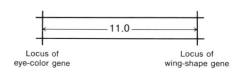

Usually the eye-color locus is referred to in shorthand as the "*pr* locus" after its most famous nonwild allele, but it means that place on any chromosome of that type.

It should be noticed that there is a strong implication here that the "distance" on a genetic map is a physical distance along a chromosome, and obviously Morgan and Sturtevant intended it that way. But we should realize that it is another example of an entity (the genetic map) constructed from a purely genetic analysis. The genetic map (or *linkage map* as it is sometimes called) could have been derived without even knowing that chromosomes existed. Furthermore, at this point, we do not know whether the "genetic distances" calculated by means of recombinant frequencies in any way reflect actual physical distances on chromosomes. In fact, it has been shown by means of a clever cytogenetic analysis of *Drosophila* that the genetic distances are largely proportional to chromosome distances.

Realize also that, given a genetic distance in map units, one can predict frequencies of progeny in different classes. For example, of the progeny from a testcross of a *pr vg*/ + + heterozygote, 11% will be *pr* + /*pr vg* or + *vg*/*pr vg*, and of the progeny from a testcross of a *pr* + / + *vg* heterozygote, 11% will be *pr vg*/*pr vg* or + +/*pr vg*.

Let's take a few other examples showing different map distances. In corn, there are two linked loci controlling seed color (*A*, colored, and *a*, colorless) and shape (*Sh*, plump, and *sh*, shrunken). In the cross *A Sh*/*a sh* × *a sh*/*a sh*, we get

A Sh	5,020
a sh	4,960
A sh	12
a Sh	8
	10,000

In this example, the number of recombinants is very small; consequently the genetic distance is short—0.2 map unit, or $[(12 + 8)/10,000] \times 100$. On the other hand, in chickens, two genes affecting feather color (*I*, white, and *i*, colored) and texture (*F*, frizzle, and *f*, normal) are less tightly linked. In the cross *i F*/*I f* × *if*/*if*, we obtain:

iF	63
If	63
IF	18
if	13
	157

In this example, the distance between the gene pairs is 19 map units, or $[(18 + 13)/157] \times 100$. In summary, Sturtevant's genetic map is derived from the frequency of recombinants detected between two linked pairs of genes.

MESSAGE

The genetic map is another example of a hypothetical entity based on genetic analysis. Its construction is not dependent on any cytological phenomenon.

How do we analyze crosses of more than two pairs of genes? In a cross that includes three loci in *Drosophila*, the nonwild-type alleles are *sc* (scute, or loss of certain thoracic bristles), *ec* (echinus, or roughened eye surface), and *vg*. If we cross *sc ec vg* flies with homozygous wild-type and then testcross the F_1 females, which are $sc/+ ec/+ vg/+$, we obtain (we have invented these numbers):

sc	*ec*	*vg*	235
+	+	+	241
sc	*ec*	+	243
+	+	*vg*	233
sc	+	*vg*	12
+	*ec*	+	14
sc	+	+	14
+	*ec*	*vg*	16
			1,008

There is considerable deviation from a $1:1:1:1:1:1:1:1$ ratio for a trihybrid cross of independently assorting genes; so there must be linkage involved. The four classes with the largest numbers must therefore consist of individuals that are not the result of a crossover between linked loci. Which loci are linked? Taking only the four largest classes, we can determine whether *sc* and *ec* are linked. Remember that by *sc* and *ec* we really mean the gene sites or loci represented by $sc/+$ and $ec/+$. Ignoring *vg* then, we get:

sc	*ec*	235
+	+	241
sc	*ec*	243
+	+	233

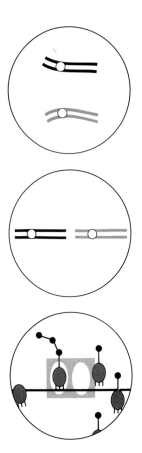

There are only two classes, *sc ec* and $+$ $+$; so we know they are linked and that these are the two parental classes. What about *vg*? Let's ask whether *ec* and *vg* are linked and ignore *sc* (since we know it is linked to *ec*). The data then are:

ec	*vg*	235
$+$	$+$	241
ec	$+$	243
$+$	*vg*	233

which is a nearly perfect $1:1:1:1$ ratio, showing that *vg* is assorting independently of *ec* (and *sc*). So we know *sc* and *ec* are on one chromosome and *vg* assorts independently. We can calculate the genetic distance between *sc* and *ec* by looking at the smaller classes, which (if we ignore *vg* because it is assorting independently) are obviously derived from crossovers.

sc	$+$	12
$+$	*ec*	14
sc	$+$	14
$+$	*ec*	16

The distance between *sc* and *ec* is $[(12 + 14 + 14 + 16)/1008] \times 100 = 5.5$ map units and we can diagram the chromosomes as

Now let's take another cross, in which the genes are *sc*, *ec*, and another nonwild-type, *cv* (crossveinless, or absence of a crossvein on the wing). We cross flies homozygous for *sc*, *ec*, and *cv* with homozygous wild-type to get *sc ec*/$+$ $+$, *cv*/$+$ females, which we then testcross to obtain (again, invented numbers):

sc	*ec*	*cv*	417
$+$	$+$	$+$	430
sc	$+$	$+$	25
$+$	*ec*	*cv*	29
sc	*ec*	$+$	44
$+$	$+$	*cv*	37
			982

First, do these genes assort independently? We see that there are two classes, *sc ec cv* and $+$ $+$ $+$, that occur much more often than the others, thereby showing that we are not getting a $1:1:1:1$ ratio of independent assortment of any two pairs. Moreover, they are reciprocal classes (that is, the alleles in one are all different from those in the other); so *sc*, *ec*, and *cv* must have been on one of the homologous chromosomes and +, +, and + on the other. We now know the three genes are linked; so we can determine how far apart they are. Looking at one pair of loci at a time, let's take *sc* and *ec* first (we can cover up the *cv* and consider only the *sc* and *ec* pair of loci).

sc	*ec*	c̸v̸	417
+	+	c̸v̸	430
sc	+	c̸v̸	25
+	*ec*	c̸v̸	29
sc	*ec*	c̸v̸	44
+	+	c̸v̸	37
			$\overline{982}$

There are $417 + 430 + 44 + 37 = 928$ parentals and $25 + 29 = 54$ recombinants. Therefore, the genetic distance between *sc* and *ec* is 5.5 map units $[(54/982) \times 100]$, which is the same distance as was found earlier for another cross involving these loci. The map can be diagrammed as:

$$\overset{sc}{\vert} \quad 5.5 \quad \overset{ec}{\vert}$$

Now let's take *ec* and *cv*; the data are:

s̸c̸	*ec*	*cv*	417
s̸c̸	+	+	430
s̸c̸	+	+	25
s̸c̸	*ec*	*cv*	29
s̸c̸	*ec*	+	44
s̸c̸	+	*cv*	37
			$\overline{982}$

There are $417 + 430 + 25 + 29 = 901$ parental and $44 + 37 = 81$ recombinants. The distance between *ec* and *cv*, then, is $(81/982) \times 100 = 8.2$ and the map is

$$\overset{ec}{\vert} \quad 8.2 \quad \overset{cv}{\vert}$$

The distance between *sc* and *ec* being 5.5 and that between *ec* and *cv* being 8.2 we can draw a map as either

$$\overset{ec}{\vert} \quad 5.5 \quad \overset{sc}{\vert} \quad \overset{cv}{\vert}$$
$$\longleftarrow \!\!\!\!\! \text{---} 8.2 \text{---} \!\!\!\!\! \longrightarrow$$

or

$$\overset{sc}{\vert} \quad 5.5 \quad \overset{ec}{\vert} \quad 8.2 \quad \overset{cv}{\vert}$$

Now let's see how far *sc* and *cv* are apart, forgetting *ec*:

sc		*cv*	417
+		+	430
sc		+	25
+		*cv*	29
sc		+	44
+		*cv*	37
			982

There are $417 + 430 = 847$ parentals and $25 + 29 + 44 + 37 = 135$ recombinants for a distance of $(135/982) \times 100 = 13.7$ map units. We know then that the map should be drawn as

Let's turn to another cross in which the genes are *cv* and two more nonwild-types, *ct* (cut, or snipped wing edges) and *v* (vermilion, or bright scarlet eye color). A homozygous *cv ct v* fly is crossed with a homozygous wild-type fly to yield a $cv/+ \ ct/+ \ v/+$ heterozygous female, which is then testcrossed to give

cv	*ct*	*v*	580
+	+	+	592
cv	+	+	45
+	*ct*	*v*	40
cv	*ct*	+	89
+	+	*v*	94
cv	+	*v*	3
+	*ct*	+	5
			1,448

Again we recognize linkage by the absence of Mendelian ratios, the *cv ct v* and $+ + +$ classes being the parental chromosomes. Unlike the previous cross, this cross yields six classes of recombinant chromosomes. We can calculate map distances between two gene pairs at a time, ignoring the third again. Let's ignore *v* and look at *cv* and *ct*, for which the recombinant classes are *cv* + and + *ct*; the genetic distance is 6.4 map units, or $[(45 + 40 + 3 + 5)/1{,}448] \times 100$. Doing the same for *ct* and *v* (and ignoring *cv*), we get a distance of 13.2 map units, or $[(89 + 94 + 3 + 5)/1{,}448] \times 100$. Now if we ignore *ct* and look at *cv* + and + *v* recombinants, we find the distance to be 18.5 map units, or $[(45 + 40 + 89 + 94)/1{,}448] \times 100$. This shows that *cv* and *v* are the farthest apart and that *ct* must be between them. However, if we add the distances between *cv* and *ct* (6.4) and between *ct* and *v* (13.2), we get a value of 19.6, which is

larger than the computed value of 18.5. Why? We see that eight of the flies
(*cv + v* and *+ ct +*) were included in calculations for both the *cv*-to-*ct* and
the *ct*-to-*v* intervals, but *not* for the calculations for the *cv*-to-*v* interval, even
though *two crossovers* had taken place between *cv* and *v* in each of these
chromosomes. The reason becomes obvious if we look at a representation of
this *double crossover* event (Figure 4-4).

FIGURE **4-4**

Although two crossovers have taken place between *cv* and *v*, the genetic result
is that *cv* and *v* remain linked in the parental combination. Only when *ct* and its
wild-type allele are present can we recognize the double-crossover chromosome.
So, in a calculation of the *cv* to *v* distance we must add in *twice* the number of
double crossovers because each double crossover consists of two exchange
events between *cv* and *v*.

In almost any crossover study of three linked genes, the double-recombinant
classes will always be the most rare. Knowing that, we could have determined
the gene order (i.e., which gene is in the middle) without calculating map
distances. We know from the noncrossover classes that the parental arrangement
of genes was *cv ct v/+ + +*. If we do not know the proper gene order, then it
could be any of *cv v ct*, *v cv ct*, and *cv ct v* (as shown in Figure 4-5). Of these

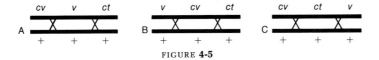

FIGURE **4-5**

sequences, only the one shown in Part C would generate the genotypes of the
double-crossover classes *cv + v* and *+ ct +*.

The detection of the double-exchange class allows us to ask another question.
Are exchanges occurring in each of two different regions of the same chromosome independently of each other? We know that we get recombinants between
cv and *ct* 6.4% of the time and between *ct* and *v* 13.2% of the time. If a crossover
in one of these regions in no way affects the probability of a crossover occurring
in the other region (i.e., they are independent), then the probability that an
individual will be produced that is recombinant in both regions at once is the
product of these individual probabilities. If they are independent, then, the
expected frequency of double-recombinant types is $0.064 \times 0.132 = 0.0084$
(0.84%). In a sample of 1,448, we would expect $1,448 \times 0.0084 = 12$ double
crossovers, whereas we actually observed 8. Had we obtained a million progeny,
the observed number of doubles would still have been lower than the expected.
What does that mean? It means that crossing over in one region is not independ-

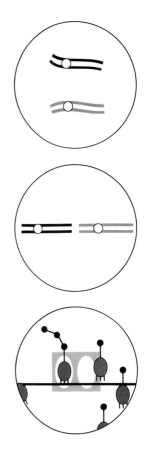

ent of crossing over in the other; in fact, when a crossover does occur in one region, it decreases the likelihood that another will occur in the other region. The effect of a crossover in one region on another area is called *interference*. This is quantified by the *coefficient of coincidence* (c.c.), which is simply the number of double recombinants observed divided by the number expected.

An interference value (I) is calculated by subtracting the c.c. from 1.

$$I = 1 - \text{c.c.} = 1 - \frac{\text{observed doubles}}{\text{expected doubles}}$$

When there are no double crossovers, the c.c. is 0, and so I is 1; therefore interference is complete. (Interference is complete in the *sc-ec-cv* interval in the example on page 86.) When there are fewer doubles than expected, the c.c. is less than 1, making I a positive number; so there is interference. When the c.c. is 1, $I = 0$; so there is no interference. Finally, when there are more doubles than expected, the c.c. is greater than 1, making I a negative number; so interference is negative. The effects of interference are further explored in Figure 4-6, which shows the redistribution of progeny types due to interference. (Note that map distances of I and II remain the same.)

A	I		B	II		C
	20 m.u.			20 m.u.		
	p(observable exchange) = 0.2			p(observable exchange) = 0.2		

	No interference	With 10% interference
Exchange in I but not II	0.2 × 0.8 = 0.16	0.164 (up 0.04)
Exchange in II but not I	0.2 × 0.8 = 0.16	0.164 (up 0.04)
Exchange in I and II	0.2 × 0.2 = 0.04	0.036 (down 0.04)
Exchange in neither I nor II	0.8 × 0.8 = 0.64	0.636 (down 0.04)
	1.00	1.000

FIGURE **4-6**

An explicatory analogy in which 1,000 people are given at random 200 buns and 200 cookies is given in Table 4-1.

The four people who give up either cookie or bun generate four additional 1-bun people and four additional 1-cookie people, but also generate four less bun-and-cookie people (their original selves) and four less people having neither (who are given one of their items).

TABLE **4-1**

	Random	A socialistic law forces 10% of those having both bun and cookie to give up one
One bun only	160	164 (up 4)
One cookie only	160	164 (up 4)
Bun and cookie	40	36 (down 4)
Neither	640	636 (down 4)

You may have noted that we were always careful to specify that, in crossover studies in *Drosophila*, heterozygous *females* were always testcrossed. When *pr vg*/ + + males are crossed with *pr vg/pr vg* females, only *pr vg*/ + + and *pr vg/pr vg* progeny are recovered. This shows that crossing over does not occur in *Drosophila* males. However, this absence of crossing over in one sex is species specific and is not the case for all males or for the heterogametic sex; in other organisms, crossing over can occur in XY males or WZ females. The reason for this sex difference is completely unknown and you will just have to remember that male *Drosophila* have this special characteristic.

The production of recombinant chromosomes is an interesting phenomenon that cries out for explanation. How are they generated? We will examine the question now, and again in greater detail in a later chapter. Morgan's explanation was that through crossing over (the actual mechanism for which he did not speculate on) a *physical exchange* of homologous chromosome parts produced the crossover chromosome. Goldschmidt, on the other hand, felt that, because chromosomes disappear during interphase, genes become more or less loose elements like unstrung beads. He suggested that during meiotic prophase the beads are restrung in the same sequence but that the members of each pair of alleles could exchange places. H. Winkler, on the other hand, suggested that when cells are heterozygous for different alleles of the same gene the two allelic states may interact in such a way that one converts the other into its own state at a low frequency. Thus in *a*/ +, sometimes *a* would be converted into + and other times + would be converted into *a*. In a heterozygote for linked genes, *a b*/ + +, *gene conversion* of the *b* allele to + would give an *a* + chromosome that would be genetically recorded as a recombinant. These three ideas are shown in Figure 4-7 in which beads represent genes, each pair of alleles consisting of a black bead and a white one.

Can we distinguish between these three possibilities? In 1931, Stern per-

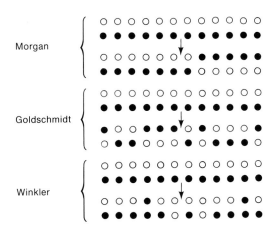

FIGURE 4-7

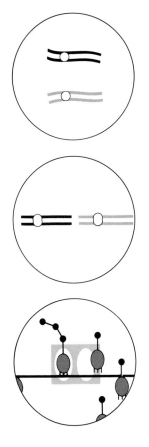

formed an elegant experiment on *Drosophila,* and Harriet Creighton and Barbara McClintock did the same on corn, to determine the method of crossover production. Stern studied two X chromosomes that differed cytologically in appearance, one having an apparent discontinuity in it

and the other carrying an extra long piece on one side of the centromere.

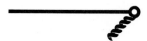

He knew (and we'll see how later in the book) that two genes, *car* (carnation or light eye color) and *B* (bar, or smaller eye), were linked near the centromere and had this appearance.

A female heterozygous for the two chromosome types and *B* and *car* would be

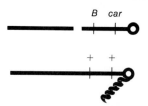

and, on testcrossing, Stern could separate crossovers from noncrossovers between *B* and *car*. He found that the parental types determined genetically (i.e., *B car,* and + +) always had parental chromosome configurations; that is, *B car* flies had

and + + flies carried

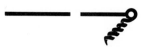

When he looked at the chromosomes of the recombinant progeny, he found new chromosome arrangements, *B* + flies having

and + *car* flies having

Thus, he showed a direct relationship between crossovers detected genetically and a physical rearrangement of the chromosome parts. This rules out both Goldschmidt's and Winkler's models, although gene conversion can be detected in specialized systems.

It is worth repeating these observations by describing the work done by Creighton and McClintock. In order to correlate crossing over with chromosome exchange, it is important to have chromosomes that are cytologically different on both sides of the region in which crossing over occurs. Creighton and McClintock were studying chromosome 9 of corn, which has two loci: one affects seed color (*C*, colored, and *c*, colorless) and the other composition (*Wx*, waxy, and *wx*, starchy). Furthermore, the chromosome carrying *C* and *Wx* was distinguished from its homolog by the presence of a large, densely staining element (called a knob) on the *C* end and by a longer piece of chromosome on the *Wx* end; thus, a heterozygote is

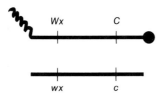

When they separated recombinants from nonrecombinants, they showed that the nonrecombinants retained the parental chromosome arrangements, whereas the recombinants were

and

again a confirmation of Morgan's suggestion that crossovers were a result of physical exchange consisting of an apparent breakage and a reunion of chromosome parts.

MESSAGE

Crossing over seems to be the result of a physical breakage and a reunion of chromosome parts.

How might the breakage and reunion be produced? Is it a "cut and paste" process like splicing sound tapes? Or is it a process that only *appears* to consist of a breakage and a reunion. An idea that favors the latter was proposed by John Belling, who studied meiosis in plant chromosomes and observed bumps along the chromosome called *chromomeres*, which he thought corresponded to genes

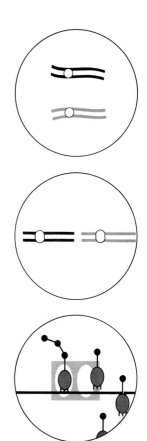

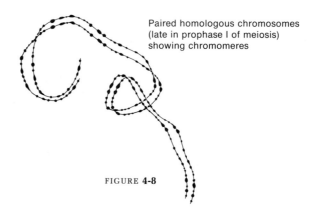

Paired homologous chromosomes
(late in prophase I of meiosis)
showing chromomeres

FIGURE **4-8**

(Figure 4-8). He visualized the genes as beads strung together with some nongenic linking substance. He reported that, during prophase of meiosis, chromomeres duplicated so that newly made chromomeres were stuck to the originals (Figure 4-9). After duplication, the newly formed chromomeres were

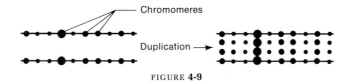

FIGURE **4-9**

hooked up together but, because all the chromomeres were tightly juxtaposed, the linking elements could switch from a newly made chromomere on one homologous chromosome to an adjacent one on the other homolog. This became known as the copy-choice, or switch, model of crossing over. You can see how it can generate a crossover chromatid that would seem to have arisen from a physical breakage and the reunion of chromosomes (Figure 4-10). You can also

FIGURE **4-10**

see that it suggests that the event generating recombinant chromosomes can take place only between newly made chromomeres (and hence newly made chromatids), so that all multiple exchanges could involve only two chromatids.

In some haploid organisms (see Chapter 5) it is possible to isolate all of the four products of a *single* meiosis. An interesting group of four is illustrated in the following diagram, in which the loci are linked in the order shown:

$$ABC \times abc$$

$$\left. \begin{array}{l} ABc \\ AbC \\ aBC \\ abc \end{array} \right\}$$ Four products
of one specific meiosis

Note there has been a recombination event between the first two (*A/a* and *B/b*) *and* the second two (*B/b* and *C/c*) loci in the same meiosis; a double-exchange event has occurred (and one of the two products, *A b C*, is a double recombinant). *But* more than two chromatids must have been involved; in fact, three were in this case (Figure 4-11). Thus, Belling's suggestion, which would

FIGURE **4-11**

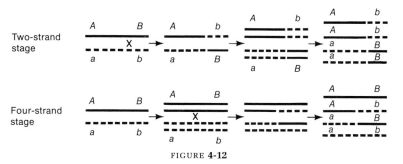

have predicted that only two chromatids could ever be involved, is most unlikely.

MESSAGE

The "cut and paste" method of chromosome crossing over wins by default.

(We'll see some *positive* evidence for breakage and reunion in Chapter 12.)

The ability to isolate the four products of meiosis in the preceding example also clears up another mystery about whether crossing over occurs at the two-chromosome (two-strand or chromatid) stage or at the four-chromosome (four-strand or chromatid) stage. If it occurs at the two-strand stage, there can never be more than two different products of a given meiosis, whereas, if it occurs at the four-strand stage, four different products of meiosis are possible (Figure 4-12). Four different products of a single meiosis are regularly observed,

Two-strand stage

Four-strand stage

FIGURE **4-12**

showing that crossing over occurs at the four-strand stage of meiosis.

One suggestion of what causes breakage of the chromatids before crossing over was put forth by C. D. Darlington. It is known that, during meiosis, paired chromosomes are tightly coiled around each other. Darlington suggested that the torsion from coiling creates tension on the chromatids that could be strong enough to break a chromatid; the unwinding resulting from the break would subject the remaining chromatids to increased stress that could induce a

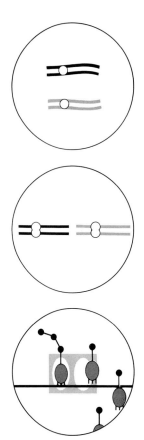

FIGURE **4-13**

second break. Reunion of broken ends of different chromatids would produce a crossover (Figure 4-13). Darlington's model, proposed in the thirties, was attractive because it provided a way of explaining positive interference by suggesting that breakage and unwinding would reduce the coiling stress along the chromosome neighboring the breaks. However, as we shall see later, crossing over consists of precise exchanges at the molecular level for which mechanically induced breaks would be far too imprecise. Moreover, the occurrence of high negative interference observed in microorganisms reduces the charm of the model.

More recent electron microscopic studies suggest that a complex apparatus, called the *synaptinemal complex,* exists in meiotic nuclei possibly for mediating crossing over. At present, however, the actual mechanism of crossing over is still an enigma although several attractive models exist and a lot is known about the sorts of chemical reactions that might ensure precision at the molecular level. But we will return to these points in Chapter 11.

Problems

1. Suppose that you have been given two strains of *Drosophila,* one strain having light yellow eyes and the other bright scarlet eyes. (Remember that wild-type *Drosophila* has deep-red eyes.) When you cross a yellow female with a scarlet male, you obtain 251 wild-type females and 248 yellow-eyed males in the F_1 generation. When you cross F_1 males and females, you obtain the following F_2 phenotypes:

    260 wild-type females

    253 yellow females

     77 wild-type males

    179 yellow males

    183 scarlet males

     80 brown males (brown is a new phenotype)

Explain these results, using diagrams where possible.

2. Suppose that you have two homozygous strains of *Drosophila*. Strain 1 has bright scarlet eyes and a wild-type thorax. Strain 2 has dark brown eyes and a humpy thorax. When you cross virgin females of Strain 2 with males of Strain 1, you obtain 232 wild-type males and 225 wild-type females in the F_1 generation. You then cross F_1 males with virgin F_1 females and obtain the following F_2 phenotypes:

  283 completely wild-type females

   78 humpy thorax, brown-eyed females

   19 humpy thorax, wild-eyed females

   20 wild thorax, brown-eyed females

    9 wild thorax, brown-eyed males

    8 humpy thorax, wild-eyed males

  139 wild thorax, scarlet-eyed males

   39 humpy thorax, brown-eyed males

   40 humpy thorax, white-eyed males

  145 completely wild-type males

   11 humpy thorax, scarlet-eyed males

   10 wild thorax, white-eyed males

Explain these results as fully as possible (using symbols wherever possible).

3. Suppose that you have a homozygous stock carrying the autosomal recessive genes a, b and c linked in that order. You cross females of this stock with males of a homozygous wild-type stock. You then cross the F_1 heterozygous males with their heterozygous sisters and obtain the following F_2 phenotypes:

+ + +	1,364
a + +	47
a + c	5
+ + c	84
+ b c	44
+ b +	4
a b +	87
a b c	365

a. What is the recombination frequency between a and b? Between b and c?

b. What is the coefficient of coincidence?

4. There is an autosomal dominant gene N in humans that causes abnormalities in nails and patellae (kneecaps), called the nail-patella syndrome. In marriages of people with the following phenotypes—nail-patella syndrome and blood type A × normal nail-patella and blood type O—some children with the nail-patella syndrome and A

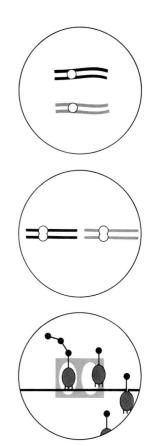

blood type are born. When marriages between those children (unrelated, of course) take place, their children are of the following types:

66% nail-patella syndrome, blood type A

16% normal nail-patella, blood type O

9% normal nail-patella, blood type A

9% nail-patella syndrome, blood type O

Analyze these data fully.

5. In a certain diploid plant, the three loci A/a, B/b, and C/c are linked as follows:

One plant is available to you (call it the parent plant) of the following constitution:

$$\frac{A \quad b \quad c}{a \quad B \quad C}$$

a. Assuming no interference, if the plant is selfed, what proportion of the progeny will be of the genotype $a\,b\,c/a\,b\,c$?
b. Again assuming no interference, if the parental plant is crossed with the $a\,b\,c/a\,b\,c$ plant, what genotypic classes will be found in the progeny, and what will their frequencies be if there are 1,000 progeny?
c. Repeat part b, this time assuming 20% interference between the regions.

6. The father of Mr. Spock, second officer of the Starship Enterprise, came from the planet Vulcan; his mother came from Earth. A Vulcanian has pointed ears (P), adrenals absent (A), and a right-sided heart (R), and all are dominant over Earth alleles. These genes are autosomal and are linked as shown in this linkage map:

$$\underset{\text{15 m.u.}}{\overset{P/p}{\vert}} \qquad \underset{\text{20 m.u.}}{\overset{A/a}{\vert}} \qquad \overset{R/r}{\vert}$$

If Mr. Spock marries an Earth woman and there is no (genetic) interference, what proportion of their children will

a. look like Vulcanians (for all three characters)?
b. look like Earth people?
c. have Vulcanian ears and a Vulcanian heart and Earth adrenals?
d. have Vulcanian ears and an Earth heart and Earth adrenals?[1]

7. Groodies are useful (but fictional) haploid organisms that are pure genetic tools. A wild-type groody has a fat body, a long tail and flagella. Nonwild-type strains are known that have thin bodies, or are tailless, or do not have flagella. Groodies can mate with each other (but they are so shy that we don't know how) and produce recombinants. A wild-type groody is crossed with a thin-bodied animal lacking a

---

[1]Problem 6 is from *Problems in Genetics* by D. Harrison, 1970, Addison-Wesley, Reading, Mass.

tail and flagella. A thousand baby groodies produced are classified as illustrated in the accompanying diagram. Assign genotypes and map the three genes.[2]

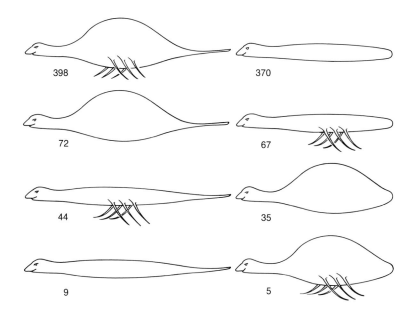

398	370
72	67
44	35
9	5

8. A geneticist studying bloops (an exotic organism found only in textbook problems) has been unable to find any examples of linkage. Suggest some explanations for this.[3]

9. A strain of *Neurospora* whose genotype is *H I* is crossed with a strain whose genotype is *h i*. Half the progeny are *H I* and half are *h i*. Explain how this is possible.

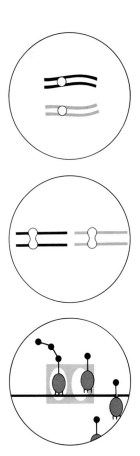

---

[2]From Burton S. Guttman, *Biological Principles,* copyright © 1971, W. A. Benjamin, Inc., Menlo Park, California.
[3]From *The Challenge of Genetics* by E. H. Simon and J. Grossfield, 1971, Addison-Wesley, Reading, Mass.

5 / Advanced Transmission Genetics

(Or: Running the chromosome theory into the ground.)

Mapping Functions

In Chapter 4, we defined a map unit as a recombination frequency (RF) of 1%. We have seen that, when this definition is used, a "reasonable" additivity of map distances results. However, when *larger* locus-to-locus intervals are being examined, the additivity becomes very shaky. This is typified in the following example:

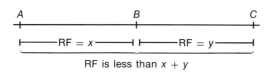

We have seen that one can *partially* compensate for this by adding in the double-crossover types (twice), but the whole point of this demonstration is that we have the notion that map distance (or the "true" distance) between loci is not linearly related to recombination frequency throughout the range of values possible in mapping experiments. We have the clue provided by double exchanges but now we have to come to grips with the notion that double exchanges are part of a much larger problem of multiple exchanges and their effect on recombination frequency. It might be pertinent to ask whether the RF value of x for the A-to-B interval in the example diagrammed is an accurate reflection of the true distance between the A and B loci. Perhaps some doubles occurred in this interval alone—we cannot detect them because we have no markers between A and B, but we intuitively feel that they should be accounted for if RF is to reflect distance at all well. This section of the chapter is about a mathematical

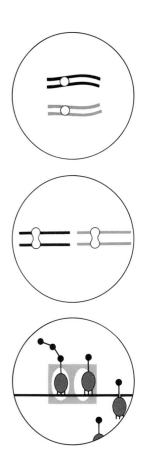

treatment of the problem that does give a way of correcting for multiple exchanges without actually seeing them.

Any relationship between one variable entity and another is called a function: the relationship between real map distance and recombination frequency is called the *mapping function*.

MESSAGE

> *The relationship between real map distance and RF is not linear.*
> *The* actual *relationship is called the mapping function.*

To calculate the mapping function we need to use a piece of mathematics that is widely used in genetic analysis because it describes many genetic phenomena well. It is called the *Poisson distribution*. A distribution is merely a device that describes the frequencies of various classes of samples. (We have already looked at one distribution—the binomial). The Poisson distribution describes the frequency of classes of samples containing $0, 1, 2, 3, 4, \ldots, n$ events when the average frequency (occurrence) of the event is small in relation to the total number of times that the event could occur. For example, the number of tadpoles one *could* get in a single dip of a net in a pond is quite large, but usually only one or two or none are found. The number of dead birds on the side of the highway is potentially very large, but in a sample kilometer the number is usually small. These kinds of samplings are described well by the Poisson distribution. Let's consider a numerical example. We'll randomly distribute 100 one-dollar bills to 100 students in a lecture room. We can distribute these dollar bills randomly by throwing them out at the students. Although the average number of bills per student will be 1.0, we might predict from past experience that many students will get one bill, a few will get two, a small number of lucky ones might get three, and so on; but many students will get *none*. Given the average (or mean) of 1.0, the frequencies of the 0, 1, 2, 3, and 4 classes can be accurately predicted by means of the Poisson distribution. It works as follows:

m = mean

e = base of natural logarithms = 2.7

i = number of events in a class

! = factorial symbol (e.g., $3! = 3 \times 2 \times 1$ and $4! = 4 \times 3 \times 2 \times 1$)

The class frequencies are then

0	1	2	3	$\cdots$	i
$\dfrac{e^{-m}m^0}{0!}$	$\dfrac{e^{-m}m^1}{1!}$	$\dfrac{e^{-m}m^2}{2!}$	$\dfrac{e^{-m}m^3}{3!}$	$\cdots$	$\dfrac{e^{-m}m^i}{i!}$

TABLE **5-1**

103

m	e^{-m}	m	e^{-m}	m	e^{-m}	m	e^{-m}
.000	1.00000	.250	.77880	.500	.60653	.750	.47237
.010	.99005	.260	.77105	.510	.60050	.760	.46767
.020	.98020	.270	.76338	.520	.59452	.770	.46301
.030	.97045	.280	.75578	.530	.58860	.780	.45841
.040	.96079	.290	.74826	.540	.58275	.790	.45384
.050	.95123	.300	.74082	.550	.57695	.800	.44933
.060	.94176	.310	.73345	.560	.57121	.810	.44486
.070	.93239	.320	.72615	.570	.56553	.820	.44043
.080	.92312	.330	.71892	.580	.55990	.830	.43605
.090	.91393	.340	.71177	.590	.55433	.840	.43171
.100	.90484	.350	.70469	.600	.54881	.850	.42741
.110	.89583	.360	.69768	.610	.54335	.860	.42316
.120	.88692	.370	.69073	.620	.53794	.870	.41895
.130	.87810	.380	.68386	.630	.53259	.880	.41478
.140	.86936	.390	.67706	.640	.52729	.890	.41066
.150	.86071	.400	.67032	.650	.52205	.900	.40657
.160	.85214	.410	.66365	.660	.51685	.910	.40252
.170	.84366	.420	.65705	.670	.51171	.920	.39852
.180	.83527	.430	.65051	.680	.50662	.930	.39455
.190	.82696	.440	.64404	.690	.50158	.940	.39063
.200	.81873	.450	.63763	.700	.49659	.950	.38674
.210	.81058	.460	.63128	.710	.49164	.960	.38289
.220	.80252	.470	.62500	.720	.48675	.970	.37908
.230	.79453	.480	.61878	.730	.48191	.980	.37531
.240	.78663	.490	.61263	.740	.47711	.990	.37158
						1.000	.36788

Source: *Statistical Tables* by F. James Rohlf and Robert R. Sokal, W. H. Freeman and Company. Copyright © 1969.

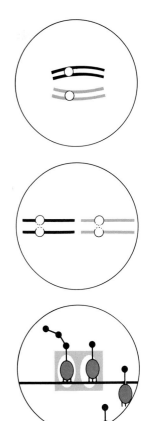

Table 5-1 gives values for e^{-m}.

How would the cash be distributed to the students at random? Plugging in an m of 1, we find that the zero class $= e^{-1} = 0.368$ (because 0! is defined as 1).

$$1 \text{ class} = \frac{e^{-1}}{1} = 0.368$$

$$2 \text{ class} = \frac{e^{-1}}{2} = 0.184$$

$$3 \text{ class} = \frac{e^{-1}}{3 \times 2} = 0.06$$

$$4 \text{ class} = \frac{e^{-1}}{4 \times 3 \times 2} = 0.02$$

A histogram of this distribution is shown in Figure 5-1. The percentage of students who get more than four dollars is negligible.

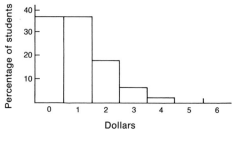

FIGURE 5-1

Similar distributions could be developed for other *m* values. Another distribution illustrated by curves rather than bars is shown in Figure 5-2.

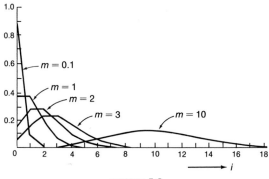

FIGURE 5-2

The occurrence of exchanges along a chromosome during meiosis can also be described by the Poisson distribution. In any given genetic region, the actual number of exchanges occurring is probably small in relation to the total number of opportunities for such an exchange in that stretch. If we knew the *mean* number of exchanges in the region per meiosis we could calculate the distribution of meioses with zero, one, two, three, four, and more multiple exchanges. This is unnecessary in the present context because, as we shall see, the only class we are really interested in is the zero class. The reason for this is that we want to correlate real distances with observable RF values, and it turns out that meioses in which there are one, two, three, four, or *any* finite number of exchanges per meiosis will produce an RF of 50% *among the products of those meioses,* whereas the meioses with no exchanges will produce an RF of 0%. Consequently, the determining force in actual RF values is the ratio of class zero to the rest!

This can be illustrated by considering meioses in which zero, one, and two exchanges occur (try three or more yourself). Figure 5-3 tells us that the only way we can get a recombinant product of meiosis is from a meiosis with at least one exchange in the marked region, and then half of the products of such meioses will *always* be recombinant.

At last we can derive the map function. Recombinants will make up one-half of the products of those meioses in which at least one exchange occurs in the

No exchanges

$RF = 0/4 = 0\%$

One exchange

(Can be between any
nonsister pair.)

$RF = 2/4 = 50\%$

Two exchanges

(Holding one constant
and varying the position
of the second produces
four equally frequent
two-exchange meioses.)

$RF = 0/4 = 0\%$

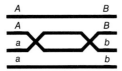

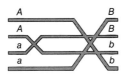

$RF = 2/4 = 50\%$

$RF = 2/4 = 50\%$

$RF = 4/4 = 100\%$

Overall two-exchange $RF = 8/16 = 50\%$

FIGURE **5-3**

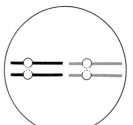
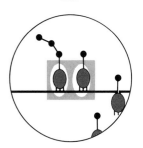

region of interest. Because products with at least one exchange have more than zero exchanges, their proportion is one minus the fraction with zero exchanges. Hence $RF = \frac{1}{2}(1 - e^{-m})$, because the zero-class frequency $= e^{-m}m^0/0! = e^{-m}$. When you think about it, m is the best measure we can have of *true* genetic distance—the *actual* average number of exchanges in a region.

If we know an RF value, then, we can calculate m by solving the equation. In plotting the function as a graph (Figure 5-4), several interesting points emerge.

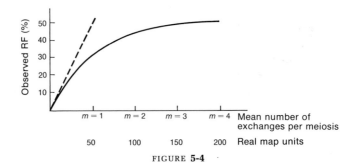

FIGURE 5-4

1. No matter how far apart two loci are on a chromosome, we will never observe an RF value of greater than 50%. Consequently, an RF value of 50% would leave us in doubt about whether two loci are linked or are on separate chromosomes. Put another way, as m gets larger, e^{-m} gets smaller, and RF tends to $\frac{1}{2}(1 - 0) = \frac{1}{2} \times 1 = 50\%$.

2. The function is linear for a certain range corresponding to very small m values (genetic distances). Therefore RF is a good measure of distance if the distance is small and no multiple exchanges are likely. In this region, the map unit defined as 1% RF would have real meaning. Then, let's use this region of the curve to define real map units. For small m, such as $m = 0.05$, $e^{-m} = 0.95$, and RF $= \frac{1}{2}(1 - 0.95) = \frac{1}{2}(0.05) = \frac{1}{2} \times m$; for $m = 0.10$, $e^{-m} = 0.90$, and RF $= \frac{1}{2}(1 - 0.90) = \frac{1}{2}(0.10) = \frac{1}{2} \times m$.

We see that RF $= m/2$, and this defines the dotted line on the graph in Figure 5-4. It allows us to translate m values into real (low-end-of-the-curve-style) map units.

So an m value of 1 is the equivalent of 50 real map units and we can represent the bottom axis in map units as indicated on the lower scale. Here we see that two loci 150 real map units apart would show only 50% RF. For regions of the graph in which the line is not horizontal we can use the function to convert the RF into map distance simply by drawing a horizontal line to the curve and dropping a perpendicular to the map-unit axis.

MESSAGE

To get good estimates of map distance, put RF values through the map function, or stick to small regions in which the graph is linear.

EXAMPLE

Suppose we get an RF of 27.5%. How many real map units does this represent?

$$0.275 = \tfrac{1}{2}(1 - e^{-m})$$
$$0.55 \;= 1 - e^{-m}$$

therefore

$$e^{-m} = 1 - 0.55 = 0.45$$

From e^{-m} tables (or by solving the hard way using logarithms) we find that $m = 0.8$, which is directly translatable into 40 real map units; so we can see that, if we had been happy to accept 27.5% RF as meaning 27.5 map units, we would have been considerably underestimating the true distance between the loci.

Tetrad Analysis

We have already hinted (in Chapter 4) at the existence of marvellous organisms in which the four products of a single meiosis are recoverable and testable. The group of four is called a tetrad, and tetrad analysis has been possible only in haploid fungi and single-celled algae, in which all the products of each meiosis are held together in a bag. There are many advantages to using haploids for genetic analysis; in addition to the convenience of tetrads, we can list a few others.

1. Because they are haploid, there is no complication of dominance. The nuclear genotype is expressed directly as the phenotype.

2. There is only one meiosis to analyze at a time (refer to the discussion of life cycles in Chapter 2), whereas, in diploids, gametes from two different meiotic events, fuse to form the zygote. In diploids, the testcross is an attempt to achieve the same end but is technically more laborious, and sometimes not possible.

3. Because the organisms are small, fast growing, and inexpensive to culture, it is possible to produce very large numbers of progeny from a cross. Thus, good statistical accuracy is possible and also very rare events at frequencies as low as 10^{-8} are detectable.

4. The structure and behavior of their chromosomes is apparently very similar to that in higher organisms.

Tetrad analysis has proved very useful for several of its own reasons.

1. It is an excellent device for students to learn formal genetic analysis.

2. It is possible to map centromeres as genetic loci.

3. Interference in crossing over between chromatids may be examined.

4. Several approaches to studying the mechanism of chromosome exchange (crossing over) are possible. We have already used it to deduce the stage at which crossing over occurs (4 strands) and to rule out Belling's copy-choice hypothesis. But perhaps the most significant use has been in the analysis of gene conversion, which will be discussed in Chapter 12.

5. It provides a unique approach to the study of abnormal chromosome sets (see Chapter 7).

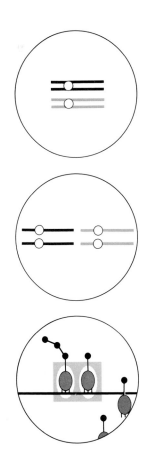

Before embarking on a study of tetrad analysis, it is worthwhile pointing out that a conventional analysis of the random products of meiosis is also possible in haploids and is much easier than in diploids. For example, the products of the cross $+ + \times a\,b$

+ +	45%
a b	45%
a +	5%
+ b	5%

allow a direct calculation of RF as 10% and 10 map units separate the a and the b loci. Note how easy it is to compare product-of-meiosis genotypes with the haploid genotypes that constituted the meiotic diploid.

CENTROMERE MAPPING

The organism we will use to illustrate tetrad analysis is the mold *Neurospora crassa*. In this fungus, not only do the products of a meiosis stay together in a gift-wrapped package called an *ascus*, but meiosis occurs in a linear way as shown in Figure 5-5, because there is no overlapping of meiotic or mitotic spindles.

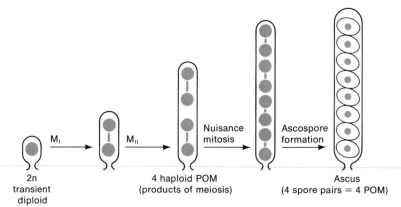

FIGURE **5-5**

This particular quirk allows us to map centromeres. Let's consider a meiosis in the cross $+ + \times a\,b$, in which the linkage arrangement is centromere–*a*–*b*. Most asci will have the ascospore arrangement of $a\,b : a\,b : + + : + +$ or $+ + : + + : a\,b : a\,b$ spore pairs. These must represent the disjunctional products of meiotic chromosomes, which can be pictured as in Figure 5-6. You

FIGURE **5-6**   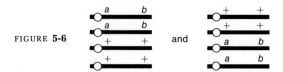   and

can see that all the separation of different alleles occurs at the first meiotic division; so this pattern is called *first-division segregation* and indicates and absence of a crossover. Let's now see what happens when a crossover occurs between *a* and *b* (Figure 5-7).

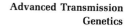

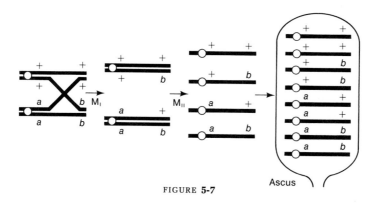

FIGURE 5-7

If we look at the distribution "pattern" of the alleles in the ascus, we can see that it is different for the two loci. Whereas the *a* locus is recovered in a $+++ +aaaa$ pattern, the *b* locus shows a $++bb++bb$ pattern. The first-division segregation of *a* indicates that the different alleles have segregated into separate nuclei at the end of meiosis I. On the other hand, segregation of the *b* alleles does not occur until the end of meiosis II and is called *second-division segregation*. No doubt you will appreciate immediately that a second division pattern can only result when a crossover occurs betwen the *b* locus and the centromere. However, because *no* crossover took place between *a* and the centromere, an M_I (first meiotic division) pattern is seen in the ascus.

Because of random spindle attachment at the M_I stage, two equally frequent M_I patterns are expected (Figure 5-8). Random spindle attachment at the M_I and the M_{II} stages of meiosis produces four equally frequent M_{II} patterns (Figure 5-9). In other words, each ascus or half ascus can be flipped over at random to simulate random spindle attachment.

As keen, budding geneticists, you should now recognize the potential for centromere mapping. The frequency of M_{II} patterns for any locus should be proportional to the distance of that locus from the centromere on that particular

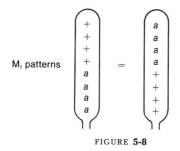

FIGURE 5-8

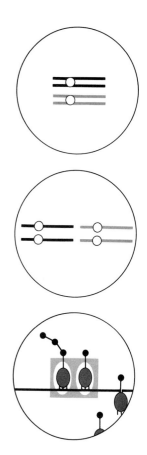

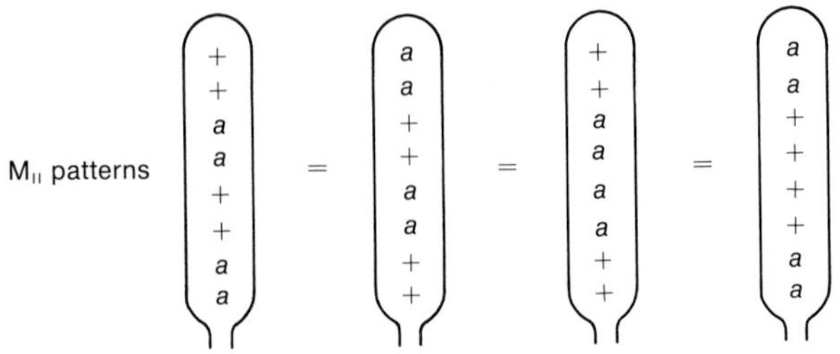

M_{II} patterns

$$\begin{pmatrix} + \\ + \\ a \\ a \\ + \\ + \\ a \\ a \end{pmatrix} = \begin{pmatrix} a \\ a \\ + \\ + \\ a \\ a \\ + \\ + \end{pmatrix} = \begin{pmatrix} + \\ + \\ a \\ a \\ a \\ a \\ + \\ + \end{pmatrix} = \begin{pmatrix} a \\ a \\ + \\ + \\ + \\ + \\ a \\ a \end{pmatrix}$$

FIGURE 5-9

chromosome. For example, we can do a linear tetrad analysis of a cross involving the locus that controls crossing ability (the so-called sex locus). We make the cross by mixing parental cultures and then isolating asci, carefully removing the ascospores and testing each of them (Figure 5-10).

Cross		A	$\times$	a			
		\multicolumn{6}{c}{Progeny asci}
		\multicolumn{2}{c}{M_I}	\multicolumn{4}{c}{M_{II}}
	1	A a		A a	A a
Spore	2	A a		a A	a A
pair	3	a A		A a	a A
	4	a A		a A	A a

Number of asci of each type: 126 132   9 11 10 12   Total: 300

FIGURE 5-10

Note that 126 approximately equals 132, reflecting random spindle attachment at M_I, and that 9, 11, 10, and 12 are approximately equal, reflecting random spindle attachment at M_{II}. The total M_{II} segregation pattern frequency $= 42/300 = 14\%$.

Can we use 14% as 14 map units for the genetic distance between the sex locus and the centromere? The answer is a resounding NO! The reason is an important one: map units were defined as the percentage of *recombinants* (or recombinant chromosomes) and our 14% is the percentage of *meioses* in which a crossover has occurred. As we saw earlier in this chapter, one crossover between two loci in a meiosis (or in this case between a gene locus and its centromere) produces only 50% recombinant chromosomes (chromatids) (Figure 5-11).

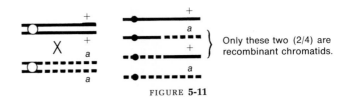

Only these two (2/4) are recombinant chromatids.

FIGURE 5-11

MESSAGE

To convert the frequency of second-division segregation into map units, simply multiply by 1/2 (or 50%). Thus, the sex locus is 7 map units from its centromere.

In this way, the centromere may be introduced into a genetic map with units consistent with the original method for constructing genetic maps.

Centromere mapping may be combined with the mapping of conventional loci, and an example is given in Figure 5-12. The cross is *nic +* × *+ ad*, in

1		2		3		4		5		6		7	
+	*ad*	+	+	+	+	+	*ad*	+	*ad*	+	+	+	+
+	*ad*	+	+	+	*ad*	*nic*	*ad*	*nic*	+	*nic*	*ad*	*nic*	*ad*
nic	+	*nic*	*ad*	*nic*	+	+	+	+	*ad*	+	+	+	*ad*
nic	+	*nic*	*ad*	*nic*	*ad*	*nic*	+	*nic*	+	*nic*	*ad*	*nic*	+
M_I	M_I	M_I	M_I	M_I	M_{II}	M_{II}	M_I	M_{II}	M_{II}	M_{II}	M_{II}	M_{II}	M_{II}
(PD)		(NPD)		(T)		(T)		(PD)		(NPD)		(T)	
808		1		90		5		90		1		5	

FIGURE **5-12**

which *nic* is an allele that causes the fungus to require nicotinic acid to grow and *ad* is an allele that confers a requirement for adenine. (Don't worry about these phenotypes for now, they are being used only as genetic markers.) In a cross involving two marker loci, only seven basic classes are possible. Note that in arriving at these seven classes the order of genotypes within the half ascus has been ignored because it simply reflects random spindle attachment; for example, class 5 also includes

nic	+		+	*ad*		*nic*	+
+	*ad*		*nic*	+		+	*ad*
+	*ad*		*nic*	+		*nic*	+
nic	+		+	*ad*		+	*ad*

The asci have also been labelled according to another classification: *parental ditypes* (PD), in which there are only two genotypes (hence *di*type) with respect to the marker loci and both are parental (classes 1 and 5); *nonparental ditype* (NPD), in which there are only two genotypes and both are nonparental or recombinant (classes 2 and 6); and *tetratype* (T) in which there are four genotypes—two parental and two nonparental (classes 3, 4 and 7).

These classifications are the result of the combinations of M_I and M_{II} segregations shown in Figure 5-13. Note that to get a tetratype, there must be

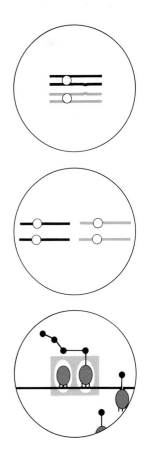

$+/nic$

	M_I	M_{II}
$+/ad$ M_I	PD NPD	T only
M_{II}	T only	T PD NPD

FIGURE 5-13

a crossover between at least one locus and its centromere. Now for the calculations.

First, we calculate the distances of each locus from the centromere. For $+/nic$, it is

$$\frac{5 + 90 + 1 + 5}{1,000} = \frac{101}{1,000} = 10.1\% \; M_{II} = 5.05 \text{ m.u.}$$

For $+/ad$, it is

$$\frac{90 + 90 + 1 + 5}{1,000} = \frac{186}{1,000} = 18.6\% \; M_{II} = 9.30 \text{ m.u.}$$

However, this still leaves us with three linkage possibilities (Figure 5-14). Most of the asci are $M_I M_I$ parental ditypes (808/1,000); therefore most genomes will

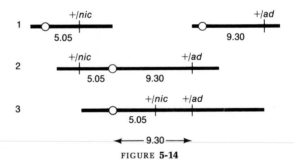

FIGURE 5-14

be parental and independent assortment cannot have occurred, and so possibility 1 can be ruled out.

If we arrange the data in a different way, then possibility 2 is also ruled out.

$+/nic$	$+/ad$	
M_I	M_I	809
M_I	M_{II}	90
M_{II}	M_I	5
M_{II}	M_{II}	96
		1,000

Looked at this way, we can clearly see that a crossover in the centromere-to-*nic* region is virtually always accompanied (96/101 times) by a crossover in the centromere-to-*ad* region; that is, it is the *same crossover*, which simultaneously generates an M_{II} pattern for +/*nic* and for +/*ad*. This is very powerful evidence in favor of possibility 3 and can be seen as:

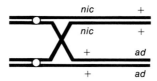

Having ascertained that possibility 3 is correct, we can calculate the recombinant frequency between +/*nic* and +/*ad*. The simple subtraction 9.30 − 5.05 does not give an accurate value for the following reason. We arrived at the figure of 9.30 m.u. by calculating the M_{II} frequency and dividing by 2. In adding up the total M_{II} frequency, we now know that we missed quite a few asci containing crossovers between the centromere and the *ad* locus. For example, class 4 was scored as M_I for +/*ad*, but we know now that *two* crossovers occurred in these asci:

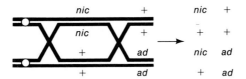

We can better calculate the +/*nic*-to-+/*ad* distance by using the formula

$$RF = \frac{NPD + \frac{1}{2}T}{total\ asci} \times 100$$

remembering that the NPD asci are full of recombinant genotypes and the T asci are "half full" of recombinant genotypes. Therefore:

$$RF = \frac{2 + \frac{1}{2}(100)}{1,000} \times 100 = 5.2\ m.u.$$

Now we can redraw the best map obtainable from these data:

```
                  nic         ad
      ─────○──────┼───────────┼──
           5.05        5.20
      ◄──────────10.25──────────►
```

(*Note:* Class 6 is a very complex ascus. Draw the crossovers needed to produce it, right now!)

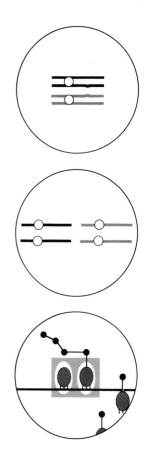

Because it is laborious to isolate linear tetrads, unordered tetrads are sometimes used as a second best but simpler way to study crossing over. The eight ascospores are still recovered clustered together, but not in linear sequence (they are shot out of the ascus by the fungus and land as a group of eight on a collection surface such as a slide.) The products can only be scored as PD, NPD, or T, and, although centromere mapping is not possible, linkage information is still available and an analysis of unordered tetrads introduces some interesting concepts.

Remember that PD asci have no recombinants, NPD are full of recombinants, and T have equal numbers of each. Therefore, a T ascus is irrelevant in deciding whether there is linkage, because its class contributes equally to the parental and recombinant genotypes. The critical test for linkage in unordered asci is therefore whether NPD is equal to or less than PD.

MESSAGE

If the number of PD asci is greater than the number of NPD, then the RF must be less than 50% and there must be linkage between the genes.

If, however, PD = NPD, then RF = 50%; this indicates independent assortment and we cannot say whether the genes are on different chromosomes or are very distantly linked loci. In this dilemma, the T class can be of some use. Let's take an example in which PD = 7, NPD = 11, and T = 2. The χ^2 test shows no significant difference between 7 and 11, yet the T frequency clearly shows that the loci are unlinked. Why is this? If they *were* linked they would have to be a very great distance apart to give a 50% RF value. We can now ask what T frequency would prevail under such conditions.

In a hypothetical cross of $a + \times + b$ in which the a and b loci are linked but very far apart, we can visualize a "maze" of crossovers at meiosis something like that shown in Figure 5-15. We can follow one chromatid—say chromatid 1—from left to right through the maze. There are so many crossovers in the maze that the probability that chromatid 1 will end up being + is equal to the probability that it will end up being b; that is, 1/2 (thus it would be $a +$ or $a b$). Once the possible outcomes for chromatid 1 have been established, the fate of

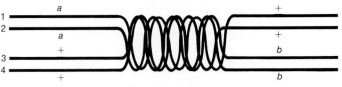

FIGURE 5-15

any other chromatid will determine the ascus type. For example, chromatid 2 now has a restricted choice: if chromatid 1 has occupied b, then there are two out of three chances that chromatid 2 will end up with $+$ and only one out of three chances that it will end up with b. In the former case, we have the two genotypes $a\,b$ and $a\,+$ already determined and of course the ascus *must* be tetratype; the probability of it is $1/2 \times 2/3 = 2/6$. In the latter case, we have the two genotypes $a\,b$ and $a\,b$ determined and the ascus must be an NPD with a probability of $1/2 \times 1/3 = 1/6$. The entire situation is given in Table 5-2.

TABLE **5-2**

Chromatid 1	Chromatid 2	Total p	Spore pairs				Ascus class
$p(+) = 1/2$ $\nearrow$	$p(+) = 1/3$	1/6	$a\ +$	$a\ +$	$+\ b$	$+\ b$	PD
$\searrow$	$p(b)\ = 2/3$	2/6	$a\ +$	$a\ b$	$++ \leftrightarrow a\ b$		T
$p(b)\ = 1/2$ $\nearrow$	$p(b)\ = 1/3$	1/6	$a\ b$	$a\ b$	$++$	$++$	NPD
$\searrow$	$p(+) = 2/3$	2/6	$a\ b$	$a\ +$	$++ \leftrightarrow a\ b$		T

$$\text{Total T} = 2/6 + 2/6 = 2/3 = 66.7\%$$
$$\text{PD} = 1/6 \qquad = 16.7\%$$
$$\text{NPD} = 1/6 \qquad = 16.7\%$$

We have thus derived certain maxima and minima for the ascus classes for distantly linked loci. In fact, we can plot the frequencies of the ascus classes against map distance for linked loci as shown in Figure 5-16. From this plot we can see that, if loci are linked and PD = NPD, then T must be 66.7%. In the example we are working on, T is, in fact, $2/20 = 10\%$; so the loci cannot be linked.

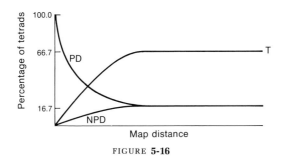

FIGURE **5-16**

At this point, we might as well calculate the equivalent curves for unlinked loci. If the loci are unlinked, a tetratype can arise only from an exchange in at least one locus-to-centromere region. The important distance, then, is the combined locus-to-centromere distance. Let's take an example in which two loci on separate chromosomes are very distant from their centromeres. The M_{II} maxima wil be 66.7%: consider the centromere to be one locus in the T

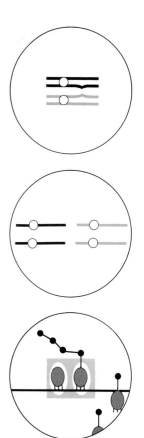

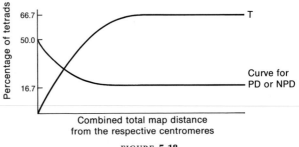

$a/+$

	$1/3\ M_I$	$2/3\ M_{II}$
$1/3\ M_I$	$1/9$ ↗ $1/2$ PD* ↘ $1/2$ NPD	$2/9 \to$ all T
$2/3\ M_{II}$	$2/9 \to$ all T	$4/9$ ↗ $1/4$ PD‡ → $1/2$ T ↘ $1/4$ NPD

$b/+$

*Because of random spindle attachment.
‡Because, if you hold one M_{II} pattern constant and draw all four possible M_{II} patterns for the other locus, the following 1:2:1 ratio for PD:T:NPD is produced.

PD	T	T	NPD
$a+$	$a\ b$	$a+$	$a\ b$
$+b$	$++$	$+b$	$++$
$a+$	$a+$	$a\ b$	$a\ b$
$+b$	$+b$	$++$	$++$

Total T $= 2/9 + 2/9 + 2/9 = 6/9 = 2/3 = 66.7\%$
PD $= 1/9 + 1/18 \qquad\qquad = 1/6 = 16.7\%$
NPD $= 1/9 + 1/18 \qquad\qquad = 1/6 = 16.7\%$

FIGURE 5-17

maximum of 66.7% to see why this is so. Thus we generate a checkerboard as shown in Figure 5-17, and we get a set of curves for unlinked loci as shown in Figure 5-18.

FIGURE 5-18

In conclusion, for linked loci, the value of NPD/T lies between 0 and 1/4. For unlinked loci, NPD/T lies between 1/4 and ∞.

CHROMATID INTERFERENCE

We can ask the question, Does the occurrence of a crossover between two nonsister chromatids affect the probability of a second crossover between the *same* chromatids? As usual, the best hypothesis to test is the null hypothesis, which is that the second crossover will occur randomly between any pair of

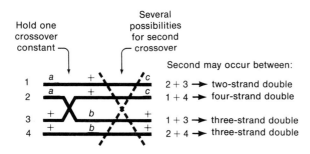

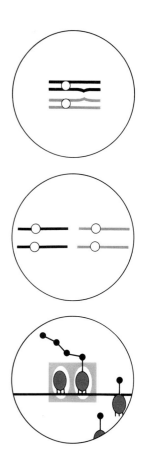

FIGURE **5-19**

(All four possibilities give separate distinguishable ascus genotypes.)

nonsister chromatids; this is illustrated in Figure 5-19. The randomized second crossover should generate a $1:2:1$ ratio of 2-strand:3-strand:4-strand doubles, and any deviation from that ratio can be considered interference between chromatids or chromatid interference. Usually there is very little or no consistent interference of this sort when the analysis of double-exchange tetrads is performed with ordered ascospores in fungi.

Mitotic Segregation and Crossing Over

We normally think of crossing over as a meiotic phenomenon, but it does occur, although less frequently, during mitosis, and can be easily demonstrated if the genetic system is appropriately constructed. (In the definition of recombination on page 80, we simply replace the word meiosis with mitosis, and meiotic diploid with mitotic diploid.)

Bridges had observed that, among *Drosophila* females that were genotypically $M/+$ (where M is a dominant gene that produces a phenotype of slender bristles and is also a recessive lethal), a few had patches of non-M or wild-type bristles. He assumed that this resulted from somatic nondisjunction (Figure 5-20). He also observed patches of mutant tissue expressed by autosomal recessives in heterozygotes, attributable to a similar mechanism.

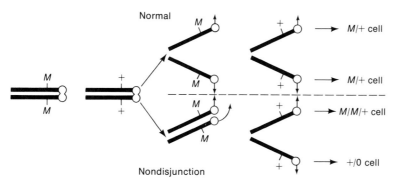

FIGURE **5-20**

Stern in 1936 showed that these somatic mosaics were not always the result of nondisjunction or somatic loss. In a cross of the sex-linked genes *y* (yellow) and *sn* (singed—short, curly bristles), $+ sn/+ sn \times y +/Y$, he obtained females that were mostly wild type but simultaneously carried adjacent patches of *y* and *sn* tissue. These are called *twin spots*. Stern reasoned that, because the occurrence of twin spots was far greater than expected from the occurrence of *y* or *sn* patches alone, they must have been the product of the same reciprocal event. It could be explained by a crossover between homologous chromosomes with subsequent normal separation of centromeres (Figure 5-21). Because there were many more

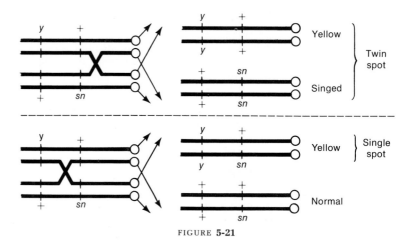

FIGURE **5-21**

twin spots than single *y* spots, most of the crossing over must occur between *sn* and the centromere.

Another widely analyzed system for studying mitotic recombination uses fungi and requires the generation of a diploid fungus. Diploids form spontaneously in many fungi: the one we will examine is *Aspergillus,* a greenish-colored mold. *Aspergillus* is a highly suitable organism for mitotic analysis for several reasons:

1. The hyphae of the fungus bud off long chains of cells called asexual spores. Asexual spores are uninucleate and the appearance or phenotype of any individual one is dependent only on the genotype of its own nucleus. This makes possible certain kinds of selective techniques, which will be discussed below.

2. If two haploid cultures are mixed, the hyphae fuse and both nuclear types are then present in a common cytoplasm. This condition is called a heterokaryon. Let's consider the following heterokaryon:

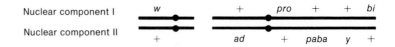

The alleles *ad, pro, paba,* and *bi* are all recessive to their wild-type alleles and each confers a requirement for certain specific chemical supplement in order for growth to occur. The alleles *w* and *y* produce white and yellow asexual spores, respectively, and are also recessive. Thus the heterokaryon does not require any supplement for growth, but, because of the phenotype autonomy described in reason 1 above, the asexual spores are either yellow or white. Thus the heterokaryon looks yellowish white.

3. In some heterokaryons, green sectors appear owing to *diploid* nuclei that have formed spontaneously. The diploid asexual spores can be isolated and diploid cultures grown for study.

4. When the diploid is fully grown, rare sectors showing either white or yellow asexual spores can be observed. Some of these spores are diploid (recognized by their large diameter) and some are haploid (small diameter). Two types are particularly suitable for illustrating the phenomena at work: (1) white sectors with white haploid spores; and (2) yellow sectors with yellow diploid spores (Figure 5-22).

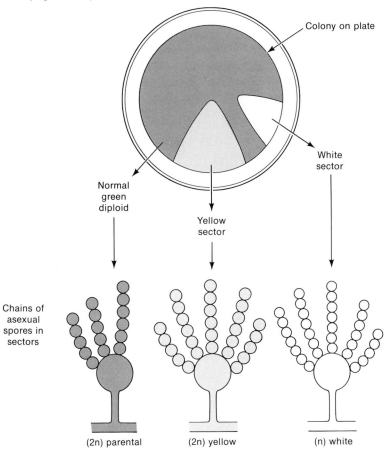

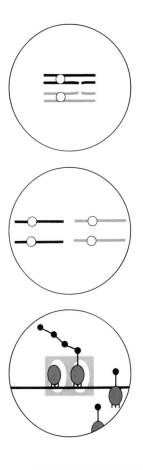

Colony on plate

White sector

Normal green diploid

Yellow sector

Chains of asexual spores in sectors

(2n) parental (2n) yellow (n) white

(Yellow haploid and white diploid are also possible.)

FIGURE **5-22**

Haploid white spores. In isolating and examining these spores, one finds that half of them carry the genotype $+ pro + + bi$ and the other half carry the genotype $ad + paba\ y +$. The diploid nucleus has *haploidized* and a kind of independent assortment has taken place in this process. Haploidization probably occurs by chromosome loss. Nevertheless, you can see that it provides a way of testing whether two loci are linked or are on separate chromosomes. The result here indicates *separate* chromosomes.

Diploid yellow spores. In isolating these spores, one usually finds that they contain recombinant chromosomes. For example, one sector type examined was yellow and also required "paba" (para-aminobenzoic acid) for growth. Mitotic crossing over explains this type as shown in Figure 5-23. Notice that we

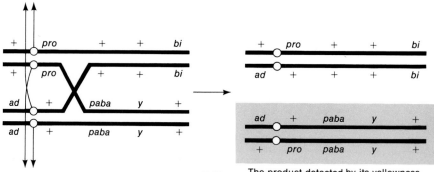

FIGURE 5-23   The product detected by its yellowness

have to follow two spindle fibers going to each pole in mitotic analysis: although one chromatid pair would normally *not* lie adjacent to its homologous chromatid pair, this *has* happened, presumably "accidentally." This yellow diploid arose from a mitotic exchange that obviously took place in the centromere-to-*paba* region; others would arise from mitotic exchange in the *paba*-to-*y* region (these would not require *paba* for growth). The relative frequencies of these two types would give a kind of mitotic linkage map for that region. Such mapping can be obtained for several fungi. As expected, the gene orders correspond to gene orders in meiotic maps, but, unexpectedly, the relative sizes of many of the intervals are very different when meiotic and mitotic maps are compared. Note that every locus from the point of crossing over to the end of the chromosome arm is made homozygous by a mitotic crossover: this can give valuable mapping information and can be a lot quicker than a meiotic analysis.

Mapping Human Chromosomes

Human beings do not submit themselves readily to traditional genetic analysis. Until recently, geneticists have had to rely on the study of family pedigrees in order to deduce linkage of various traits by inference from these (from the geneticist's standpoint) hopelessly inadequate data. Most of the analyses are concerned with the X chromosome where sex-linkage is relatively easily shown, and test crosses are not necessary since the male's phenotype reflects his X chromosome genotype. (Another approach is illustrated in problem 4 of Chapter 4.) Recently, however, a technique has been developed that has great promise for locating the chromosome position of human genes both X-linked and autosomal.

There is a virus called Sendai that has a useful property. Normally, viruses have a specific point for attachment and penetration of a host cell. Each Sendai virus has several points of attachment so that it can simultaneously attach to two different cells if they happen to be close together. A virus, though, is very small in comparison with a cell (similar to the comparison between the planet Earth and the sun) so that the two cells to which it is attached are pulled very close together indeed. In fact, in many cases, the membranes of the two cells fuse together and the two cells become one, a binucleate heterokaryon (Figure 5-24).

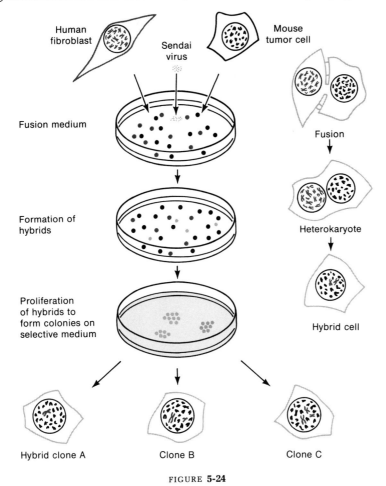

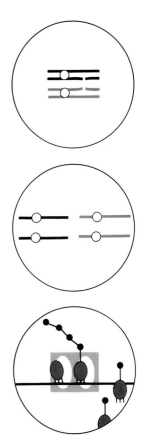

FIGURE 5-24

If suspensions of human and mouse cells are mixed together in the presence of Sendai virus (which has been inactivated by ultraviolet light), the virus can mediate fusion of the two kinds of cells. Once the cells fuse, the nuclei can fuse to form a uninucleate cell line. Because the mouse and human chromosomes are very different in number and shape, the hybrid cells can be readily recognized. However, for some unknown reason, the human chromosomes are gradually eliminated from the hybrid as the cells divide (perhaps this is analogous to haploidization in *Aspergillus*). This process can be arrested to encourage the formation of a stable partial hybrid in the following way. The mouse cells that are used can be made genetically deficient in some function (usually a nutritional one) so that, for growth of cells to occur, the function must be supplied by the human genome. This selective technique usually results in the maintenance of hybrid cells that have a complete set of mouse chromosomes and a small number of human chromosomes, which vary in number and type from hybrid to hybrid.

Luckily, this process can be followed under the microscope because mouse chromosomes can be easily distinguished from human chromosomes. Recently, this has been made a lot easier by the development of fluorescent stains (e.g.,

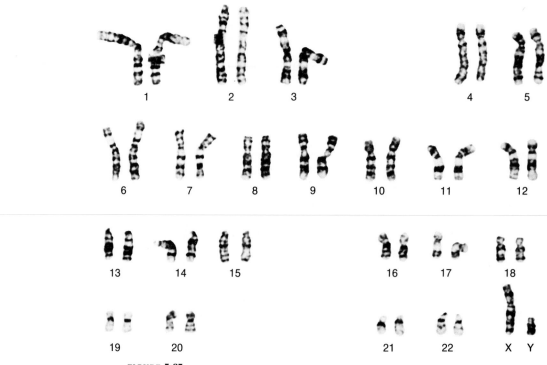

FIGURE 5-25
Karyotype of a human male, showing how size, centromere position, and banding pattern (produced by trypsin-giemsa treatment) can be used to recognize specific human chromosomes.

quinacrine and giemsa) that reveal a pattern of "banding" within the chromosomes. The size and the position of these bands vary from chromosome to chromosome, but the banding patterns are highly specific and constant for each chromosome. Thus, for any hybrid, it is relatively easy to identify the human chromosomes that are present (Figure 5-25).

The mapping technique works as follows: if the human chromosome set contains a genetic *marker* (such as a gene that controls a specific cell-surface antigen, drug resistance, a nutritional requirement, or a protein variant), then the presence or absence of the genetic marker in each line of hybrid cells can be correlated with the presence or absence of certain human chromosomes in each line as shown in Table 5-3. We can see that in the different hybrid cell lines,

TABLE **5-3**

		Hybrid cell lines				
		A	B	C	D	E
Human genes	1	+	−	−	+	−
	2	−	+	−	+	−
	3	+	−	−	+	−
	4	+	+	+	−	−
Human chromosomes	1	−	+	−	+	−
	2	+	−	−	+	−
	3	−	−	−	+	+

genes 1 and 3 are always present or absent together. We conclude then that they are linked. Furthermore, the presence or absence of genes 1 and 3 is directly correlated with the presence or absence of chromosome 2 on which we assume they are located. By the same reasoning, gene 2 must be on chromosome 1, but the location of gene 4 cannot be assigned.

Large numbers of human genes have now been localized to specific chromosomes in this way but of course we cannot derive a linkage map showing the order and distances between genes. Other tricks are needed—for example, the loss or gain of variously sized bits of a specific chromosome might be correlated with the presence or absence of genetic markers. A problem at the end of Chapter 6 will enable you to think through the kind of logic involved. However, the results of this kind of intrachromosomal mapping are so far not nearly as extensive as those on simple chromosome location (Figure 5-26).

MESSAGE

Mitotic as well as meiotic phenomena can give information on gene location in fruit flies, fungi, and man.

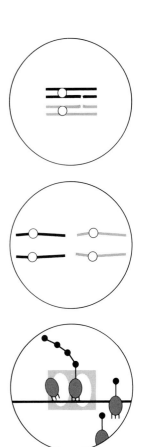

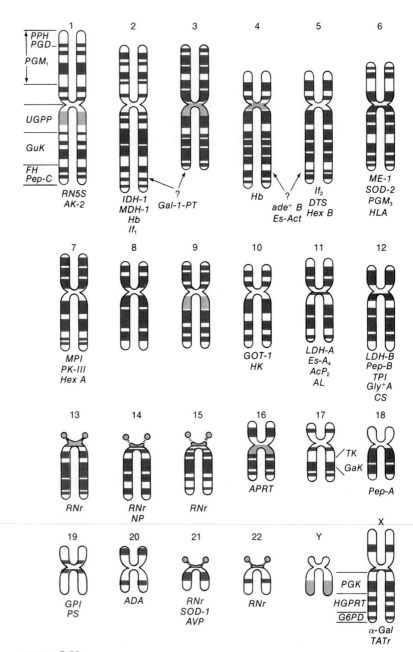

FIGURE 5-26
Genes that have been assigned to specific human chromosomes by the cell hybridization technique. Characteristic quinacrine banding patterns of each chromosome are shown. The shorthand gene designations relate to their phenotypes. Note that only a few approximate loci have been assigned.

In this chapter we have explored the mechanics of the chromosome theory of heredity and have found a perfect internal consistency. We can consider the chromosome theory proved: it is now an accepted part of the day-to-day arsenal of weapons at the disposal of the genetic analyst.

Problems

1. In *Drosophila melanogaster*, two entire X chromosomes can be attached to the same centromere:

The two arms now behave as a single chromosome called an attached-X. Work out the inheritance of sex chromosomes in crosses of attached-X-bearing females with normal males. During meiosis, the duplicated attached-X chromosome segregates from its sister chromatids as:

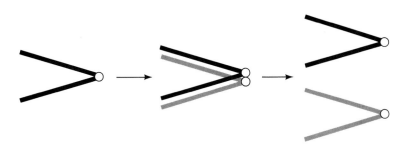

Crossing over can take place between nonsister chromatids of attached-X chromosomes. Suppose you have an attached-X chromosome of the following genotype:

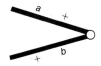

Diagram all of the possible genotypic products from:
a. Single crossovers between *a* and *b* and between *b* and the centromere.
b. Double exchanges between *a* and *b* and *b* and the centromere.
Can you do a tetrad analysis of the data? Try.

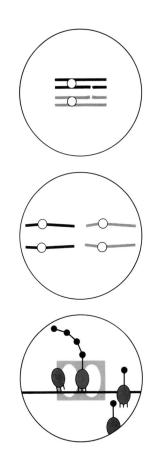

2. An *Aspergillus* diploid that was $+ + + +/w\,a\,b\,c$ was haploidized and many white haploids were scored for *a*, *b*, and *c*. The results were

w a b c	25%
w + + +	25%
w a + *c*	25%
w + *b* +	25%

What linkage relations can you deduce from these frequencies? Sketch your conclusions.

3. In *Neurospora*, the *a* locus is 5 map units from the centromere on chromosome 1. The *b* locus is 10 map units from the centromere on chromosome 7. From the cross $a + \times + b$, what will be the frequency of:

a. Parental ditype asci?
b. Nonparental ditype asci?
c. Tetratype asci?
d. Recombinant ascospores?
e. Colonies growing from ascospores plated on minimal medium if *a* and *b* represent nutritional requirements?

(*Note:* Don't bother with map function complications in this problem).

4. In *Neurospora* crosses $a\,b \times + +$ (in which *a* and *b* represent different loci in each cross), 100 linear asci are analyzed from each cross to give the following results:

	Ascus types						
	a b	a +	a b	a b	a b	a +	a +
	a b	a +	a +	+ b	+ +	+ b	+ b
	+ +	+ b	+ +	+ +	+ +	+ b	+ +
	+ +	+ b	+ b	a +	a b	a +	a b
Cross	Number of each ascus type						
1	34	34	32	0	0	0	0
2	84	1	15	0	0	0	0
3	55	3	40	0	2	0	0
4	71	1	18	1	8	0	1
5	9	6	24	22	8	10	20
6	31	0	1	3	61	0	4
7	95	0	3	2	0	0	0
8	6	7	20	22	12	11	22
9	69	0	10	18	0	1	2
10	16	14	2	60	1	2	5
11	51	49	0	0	0	0	0

For each cross, map the genes in relation to each other and to their respective centromere(s).

5. a. By using the map function, calculate how many real map units are indicated by a recombinant frequency of 20%. Remember, a mean of 1 equals 50 real map units.

   b. If you obtained a RF value of 45% in one experiment, what can you say about linkage? (Actual figures are 58, 52, 47, and 43 out of 200 progeny.)

6. In *Neurospora*, the cross $+ + + + \times a\,b\,c\,d$ is made in which *a*, *b*, *c*, and *d* are linked in the order written. Draw crossover diagrams to illustrate how the following unordered (nonlinear) ascus patterns could arise:

$+ + c +$	$+ b\,c\,d$	$+ + c +$	$+ + c +$	$+ b + d$	$+ b\,c\,d$	$+ b + d$	$+ b + +$	$+ b + d$
$a\ b\,c +$	$+ + + d$	$+ b + d$	$+ b + d$	$+ b + d$	$a\ b\,c +$	$+ + + d$	$+ b + +$	$a + c +$
$+ b + d$	$a\,b\,c +$	$a + c +$	$a + c +$	$a + c +$	$+ + + d$	$a\,b\,c +$	$a + c\,d$	$+ b + d$
$a + + d$	$a + + +$	$a\,b + d$	$a\,b + d$	$a + c +$	$a + + +$	$a + c +$	$a + c\,d$	$a + c +$

7. Complete the table for a situation in *Neurospora* in which the *mean* number of crossovers between the *a* locus $(a/+)$ and its centromere is equal to one per meiosis.

	Number of exchanges				
	0	1	2	3	4
Probability of this kind of meiosis?					
What proportion of each of these kinds of meiosis will result in an M_{II} pattern for $a/+$?					
What proportion of all asci from this cross will show an M_{II} pattern as a result of each of these kinds of meiosis?					

   a. What is the total M_{II} frequency if the mean $= 1$?

   b. Complete similar tables for means of 0.5, 2.0, and 4.0, and hence draw a mapping function for M_{II} frequency (i.e., plot the total M_{II} frequency against mean crossover frequency per meiosis).

   c. Why does the curve bend downward? How would you correct this?

8. In an *Aspergillus* diploid $+ + / y\ ribo$, the two loci are linked, but the order with respect to the centromere is not known. Yellow diploid segregants were obtained: 80% of them were $ribo^+$ and 20% were *ribo*-requiring. What is the most likely order?

9. In several nonlinear tetrad analyses, in which each experiment consists of a cross between different pairs of marked loci, the following results were obtained. Deduce in each case if there is any indication of linkage (derive as much information as possible).

Cross	PD	NPD	T
1	22	3	21
2	1	2	0
3	1	1	3
4	10	10	9
5	10	11	1
6	18	15	50

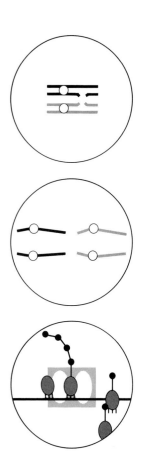

10. An *Aspergillus* diploid was + + +/*pro fpa paba*, in which *pro* is a recessive allele for proline requirement, *fpa* is a recessive allele for fluorophenylalanine resistance, and *paba* is a recessive allele for *para*-aminobenzoic acid requirement. By plating asexual spores on fluorophenylalanine, selection for resistant colonies can be made. Of 154 *diploid* resistant colonies, 35 required neither pro nor paba, 110 required paba, and 9 required both.

    a. What do these figures indicate?

    b. Sketch your conclusions in the form of a map.

    c. Some resistant colonies (not the ones above) were haploid. What would you predict their genotype would be?

11. The following diagram shows three man/mouse hybrid clones and the only human chromosomes they contain.

Human chromosomes
(+ means present)

		1	2	3	4	5	6	7	8
Hybrid clones	A	+	+	+	+	−	−	−	−
	B	+	+	−	−	+	+	−	−
	C	+	−	+	−	+	−	+	−

Five enzymes, α, β, γ, δ, and ε were tested in each of the clones and the results were as follows:

α: activity only in clone C

β: activity in all three clones

γ: activity only in clones B and C

δ: activity only in clone B

ε: no activity in any clone

What can you say about the location of the genes responsible for these enzyme activities?

12. Four histidine loci are known in *Neurospora*. As shown below, each of the four loci is located on a different chromosome.

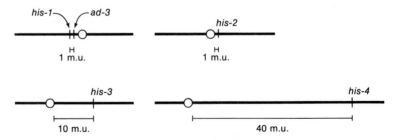

The *his*-1 locus is closely linked (1 m.u.) to both *ad*-3 and the centromere; *his*-2, -3, and -4 are 1, 10, and 40 m.u., respectively, from their centromeres. In your experiment you began with an *ad-3* strain from which you recover a cell that also requires histidine. You wish now to determine which of the four histidine loci was involved. You cross the *ad-3 his-?* strain with wild type (+ +) and analyse ten unordered tetrads: two are PD, six are T, and two are NPD. From this result, which of the four *his* loci was most probably the one that changed from *his+* to *his?*[1]

13. The accompanying drawing shows a germinated teliospore of *Ustilago hordei* that has just undergone meiosis. Each haploid cell of the promycelium has undergone

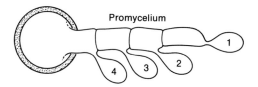

Promycelium

1

2

3

4

mitosis to form a sporidium. The four haploid sporidia are numbered in sequence, beginning with the terminal cell. The sporidia can be removed, in sequence, to establish haploid colonies, which are easy to maintain in the laboratory. Haploid colonies can be subcultured and combined in pairs and, when this is done, compatible combinations result in the formation of dikaryons; incompatible combinations do not. For one such set of haploid cultures the following result was obtained:

Mating

$\left.\begin{array}{l} 1 \times 2 \\ 1 \times 4 \\ 3 \times 2 \\ 3 \times 4 \end{array}\right\}$ Dikaryons were formed

$\left.\begin{array}{l} 2 \times 4 \\ 1 \times 3 \end{array}\right\}$ Dikaryons were not formed

a. What can you conclude about the genetic determination of compatibility? Dikaryons of *U. hordei* are parasitic on cultivated barley. The four dikaryons listed above, when inoculated into three different barley cultivars (varieties of barley), gave the following results:

	Disease on cultivar		
Dikaryon	A	B	C
1 × 2	None	Severe	None
1 × 4	Severe	None	None
3 × 2	Severe	None	None
3 × 4	Severe	None	Severe

[1] Problem 12 courtesy of Luke deLange.

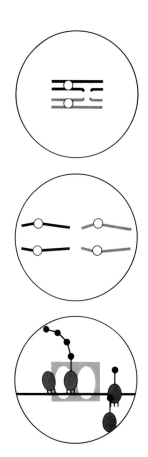

b. What can you conclude about the genetic determination of virulence (i.e., of the capacity of dikaryons to incite a severe disease reaction) on these cultivars?

c. Using your own system of symbols, assign genotypes to each of the four haploid cultures.[2]

---

[2]Problem 13 courtesy of Clayton Person.

6 / Recombination in Bacteria and Their Viruses

(Or: The extension of recombination analysis to some exotic life cycles.)

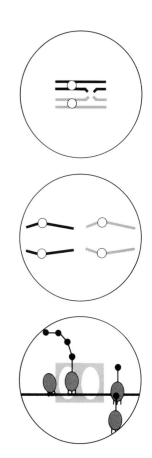

So far, we have been dealing exclusively with eukaryotic organisms; that is, those with their genes packed into chromosomes and contained within a nucleus. However, a very large part of the history of genetics and current genetic analysis (especially molecular genetics) is concerned with prokaryotic organisms and viruses. Viruses are a problem for biologists because they aren't cells and can't grow and multiply alone; they must parasitize living cells, using the cell's metabolic machinery to reproduce. Nevertheless, as we shall see, they have hereditary properties and can be used for genetic analysis. We may refer to viruses as organisms, but always bear in mind that such free use of the word rattles a lot of biologists and is used only to simplify discussion.

Prokaryotic organisms and viruses have very simple chromosomes (compared with those of eukaryotes) that are not contained in a nuclear membrane. Nor do they undergo meiosis and, although the ways in which they undergo reproduction may seem strange, they still have stages that are analogous to meiosis. The genetic analysis of prokaryotic recombination is surprisingly similar to that for eukaryotes, and, although treated in different chapters in this book, there is no fundamental difference between the two systems.

The prokaryotes are bacteria and blue-green algae, but only the bacteria have been extensively used in genetics. The best-studied viruses are those that parasitize bacteria and are called bacterial viruses or *bacteriophages*.

The early history of genetics is dominated by the stunning success in working out the basic features of Mendelian genetics and linkage in *Drosophila*. Bacteria are biologically important not only from a medical standpoint, but because of their great evolutionary success, as measured by the very large proportion of the biomass on Earth that they compose. Also, their very simplicity as single-celled organisms makes them attractive for the study of basic cellular processes. But how is it possible to study inheritance in organisms too small to be seen without a microscope?

Bacteria can be grown in two ways—in a liquid medium and on a solid surface such as agar—so long as basic nutritive ingredients are supplied. In liquid medium, the bacteria divide by binary fission and multiply geometrically until the nutrients are exhausted or toxic factors accumulate. A small amount of the liquid culture can be pipetted onto a petri plate containing an agar medium and spread evenly on the surface with a sterile glass rod. Each cell will then reproduce by fission, but because they are immobilized all of the daughter cells will remain together in a clump and soon this mass of more than 10^7 cells will become visible to the naked eye as a *colony*. Note that, if the sample initially *plated* consists of a small number of cells, each distinct colony will derive from a single, original cell (Figure 6-1).

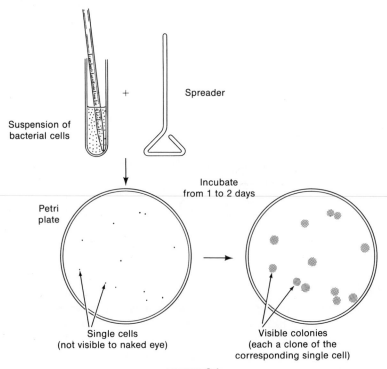

FIGURE **6-1**

Can we show heredity in bacteria? The gut bacterium *Escherichia coli* (hereafter *E. coli*) has been very popular among geneticists, probably because it is not a serious pathogen in humans and can be grown on a very simple medium of salts and glucose. In the early 1940s, a number of different strains of *E. coli* were selected (we'll tell you how in a later chapter). Some of them were resistant to antibiotics and others could no longer grow on the basic medium and required supplements of specific organic molecules such as amino acids. Thus, for example, wild-type *E. coli* thrives on the basic medium (we'll call it a *minimal medium*), whereas a *thr⁻* strain will grow only if the minimal medium is supplemented with the amino acid threonine. Both strains "breed true" in that progeny cells have the same growth requirements as the parental cells. Once we know what the genetic differences between the strains are, we can ask whether "crosses" are possible and whether exchanges of genes can occur.

Joshua Lederberg and Edward Tatum studied two strains of *E. coli* in a line called K12. They had the following phenotypes:

1. Strain A will grow on minimal medium if supplemented with methionine and biotin.
2. Strain B will grow on minimal medium if supplemented with threonine, leucine, and thiamine.

We can designate strain A as *met⁻ bio⁻ thr⁺ leu⁺ thi⁺* and strain B *met⁺ bio⁺ thr⁻ leu⁻ thi⁻*. When millions of cells from either strain A or strain B are plated on minimal medium, no colonies are ever seen. Lederberg and Tatum then did the following experiment. Cells of strains A and B were mixed together in a liquid medium containing all five growth factors and incubated for several hours; then the culture was centrifuged so that the cells were forced to the bottom of the test tube in a "pellet." The liquid supernatant was removed, the cells were washed and then suspended (by vigorous stirring) in minimal medium, and an aliquot (a sample) was plated on minimal medium. From the suspension of cells of both strains colonies were recovered at a frequency of about 1 for every 10⁷ cells (or, we say, 1×10^{-7}). That is a rather low frequency compared with the recombination values obtained from fruit flies, but it still suggests that bacteria can exchange genetic information to give *met⁺ bio⁺ thr⁺ leu⁺ thi⁺*. (Notice how much labor is saved using bacteria. One cell can be picked out of ten million because it is the only one that survives on minimal medium. Geneticists are quite lazy; they love such selective techniques.)

It could be suggested that the cells of the two strains do not really exchange genes but leak substances that the other cells can absorb and use for growing (this won't seem so farfetched in later chapters). This possibility of "cross feeding" was ruled out by Bernard Davis who constructed a U tube in which the two arms were separated by a fine filter. The pores of the filter were too small to allow bacteria through, but the medium percolated through easily enough

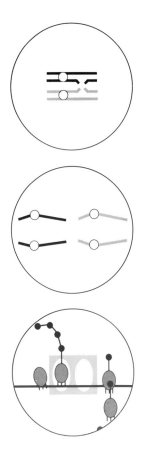

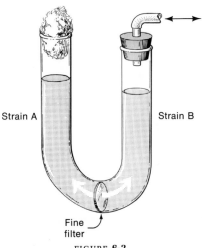

Strain A   Strain B

Fine
filter

FIGURE 6-2

(Figure 6-2). Strain A was put in one arm and strain B in the other; after they had been incubated for a while, the content of each arm was tested for cells able to grow on minimal medium, and none was found. In other words, *physical contact* between the two strains was needed for wild-type cells to form. It looked as though "sex" was involved.

Lederberg and Tatum assumed that a process similar to meiosis was occurring and therefore that linkage might be established. How could this be done? If the cross between strain A and strain B is repeated but the cells are then plated on minimal medium plus *one* of the nutritional requirements, then wild-type alleles for all but one of the genes are needed for growth. For example, if leucine is added to the medium, then both bio^+ met^+ thr^+ leu^+ thi^+ and bio^+ met^+ thr^+ leu^- thi^+ will grow. The allelic state of the leucine gene can be determined later. So, if leu^+ is linked to either thr^+ or thi^+, then most of the survivors on minimal medium should also be leu^+, because they come in from strain B cells. The ratio of the $-$ to $+$ alleles of the "unselected" gene should indicate no linkage if it is 1.0, linkage to a wild-type gene if it is less than 1.0, and linkage to a nonwild type if it is greater than 1.0. The results obtained were:

Medium supplement	Ratio of $-/+$ alleles of unselected gene
Biotin	0.170
Threonine	0.240
Leucine	0.096
Thiamine	9.880

This suggests that there is linkage. For example, there is an excess of leu^+ among the survivors, thereby showing that selection for the thr^+ and thi^+ also brings *leu⁺*. The value for the thiamine gene shows that *thi⁻* tends to be recovered among survivors, thereby showing that it is included when *bio⁺* and *met⁺* are recovered.

This is a complicated way to do a genetic analysis and, in fact, Lederberg found it increasingly difficult to interpret his data. William Hayes then made a remarkable discovery. He made a cross similar to that made by Lederberg and Tatum:

Strain A   Strain B

met⁻ thr⁺ leu⁺ thi⁺ × *met⁺ thr⁻ leu⁻ thi⁻*

However, if he treated either strain A or strain B with the antibiotic streptomycin (which does not kill immediately but which prevents cell division) and then mixed in the other strain, different results were obtained. Treatment of strain B resulted in no survivors on minimal medium. But treatment of strain A before mating gave the same frequency of survivors as untreated controls.

What does that mean? One interpretation would be that:

MESSAGE

The exchange of genetic material in E. coli *is not reciprocal; in fact one cell (strain B) acts as the* recipient, *whereas the other (strain A) is a* donor.

The donor can still transmit genes after exposure to streptomycin; this kind of unidirectional exchange of genes could be analogized to a sexual difference, the donor being male and the recipient female (Figure 6-4 on page 136).

By accident, Hayes discovered a variant of his original A strain (male) that, on crossing with the B strain (female), no longer gave recombinants. Had the A males changed into females or perhaps even into homosexuals? The original male strain was sensitive to streptomycin (*strˢ*) and Hayes was able to recover a streptomycin resistant strain (*strʳ*) in the sterile A variants. When he mixed the sterile A *strʳ* cells with the fertile A male *strˢ* cells and then plated on a medium containing streptomycin, he found that as many as one-third of the A *strʳ* cells were fertile when crossed with B females (Figure 6-3). Hayes

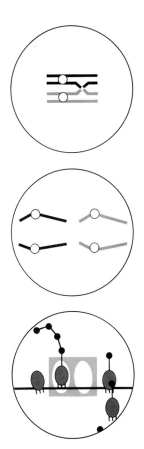

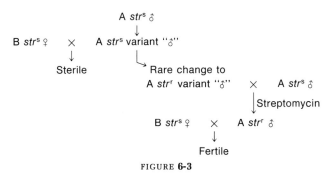

A *strˢ* ♂
↓
B *strˢ* ♀   ×   A *strˢ* variant "♂"
↓   ↓
Sterile   Rare change to
                  A *strʳ* variant "♂"   ×   A *strˢ* ♂
                                              │ Streptomycin
                                              ↓
                  B *strˢ* ♀   ×   A *strʳ* ♂
                            ↓
                         Fertile

FIGURE **6-3**

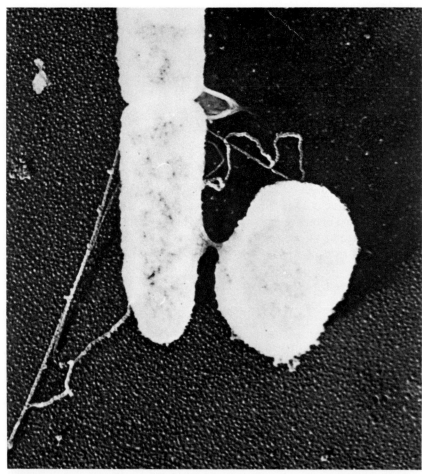

FIGURE **6-4**
Electron micrograph of a pair of cells from a cross in *E. coli*. The long cell of the
male strain and the oval cell of the female strain are connected by a bridge, which is
the probable route through which gene exchange occurs.

explained these bizarre results by suggesting that maleness or donor ability
is itself a hereditary state imposed by a *fertility factor,* or F. Females lack F
and are therefore recipients. Thus, females can be designated F^- and males F^+.
The A strain variant must have lost F and become F^-. Note that, although
recombination is rare, F appears to be transmitted very effectively by physical
contact or conjugation that can be seen with the electron microscope. The
physical basis for the genetically inferred fertility factor was found years later.

Hayes found that, in $F^+ \times F^-$ crosses, most of the genes in recombinants
were from the F^- parent. For example, in a generalized cross, $F^+ \ a^+ \ b^+ \ c^- \ d^+ \times$
$F^- \ a^- \ b^- \ c^+ \ d^-$, if only $b^+ c^+$ recombinants are selected initially and then the
a and d genes scored, most of such recombinants will be $a^- \ d^-$ rather than $a^+ \ d^+$.

This is quite different from any kind of meiotic process and seems to indicate that only part of the F$^+$ genes are transferred to the F$^-$.

The big break in trying to figure out the processes taking place came when Luca Cavalli-Sforza recovered from the A male strain a new strain that, on crossing with B females, gave a thousand times as many recombinants as an A F$^+$ × B F$^-$ cross. He called this derivative a *high frequency of recombination* or Hfr strain. It became known as Hfr C (for Cavalli-Sforza). Hayes later found a similar type of derivative, Hfr H. In F$^+$ × F$^-$ crosses, a large proportion of the originally F$^-$ genotypes recovered (called exconjugants) was found to have been converted into F$^+$, but in Hfr × F$^-$ crosses, none of the F$^-$ exconjugants had been converted. If all this seems weird to you, you know why geneticists until the late 1950s tried to ignore bacterial genetics.

It all began to fit together when Ellie Wollman and François Jacob asked the question, When are the genes of an Hfr transferred to an F$^-$ during mating? They conducted the following experiment. They crossed Hfr *str^s a$^+$ b$^+$ c$^+$ d$^+$* with F$^-$ *str^r a$^-$ b$^-$ c$^-$ d$^-$* and, at specific time intervals after the cross, they removed a sample, which was put in a Waring Blendor for a few seconds to disrupt mating and then plated onto a medium containing streptomycin to select for exconjugants. This is called an "interrupted mating" experiment. The exconjugants were then tested for the presence of donor markers. The results obtained can be plotted as shown in Figure 6-5, in which *azi^r*, *ton^r*, *lac$^+$*, and *gal$^+$* correspond to the *a$^+$ b$^+$ c$^+$ d$^+$* above.

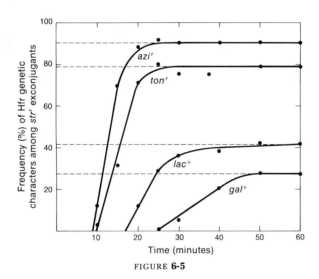

FIGURE **6-5**

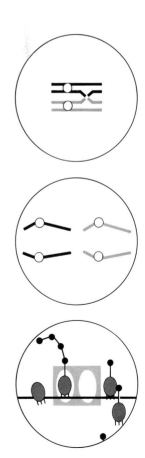

How can we interpret these results? The most striking thing is that each donor gene first appears in the F$^-$ recipients at a specific time after mating begins. Furthermore, the donor genes appear in a specific sequence. Finally, the maximal yield of recombinants is lower for the donor markers that enter later. Putting it all together, it suggests:

MESSAGE

The Hfr chromosome is transferred to the F⁻ in a linear fashion, beginning at a specific point (called the origin or O). The further a gene is from O, the later it is transferred to the F⁻. For later genes, it is likely that the transfer process will stop before they are transferred (hence the decreased slope and maximum for later genes).

Thus, Wollman and Jacob realized that linkage maps could be easily constructed from interrupted-mating studies using the time at which the first recombinants appear after mating. Here the units of distance are in minutes; that is, if *b* begins to enter the F⁻ 10 minutes after *a* began to enter, then *a* and *b* are 10 units apart. Again, this abstract linkage map was purely a genetic construction and had no known physical basis (Figure 6-6).

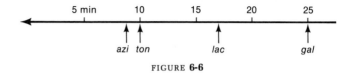

FIGURE 6-6

When Wollman and Jacob allowed Hfr × F⁻ crosses to continue for as long as two hours before blending, they found that some of the F⁻ recipients were converted into Hfr. In other words, the fertility factor conferring maleness or donor ability was eventually transmitted, but at a very low efficiency and as the last element of the linear "chromosome." We now have the picture

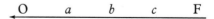

However, when several different Hfr linkage maps were derived by interrupted-mating and time-of-entry studies, they were found to differ from strain to strain:

Hfr H	O	thr	pro	lac	pur	gal	his	gly	thi
1	O	thr	thi	gly	his	gal	pur	lac	pro
2	O	pro	thr	thi	gly	his	gal	pur	lac
3	O	pur	lac	pro	thr	thi	gly	his	gal
AB 312	O	thi	thr	pro	lac	pur	gal	his	gly

At first glance this does not make sense but, if you compare the order, you will see that a pattern does emerge. The genes are not thrown together at random in each strain. For example, all *his* genes have *gal* on one side and *gly* on the other. The same applies to the other genes unless they are on opposite ends of the linkage map. The order in which the genes are transferred is not constant, for example, in two strains the order is O *his gly* and in three, it is O *gly his*.

The fertility factor is always at the end opposite the origin. Can you see how the different Hfr linkage maps could be derived? Suppose that, in F$^+$ males, F is a small cytoplasmic element (and therefore easily transferred to an F$^-$ on conjugation). If the "chromosome" of the F$^+$ male is a *ring,* then any of the linear Hfr chromosomes could be generated simply by inserting F at different places in the ring (Figure 6-7).

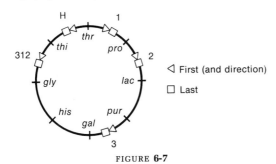

FIGURE 6-7

Again, chromosome circularity was a wildly implausible concept that was inferred from genetic data and was not confirmed physically until years later. The insertion of F seems to determine polarity, with the end opposite F being the origin. How might we account for F attachment? Wollman and Jacob suggested that some kind of recombination event between F and the chromosome might generate the Hfr chromosome. Alan Campbell then made a brilliant suggestion that provided a genetic model.

As we shall see in a later chapter, a single crossover between two circular chromosomes results in the production of a single larger ring. Thus, Campbell proposed that F, like the chromosome, is circular. Hence, a crossover between the two elements

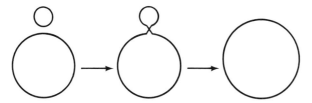

would produce a single ring. If we suggest that F consists of three different regions

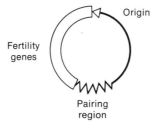

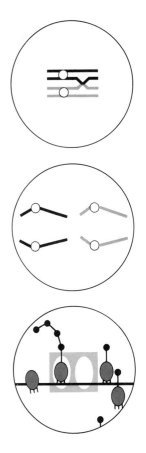

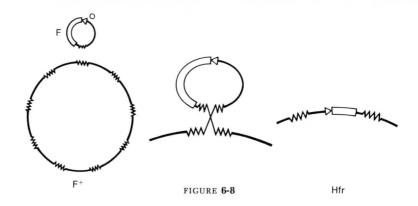

F⁺   FIGURE **6-8**   Hfr

and that the bacterial chromosome has regions of pairing homology with F, Hfr chromosomes can be generated easily by a crossover (Figure 6-8). Probably *all* recombinants in an F⁺ × F⁻ cross are attributable to infrequent insertion of the F factor in a few cells.

The fertility factor is seen to exist in two states—as a free cytoplasmic element that is easily transferred to F⁻ recipients and as an integrated part of a circular chromosome that is transmitted only very late in conjugation. Genetic factors that have these two states are called *episomes*.

MESSAGE

Episomes are genetic factors in bacteria that can exist either as elements in the cytoplasm or as integrated parts of a chromosome.

Does an Hfr die on conjugation? The answer is no, and there is evidence that shortly before or during conjugation, the Hfr chromosome replicates, thereby ensuring a complete chromosome for the donor after mating. Finally, because F⁻ is readily converted into F⁺ from which Hfr can be derived, we assume the F⁻ chromosome is also circular.

We can now summarize the sequence of sexual events in *E. coli* (Figure 6-9).

Up to this point we have discussed only the process of transfer of genetic information between individuals in a cross. This was inferred from recombinants produced from the cross. However, before a recombinant is recovered, it is necessary for the transferred genes to be "integrated" or *incorporated* into the host's genome by an exchange mechanism. We will now consider some of the special properties of this exchange event. Genetic exchange in prokaryotes does not take place between two whole genomes (as it does in eukaryotes); rather it takes place between a complete genome (called the F⁻ *endogenote*) and an incomplete one (called the donor *exogenote*). What we have in fact is a partial diploid or *merozygote*. Bacterial genetics is merozygote genetics.

Diagrammatically, the merozygote looks like this:

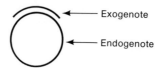

Exogenote

Endogenote

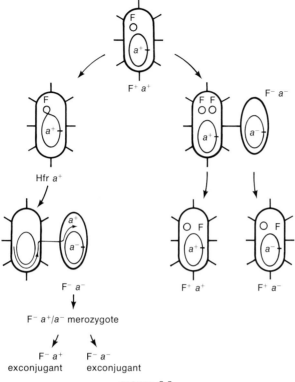

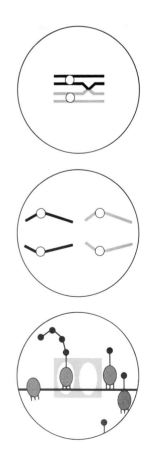

FIGURE **6-9**

It is obvious that a single "crossover" would not be very useful in generating viable recombinants because the ring is broken and a strange partially diploid linear chromosome is produced.

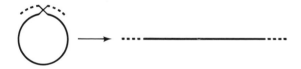

An even number is needed to maintain the F⁻ intact.

The fragment is only a partial genome and usually gets lost during subsequent cell growth. (We say "usually" because there are ways to maintain "stable partial diploids.") Hence, it is obvious that reciprocal products of recombination do not survive—only one does. A further unique property of bacterial exchange, then, is that we have to forget about reciprocal events.

Recombination Mapping

We can use recombination mapping to calculate precise distances if the unit of transfer between two genes is less than about two minutes, in which case the time-unit method is not too reliable. With respect to two closely linked genes—such as *leu* and *ade*, for example—we discover by interrupted-mating tests that, for certain Hfr donors, *ade* enters last. So, if we select for ade^+ F^- exconjugants (by plating on medium having no adenine but all the other requirements), we can recover recombinant products of an event that must have originated from a merozygote that received *both* ade^+ and leu^+ from the donor (because ade^+ enters last). We can then ask the question, What proportion of ade^+ F^- recombinants is also leu^+? knowing that the fraction that is ade^+ leu^- must have had a recombination event *between* the *ade* and *leu* loci.

Diagrammatically then, the incorporation of ade^+ could have resulted from two types of double crossover events.

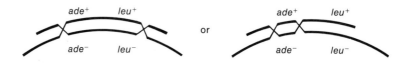

Thus the fraction

$$\frac{ade^+ \ leu^-}{(ade^+ \ leu^+) + (ade^+ \ leu^-)} \times 100$$

is an index of the separation distance between the two loci and can be called the recombination frequency. For two loci (such as *ade* and *leu*) separated by about one minute (time unit), a recombination frequency of about 20% will be obtained. However, our point is that, for two minutes or less, the time measurement is unreliable and the RF measurement more constant.

MAPPING SITES THAT ARE CLOSE TOGETHER

As we shall see in the next chapter, it often happens that we want an estimate of map distance between sites that are very close together, which poses two problems.

First, it can be difficult to make a double mutant to be used as a recipient. (Why is this?) So we cannot make the desired cross $a^+ b^+ \male \times a^- b^- \female$ and we have to cross $a^+ b^- \times a^- b^+$. This produces a

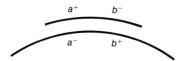

merozygote. (In this situation the sites are so close that we do not need to worry about selecting the marker that enters last because the exogenote will nearly always carry both donor alleles). We could select for recombinants by plating F^- exconjugants on minimal medium and counting the colonies that will be $a^+ b^+$ and will have come from this kind of exchange.

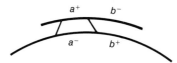

Having counted them, we have the numerator in a possible RF ratio, but we are still left with a dilemma, which is the second problem. We have no value to use as denominator. We have no idea how many merozygotes these wild-type cells come from. One might think that the number of conjugation partners put into the mating mixture would be a possibility, but in practice this does not work because the number of them that actually enter into mating events is greatly dependent on experimental conditions.

One way around this dilemma is to use the incorporation of another marker as a standard. For example, the *trp A* and *trp B* loci are very close together and mapping the distance between them requires the use of another locus called *his*. Two merozygotes are of interest.

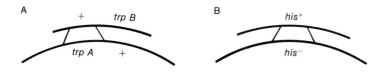

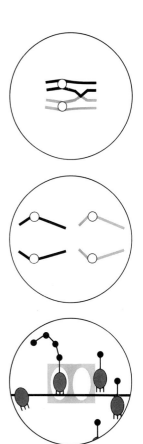

It can be seen that the double exchange needed to recombine *trp A*$^+$ into the F^- requires one highly specific exchange between the loci, whereas, to get the *his*$^+$ in, unspecific exchange on each side is needed. We can use the number of *his*$^+$ recombinants as an index of "general" double exchange in this mating pair, and the number of *trp A*$^+$ *trp B*$^+$ cells for the specific exchange whose frequency depends on the distance between the sites we are interested in.

For example, a good index of the relative distance between *trp A* and *trp B* would be

$$\frac{trp\ A^+\ trp\ B^+\ \text{recombinants}}{his^+\ \text{recombinants}}$$

Most of the merozygote systems in bacteria can use this technique. Such ratios remain very constant and independent of experimental conditions. We hope you are beginning to appreciate that bacteria do not do things the way eukaryotes do. Yet, although their analysis bears little superficial resemblance to that of eukaryotes, the principles of mapping are very similar. The geneticist's approach is similar, whether studying humans, peas, flies, or bacteria.

DERIVING GENE ORDER

Very often, loci, or *sites*, are so close together that it becomes difficult to order them in relation to a third locus. For example, when three loci are linked thus,

the RF between *a* and *b* under most experimental conditions will be more or less the same as that from *a* to *c*. Is the order *a–b–c* or *a–c–b*? Once again a trick is needed to resolve the question. We need to make a pair of *reciprocal crosses*. Depending on the order, the crossover events needed to generate prototrophs from the cross *a b c⁺* × *a⁺ b⁺ c* are shown in Figure 6-10.

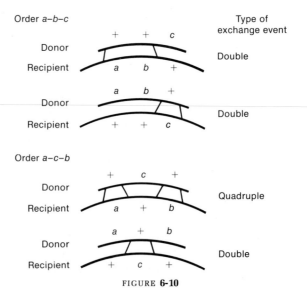

FIGURE **6-10**

Thus, if the reciprocal crosses give dramatically different frequencies of wild-type survivors on minimal medium, we know the order is *a–c–b*. If no difference between the reciprocal crosses is observed, the order must be *a–b–c*.

Sexduction

The initially confusing results of recombination work in *E. coli* have been very nicely explained now that we understand the F⁻, F⁺, and Hfr states. Knowing all this, Edward Adelberg began to do recombination experiments with an Hfr strain in 1959 but found that the particular Hfr strain kept producing F⁺ cells and therefore recombination frequencies were not all that high. He called this particular fertility factor F′ to signify a difference from the normal F for the following reasons:

1. The F′-bearing F⁺ strain reverted to an Hfr state quite often.
2. F′ always integrated at the *same place* to give back the original Hfr chromosome (remember, randomly selected Hfr derivatives from F⁺ males have origins at many different positions).

How could these properties of F′ be explained? The answer came from the recovery of a new F′ from an Hfr strain in which the *lac⁺* locus was near the end of the Hfr chromosome (i.e., was transferred very late). Jacob and Adelberg found an F⁺ derivative that transferred *lac⁺* to F⁻ *lac⁻* recipients at a very high frequency. Furthermore, the recipients, which became F⁺ *lac⁺* phenotypically, occasionally (1×10^{-3}) produced F⁺ *lac⁻* daughter cells. Thus, the genotype of the recipients appears to be F′ *lac⁺/lac⁻*. Now we have the clue: F′ is a cytoplasmic element that carries a part of the bacterial chromosome. Its origin and reintegration can be visualized as shown in Figure 6-11. This F′ is called F-*lac*. Because F′ *lac⁺/lac⁻* cells are *lac⁺* in phenotype, we know *lac⁺* is dominant over *lac⁻*. As we will see later, the dominance-recessive relationship between alleles can be a very useful bit of information in interpretations of gene function. Partial diploidy for specific segments of the genome can be made with an array of F′ derivatives from Hfr strains. The F′ cells can be selected by looking for early transfer of normally late genes in a specific Hfr strain. Try working out a screen using a lawn of F⁻ *strʳ alaˉ* cells on minimal medium containing streptomycin onto which are put Hfr *strˢ ala⁺* cells (in which *ala⁺* is near the F end).

The use of F′ elements to create partial diploids is called sexduction or F-duction. Some F′ strains can carry very large parts of the bacterial chromosome and, if marked properly, recombination studies can be done in the merozygotes.

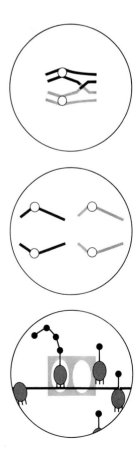

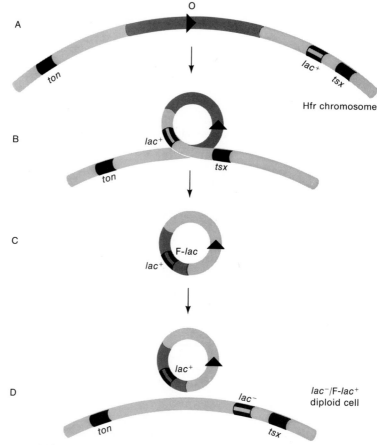

FIGURE **6-11**
A. F factor inserted in an Hfr strain between the *ton* and *lac*⁺ alleles.
B. Abnormal outlooping to include *lac* locus.
C. F-*lac*⁺ particle (= F′).
D. Constitution of an F′ *lac*⁺/*lac*⁻ partial diploid produced by the transfer of the F-*lac*⁺ particle into an F⁻ *lac*⁻ recipient.

Phage Genetics

Most bacteria are susceptible to attack by bacteriophages (translated literally as eaters of bacteria), which is abbreviated to phages. The phages most extensively studied consist of nucleic acid surrounded by a coat of protein molecules. The complicated structure of a phage belonging to the class of so-called T-even (T2, T4, etc.) phages is shown in Figure 6-12.

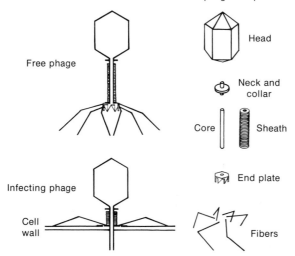

T4 phage components

Head

Neck and
collar

Core   Sheath

End plate

Free phage

Infecting phage

Cell
wall

Fibers

FIGURE **6-12**
Phage T4, shown in its free state and in the process of infecting a cell of *E. coli*. On the right, a phage has been diagrammatically exploded to show its highly ordered structure
in three dimensions.

VIRULENT PHAGES

Virulent phages attach to a bacterium and inject their genetic material into the
bacterial cytoplasm. The phage information then takes over the machinery of
the bacterial cell by turning off the synthesis of bacterial components and
redirecting the bacterial synthetic machinery to make more phage components.
Ultimately, many phage descendants are released when the bacterial cell wall
breaks open in a process called *lysis*.

But we can ask again, how does one study inheritance in phages when they are
too small to be seen except with an electron microscope? One thing we can
examine is *plaque morphology*. When a phage lyses a bacterium, progeny phage
infect neighboring bacteria and these in turn lyse and infect other bacteria. This
is an algebraically explosive phenomenon (exponential increase) and very soon
after starting an experiment of this type (overnight) the effects can be seen with
the naked eye—a *clear* area, or *plaque*, is present on the opaque lawn of bacteria
on the surface of a dish of solid medium (Figure 6-13). Depending on the phage
genotype, such plaques can be large or small, fuzzy or sharp, and so forth.
Another phage phenotype that can be analyzed genetically is *host range*. Certain
strains of bacteria are immune to adsorption (attachment) or injection by phages.
Phages, in turn, may differ in the spectra of bacterial strains they can infect and
lyse.

To illustrate a phage cross, we shall look at a cross of T2 phages (virulent
phages of *E. coli*), $h\,r^+ \times h^+\,r$, originally done by Alfred Hershey. The alleles

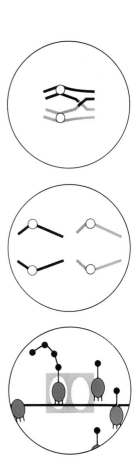

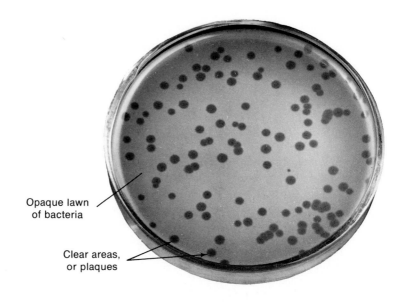

Opaque lawn
of bacteria

Clear areas,
or plaques

FIGURE **6-13**

have the following meanings: h can infect two different *E. coli* strains (which we will call 1 and 2); h^+ can infect only strain 1; r rapidly lyses cells, thereby producing large plaques; and r^+ slowly lyses cells, producing small plaques. The cross is made by infecting strain 1 with both parental T2 phage genotypes at a concentration (called *multiplicity of infection,* which is a ratio of phages to bacteria) sufficiently high to ensure a high proportion of cells that are simultaneously infected by both phage types (this is called a *mixed* or *double infection*).

The phage lysate (progeny phage) is then analyzed by being spread onto a bacterial lawn composed of a mixture of both 1 and 2. Four plaque types are distinguishable (Figure 6-14 and Table 6-1). These four genotypes can easily be scored as parental (the top two) and recombinant, and an RF can be calculated.

$$RF = \frac{(h^+ \, r^+) + (h \, r)}{\text{Total plaques}}$$

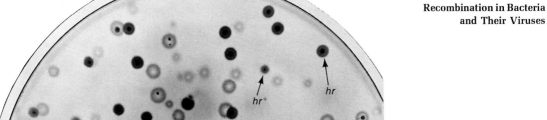

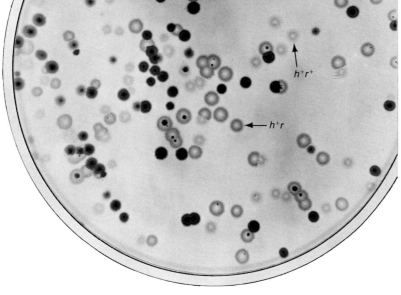

FIGURE **6-14**

TABLE **6-1**
Phage Plaques

Phenotypes	Inferred genotypes
Clear and small	$h\,r^+$
Cloudy and large	h^+r
Cloudy and small	h^+r^+
Clear and large	$h\,r$

Note: Clearness is produced by the *h* allele, which allows infection of *both* bacterial types in the lawn, and cloudiness is produced by the h^+ allele, which limits infection to the strain 1 cells.

If we assume that entire phage genomes are recombining, then we do not have a merozygotic situation as in a bacterial cross. Presumably, then, single exchanges can occur and produce viable reciprocal products. However, phage crosses are subject to several complications, which will only be mentioned here. The first is that several *rounds of exchange* can potentially occur within the host. Because hundreds of phages can be released from a single infected cell, each parental infecting phage can be duplicated many times. If recombination does not occur at one specific time in the lytic cycle, then a recombinant produced shortly after infection might undergo further exchange at later times. Second, recombination between genetically similar phages as well as between different types can occur. Thus $P_1 \times P_1$ and $P_2 \times P_2$ crosses occur in addition to $P_1 \times P_2$. For both of these reasons, recombinants from phage crosses are a consequence of a *population* of events rather than defined, single-step exchange events.

Nevertheless, *other things being equal*, the preceding RF calculation represents a valid index of map distance in phages. Hershey recovered several different T2 strains with the rapid lysis phenotype and designated them as *r1, r2*, and so forth, in the order he picked them up. Let's indicate three different *r* strains as r_a, r_b, and r_c and make the cross $r_x h^+ \times r^+ h$ (in which r_x represents one of the *r* genes). The results obtained are given in Table 6-2.

TABLE **6-2**

Cross	Percentage of each genotype			
	rh^+	r^+h	r^+h^+	rh
$r_a h^+ \times r^+ h$	34.0	42.0	12.0	12.0
$r_b h^+ \times r^+ h$	32.0	56.0	5.9	6.4
$r_c h^+ \times r^+ h$	39.0	59.0	0.7	0.9

You can see that we can construct a linkage map just as Sturtevant did with *Drosophila*. The parental types ($r h^+$ and $r^+ h$) occur with the highest frequency, although not with equal frequency. The two recombinant classes, however, are equally frequent. We can construct linkage maps for each cross:

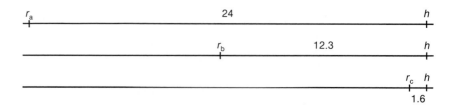

These different recombination values suggest that the loci for the three *r* genes are different. Thus, there are four possible linkage maps (Figure 6-15).

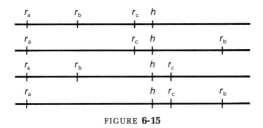

FIGURE **6-15**

Can we distinguish between these alternatives? First, let's take only r_b, r_c, and h and ask whether the order is r_c-h-r_b or h-r_c-r_b. We can make the cross $r_c r_b^+ \times r_c^+ r_b$ and compare the RF with the value of 12.3 obtained for the r_b-h interval. From this comparison h is found to be located between r_b and r_c (r_c-h-r_b).

Now the question is, On which side of h does r_a lie? Next to r_b or r_c? This cannot be answered simply by crossing r_a with r_b and r_c because the data are not clear-cut. Only when many different strains of T2 are recovered, which allow intensive genetic mapping, do we discover that *both* maps are correct. How can both r_a-r_c-h-r_b and r_c-h-r_b-r_a be right? It turns out that the linkage map of the T2 phage is also circular.

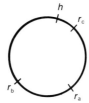

In fact, the total genetic length of the T2 linkage map is about 1,500 map units.

The important message from both the bacterial and phage experiments is that, once the novelty of the conditions for "crossing" is understood, recombination analyses and linkage maps are fairly straightforward.

LYSOGENY

In the 1920s, long before *E. coli* became the darling of microbial geneticists, it was reported that some strains of the bacterium, although resistant to certain phage infection, could cause lysis of other bacteria sensitive to those phages. Thus, the phage carriers were said to be "lysogenic" (i.e., capable of inducing lysis in other cells).

André Lwoff, in the mid-1940s, looked at lysogenic strains of *Bacillus megaterium* by physically separating newly formed daughter cells for nineteen consecutive divisions. After each division, one cell was put into a culture while

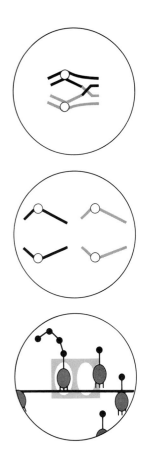

the other was being observed for separation when it divided. All nineteen cultures were lysogenic. Occasionally, Lwoff saw a cell spontaneously lyse: he spread the culture medium on nonlysogenic cells and observed plaque formation. These results suggest that the lysogenic state is inherited from cell to cell and that such cells contain a noninfective factor that nevertheless may dictate the production of an infective phage. This noninfective agent was called a *prophage* and was assumed to be "induced" into a lytic cycle spontaneously in a small fraction of cells. Indeed, later studies showed that free phage could be induced by a variety of agents such as ultraviolet light and certain chemicals.

Earlier, it had been found that if nonlysogenic cells were infected with phages from a lysogenic strain, a small fraction of the infected cells did not lyse but instead became lysogenic. Thus, the bacterial hosts could exist in either of two states—lysogenic or nonlysogenic—in which the phages were present either as active phages or as prophages, respectively. Obviously, the lysogenic cells must be resistant to *superinfection* from phages or else the phages produced by spontaneous induction would kill them.

Thus we can summarize the two types of phages: those for which there are no lysogenic bacteria are *virulent* (although there may be resistant bacterial mutants, it is not because of lysogeny); and those capable of lysogenizing bacteria are *temperate* (resistance to superinfection is in fact an "immunity" conferred by the presence of the prophage). Figure 6-16 diagrams a typical temperate phage life cycle.

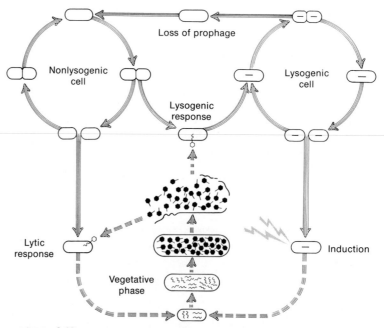

FIGURE **6-16**
Alternative cell cycles associated with a temperate phage and its host.

What is the nature of the prophage? Because the prophage is capable of producing a complete, mature phage upon induction, all of the phage genome must be present. But are there many copies in the cytoplasm or is it associated with the bacterial genome? It is very fortuitous that the original strain of *E. coli* used by Lederberg and Tatum was found to be lysogenic for a temperate phage called lambda (λ). Lambda has become the most intensively studied and best-characterized phage (Figure 6-17). Interestingly, if crosses were made

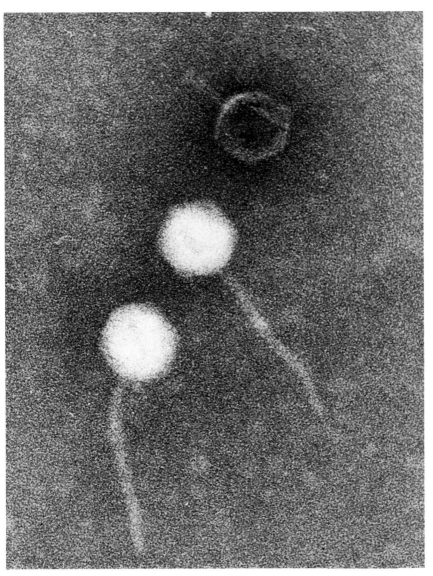

FIGURE **6-17**

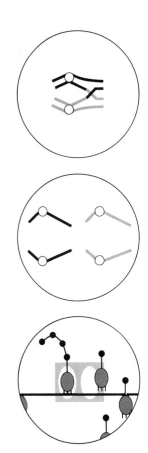

between F^+ and F^- cells, it was found that $F^+ \times F^- (\lambda)$ crosses yielded recombinant lysogenic recipients, whereas the reciprocal cross $F^+ (\lambda) \times F^-$ almost never gave lysogenic recombinants.

These results were made understandable when Hfr strains came on the scene. In the cross Hfr $\times F^- (\lambda)$, lysogenic F^- recombinants with Hfr genes were readily recovered. However, in the reciprocal cross Hfr $(\lambda) \times F^-$, early genes from the Hfr strain were recovered in the F^- recipients but, after a certain time in mating, recombinants for late markers were *not* recovered. How do we explain this? In the cross of a lysogenic Hfr with a nonlysogenic F^- recipient, the entry of the λ prophage into the cytoplasm of a nonimmune cell could be enough to trigger the prophage into the lytic cycle. This is called *zygotic induction* and, in interrupted-mating experiments, the λ prophage can be shown to enter the F^- at a specific time, closely linked to the *gal* locus. Thus, the prophage is closely associated with a specific position in the bacterial chromosome.

Zygotic induction suggests one other point. In the cross Hfr $(\lambda) \times F^- (\lambda)$, any recombinants are readily recovered (i.e., there is no phage induction). Because entry of the λ prophage into the F^- cell immediately induces the lytic cycle, it seems that the cytoplasmic state of the F^- is responsible. Immunity and lysogeny might be explicable if a cytoplasmic factor specified by the prophage somehow represses the multiplication of the virus. Entry of the prophage into a nonlysogenic environment immediately dilutes this repressing factor and therefore the virus reproduces. But if the virus specifies the repressing factor, why doesn't it shut itself off again? Perhaps it does, because a fraction of infected cells becomes lysogenic: there may be a race between the λ gene products signalling reproduction and those specifying a shutdown. The model of a phage-directed cytoplasmic repressor would nicely explain immunity because any superinfecting phage would immediately encounter a repressor and be inactivated.

How is the prophage attached to the bacterial genome? In the days before chromosome circularity was known, there seemed to be two possible models (Figure 6-18).

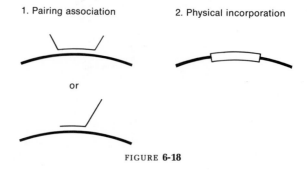

FIGURE **6-18**

A λ mutant that had a large section of its chromosome deleted and could reproduce, although it could not lysogenize, supplied the clue. The deleted segment, though not containing genes for λ production, could be a region of homology with the bacterial chromosome. Crossing over between the λ and *E. coli* chromosomes could incorporate the entire λ genome at a specific point and as a continuous part of the *E. coli* chromosome (the Campbell model again) as shown in Figure 6-19.

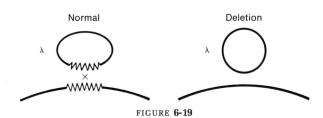

FIGURE **6-19**

Can this simple model of lysogeny be tested? The attraction of Campbell's proposal is that it does make testable predictions, and λ gives us a chance to test these predictions.

1. Physical integration of the prophage should increase the genetic distance between flanking bacterial markers (Figure 6-20). (Time-of-entry or recom-

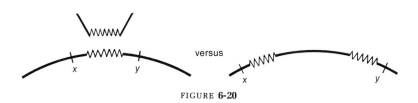

FIGURE **6-20**

bination distances between the bacterial genes *are* increased by lysogeny.)

2. Some deletions of bacterial segments adjacent to the prophage site should delete phage genes (Figure 6-21). (This is also found.)

Extent of various host deletions

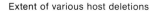

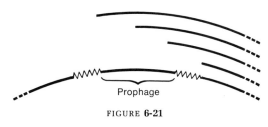

Prophage

FIGURE **6-21**

The phenomenon of lysogeny is a very clever way for a temperate phage to keep from eating itself out-of-house-and-home. Lysogenic cells can perpetuate and carry the phages around. Amazingly, the study of this biologically interesting but seemingly esoteric phenomenon has burst forth as a very relevant model for cancer. Many animal cancers are caused by viruses that may exist in two states, a quiescent state (comparable to prophage) and an active (cancer-forming) one.

Transduction

In 1951, Joshua Lederberg and Norton Zinder were testing for recombination in another bacterium, *Salmonella typhimurium,* using the same techniques as had been successful with *E. coli.* They had two different strains: one was *phe⁻ trp⁻ tyr⁻* and the other was *met⁻ his⁻*. (Let's not worry about the symbols except to say that the nonwild-type alleles conferred nutritional requirements.) When either strain was plated on minimal medium, no wild-type cells were observed. However, on mixing the two strains, about 1 in 10^5 wild-type cells were found. So far, it is reminiscent of recombination in *E. coli.*

However, when they placed the two strains into different arms of a U tube in which cell contact was prevented by a filter, wild-type cells were again recovered. By varying the size of the pores in the filter separating the two arms, they found that the filterable agent responsible for recombination had the approximate size of the virus P22, a known temperate phage of *Salmonella.* The suggestion that the vector of recombination was P22 was supported by their identical properties: size, sensitivity to antiserum, and immunity to hydrolytic enzymes. Thus, rather than confirming conjugation in *Salmonella,* Lederberg and Zinder had discovered a new type of gene transfer mediated by a virus. This was called *transduction.* Somehow, some virus particles during the lytic cycle pick up bacterial genes, which are then transferred to another host where the unwitting virus inserts its contents.

The general method for transduction studies is to infect nonlysogenic donor strains with the virus and to harvest the phage lysate. This is then used to infect either lysogenic or nonlysogenic recipients that are plated on a medium allowing selection for recombinants.

When P22 infects a bacterial cell, the bacterial chromosome is broken into smaller pieces. In 1965, K. Ikeda and J. Tomizawa found that occasionally the forming phage particles mistakenly incorporate a length of bacterial genome rather than phage material into the head. (You might consider ways that this could be shown experimentally.) Because it is the phage coat proteins that determine their ability to attack a cell, such viruses or transducing particles can bind to a bacterial cell and inject their contents, which now happen to be donor bacterial genes. When the contents of a transducing phage are injected into a

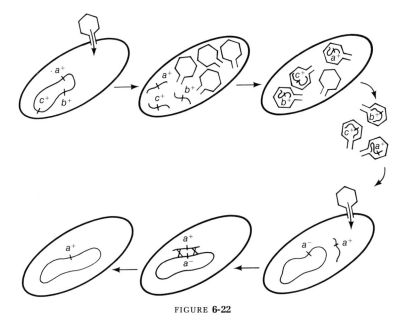

FIGURE **6-22**

recipient cell, a merodiploid situation is set up in which the transduced genes can be incorporated by recombination (Figure 6-22).

We now have a merozygote from which linkage information about bacterial genes can be derived. Transduction from an $a^+ b^+$ donor to an $a^- b^-$ recipient produces various transductants for a^+ and b^+. The ratio of

$$\frac{(a^+ b^-) + (a^- b^+)}{(a^+ b^-) + (a^- b^+) + (a^+ b^+)}$$

transductants is capable of yielding linkage information. This is because presumably the chance of a^+ or b^+ being included individually in the transducing phage is proportional to their distance apart. If they are close together, they will be picked up and transduced by the phage together. Of course, the method only gives linkage values if the genes under test *are* close enough for *both* to be included in the transducing phage and hence capable of forming an $a^+ b^+$ transductant (a *co*transductant). But the failure to find cotransductants does provide some linkage information in a negative sense.

An idea of the size of the piece of host chromosome that a phage can pick up can be estimated from the following type of experiment using the P1 phage of *E. coli.*

Donor *leu⁺ thr⁺ azi^r* → Recipient *leu⁻ thr⁻ azi^s*

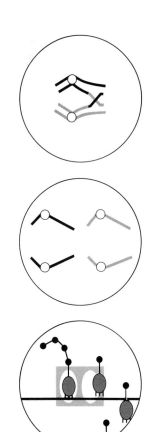

We can select for one or more donor markers in the recipient and then, in true merozygote genetics style, look for the presence of the other unselected markers (Table 6-3). Experiment 1 tells us that *leu* is relatively close to *azi* and distant

TABLE 6-3

Experiment	Selected marker(s)	Unselected
1	*leu*$^+$	50% are *azi*^{r}; 2% are *thr*$^+$
2	*thr*$^+$	3% are *leu*$^+$; 0% are *azi*^{r}
3	*leu*$^+$ and *thr*$^+$	0% are *azi*^{r}

from *thr,* but does not tell us whether the order is

or

thr		leu azi

thr		azi leu

Experiment 2 tells us that *leu* is closer to *thr* than is *azi:*

thr		leu azi

By selecting for *thr*$^+$ and *leu*$^+$ in the transducing phage in experiment 3, we see that the transduced piece of genetic material never includes the *azi* locus. If enough markers were studied to flesh out a linkage map, the size of a transduced segment could be estimated.

The phage P22 belongs to a group that shows *generalized* transduction; that is, the transfer of many different parts of the bacterial chromosome. As we shall see now, another class called *specialized* or restricted transducing phages carry only restricted parts of the bacterial chromosome.

Lambda phage is a good example of a restricted transducer. A lysogenic strain K12 (λ) can be induced by ultraviolet light and used to attempt to transduce various recipient mutants. The only successful transductions involve the *gal* locus. As you may recall, λ always attaches next to the *gal* locus. When *gal*$^+$ transductants are tested, some interesting features emerge.

1. All the *gal*$^+$ transductants are immune from superinfection by λ phage. Thus they have probably acquired part of a λ genome as a prophage. But these transduced strains cannot produce λ in lysis—they are λdg, or λ-*defective gal*$^+$.

2. λdg cells are unstable and produce some *gal*$^-$ cells. Thus they must be λdg *gal*$^+$ *gal*$^-$.

These two points may be explained by the model in Figure 6-23. A very efficient high-frequency transducing strain can be produced by doubly infecting with

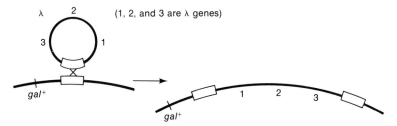

λ 2 (1, 2, and 3 are λ genes)

A. Production of lysogenic K12(λ)

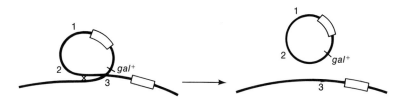

B. Production of transducing phage
   by inaccurate reversal of part A

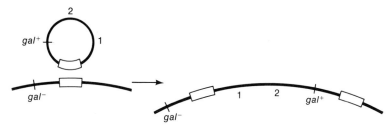

C. Transduction to produce λ*dg gal⁺ gal⁻*

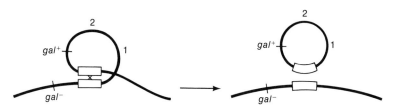

D. Production of *gal⁻* cells by λ*dg gal⁺ gal⁻*
   by *accurate* reversal of part C

FIGURE **6-23**

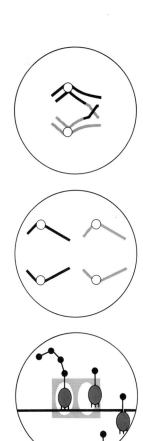

2 lines short

transducing phage *and* a normal λ. The normal λ supplies complete lytic functions. The genotype is described as λλ*dg gal⁺ gal⁻*. Can you draw its origin in the style of the above diagrams?

MESSAGE

Transduction is due to the accidental pickup of host genes by a bacteriophage. General transducing phages accidentally incorporate bacterial genes during the packaging of new phage. Restricted transducers acquire genes by the faulty outlooping of a prophage.

The elucidation of the genetic features of recombination in bacteria and their viruses is a fascinating chapter in genetic analysis. The now completed picture is satisfying because each clue in the mystery can be laid out in a logical sequence until the solution emerges. It was not always so clear when the answers were not yet apparent, but the application of clever techniques and the posing of appropriate questions illustrate genetic analysis at its best.

MESSAGE

The elucidation of the cryptic life cycles of bacteria and their phages was made possible by the application of the standard genetic techniques of recombination mapping.

Problems

1. Linkage maps in an Hfr bacterial strain are calculated in units of minutes, the number of minutes between genes indicating the length of time it takes for the second gene to follow the first after conjugation. When such maps are made, it is assumed by microbial geneticists that the bacterial chromosome is transferred from Hfr to F⁻ at a constant rate. Thus, two genes separated by 10 minutes near the origin end are presumed to be the same *physical* distance apart as two genes 10 minutes apart on the F-attachment end. Can you suggest a critical experiment to test whether this assumption is correct?

2. Suppose that you are given two strains of *E. coli*. One is an Hfr strain and is *arg⁺ ala⁺ glu⁺ pro⁺ leu⁺ Tˢ*, the other is F⁻ and is *arg⁻ ala⁻ glu⁻ pro⁻ leu⁻ Tʳ*. The markers are all nutritional except *Tˢ* and *Tʳ*, which are concerned with T1 phage sensitivity and resistance. The order of entry is as given, with *arg⁺* entering the recipient first and *Tˢ* last. You find that the F⁻ strain dies when exposed to penicillin (*penˢ*) but the Hfr does not (*penʳ*). How would you locate the gene for penicillin resistance on the bacterial chromosome with respect to *arg, ala, glu, pro,* and *leu?* Formulate your answer in logical, well-explained steps, illustrated with explicit diagrams if possible.

3. In *E. coli,* the following Hfr strains donate the markers shown in the order given:

Hfr strain	Order				
1	Q	W	D	M	T
2	A	X	P	T	M
3	B	N	C	A	X
4	B	Q	W	D	M

All these Hfr strains were derived from the same F^+ strain. What was the order of these markers on the original circular F^+ chromosome?

4. Suppose that you make this *E. coli* cross: Hfr Z_1^- *ade*$^+$ *str*^{s} × F$^-$ Z_2^- *ade*$^-$ *str*^{r}, in which *str* stands for streptomycin resistance or sensitivity, *ade*$^-$ stands for adenine requirement for growth, and Z_1^- and Z_2^- are two very close sites that result in an inability to use lactose as an energy source. After about an hour, the mixture is plated on medium containing streptomycin, with glucose as the energy source. Many of the *ade*$^+$ colonies growing were found to be capable of using lactose. However, hardly any *ade*$^+$ colonies from the reciprocal cross Hfr Z_2^- *ade*$^+$ *str*^{s} × F$^-$ Z_1^- *ade*$^-$ *str*^{r} were found to be capable of using lactose. What is the order of the Z_1 and Z_2 sites in relation to the *ade* locus? (Note that the *str* locus is terminal.)

5. In the cross Hfr *aro*$^+$ *arg*$^+$ *ery*^{r} *str*^{s} × F$^-$ *aro*$^-$ *arg*$^-$ *ery*^{s} *str*^{r}, the markers were transferred in the order given, but the first three were very close together. Exconjugants were plated on medium containing streptomycin (to contraselect Hfr cells), erythromycin, and arginine, and aromatic amino acids. Three hundred colonies from these plates were isolated and tested for growth on various media. The results were as follows:

On erythromycin only	263 strains grew
On erythromycin plus arginine	264 strains grew
On erythromycin plus aromatics	290 strains grew
On erythromycin plus arginine plus aromatics	300 strains grew

From these data:

a. Draw up a list of genotypes and indicate how many of each there are.
b. Calculate the recombination frequencies.
c. Calculate the ratio of the size of the *arg*-to-*aro* region to the size of the *ery*-to-*arg* region.

6. Suppose that the gene map in *E. coli* is

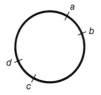

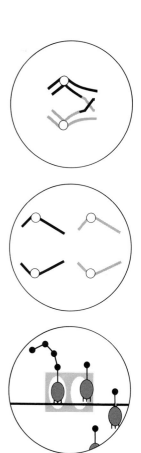

and that you have two different Hfr strains with the following genetic order:

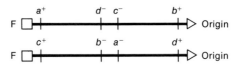

You then convert one strain into a "pseudofemale" by stripping off its pili* chemically and cross it with the other Hfr strain. After allowing sufficient time for the entire Hfr chromosome to be transferred, you select for $a^+ b^+ c^+ d^+$ Hfr recombinants (assume that the diploid recipient state is unstable and reverts to the haploid state). Diagram the recombinants selected. How might they transmit their chromosomes to F$^-$ recipients? Design a test to prove which method is correct.

7. Suppose that you have three strains of *E. coli*. Strain A is F$'$ *cys$^+$ trp1/cys$^+$ trp1* (the F$'$ and the chromosome carry *trp1*, an allele for tryptophan requirement, and *cys$^+$*). Strain B is F$^-$ *cys trp2 Z* (the F$^-$ requires cysteine for growth and carries *trp2*, which also causes a requirement for tryptophan; the strain is also lysogenic for the generalized transducing phage Z). Strain C is F$^-$ *cys$^+$ trp1* (an F$^-$ derivative of strain A that has lost the F$'$).

   a. How would you determine whether *trp1* and *trp2* are alleles (describe the crosses and the results expected).
   b. Suppose *trp1* and *trp2* are alleles and that the *cys* locus is cotransduced with the *trp* locus. Using phage *Z* to transduce genes from strain C to strain B, how would you determine the genetic order of *cys, trp1,* and *trp2*?

8. In the bacteriophage T4, gene *a* is 1.0 map unit from gene *b*, which is 0.2 map unit from gene *c*. The gene order is *a–b–c*. In a recombination experiment, you recover five double crossovers between *a* and *c* out of 100,000 progeny viruses. Interference is negative. True or false?

9. Suppose that you have infected *E. coli* cells with two strains of T4 virus: one that was minute (*m*), rapid lysis (*r*), and turbid (*tu*), and one that was wild type for all three markers. The lytic products of this infection were plated and classified as follows:

Genotype	Number of plaques
m r tu	3,467
+ + +	3,729
m r +	853
m + *tu*	162
m + +	520
+ *r tu*	474
+ *r* +	172
+ + *tu*	965
	10,342

   a. Determine the linkage distances between *m* and *r*, *r* and *tu*, and *m* and *tu*.
   b. What linkage order would you suggest for the three genes?
   c. What is the coefficient of coincidence in this cross, and what does it signify?[1]

---

*Pili are tubelike appendages on male cells and are presumed to be the route of genetic transfer.
[1] Problem 9 is reprinted with the permission of Macmillan Publishing Co., Inc., from *Genetics* by Monroe W. Strickberger. Copyright ©, Monroe W. Strickberger, 1968.

10. In a P_1 phage transduction experiment involving three very closely linked *arg* loci (*arg-1*, *arg-2*, and *arg-3*) and another remote locus *pro*, the following data were obtained:

	arg-1 Number of transductants		arg-2 Number of transductants	
		pro⁻ arg⁻ recipient		
Donor (pro⁺ arg⁻)	arg⁺	pro⁺	arg⁺	pro⁺
arg-1	—	—	492	14,959
arg-2	403	12,358	—	—
arg-3	55	2,978	996	18,239

Draw a map of the loci represented by *arg-1*, *arg-2*, and *arg-3*.

11. Suppose that you have two strains of phage λ that can lysogenize *E. coli*:

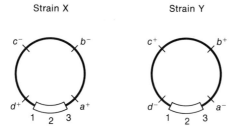

Strain X   Strain Y

The segment shown at the bottom of the chromosome and designated 1–2–3 is the region responsible for pairing and crossing over with the *E. coli* chromosome. (Keep the markers on all your drawings.)

a. Diagram the way in which strain X is inserted into the *E. coli* chromosome (so that the *E. coli* is lysogenized).

b. It is possible to superinfect the bacteria lysogenic for strain X with strain Y. A certain percentage of the time, the bacteria can become "doubly" lysogenic; that is, lysogenic for both strains. Diagram how this will occur. (Don't worry about how double lysogens are detected.)

c. Diagram how the two λ prophages can pair.

d. It is possible to recover crossover products between the two prophages. Diagram a crossover event and the consequences.

12. Jacob selected eight *lac⁻* strains and then attempted to order the markers with respect to the outside markers *pro* (proline) and *ade* (adenine) by performing a pair of reciprocal crosses for each pair of *lac* markers:

Hfr *pro⁻ lac-x ade⁺* × F⁻ *pro⁺ lac-y ade⁻*
Hfr *pro⁻ lac-y ade⁺* × F⁻ *pro⁺ lac-x ade⁻*

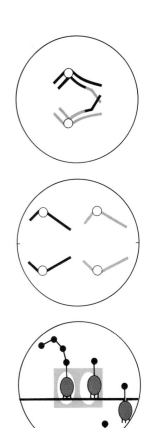

In all cases, prototrophs are selected by plating on minimal medium with lactose as the only carbon source. The following data give the number of colonies in the two crosses for each pair of markers; for example; x/y 100 and y/x 50. Determine the relative order of the markers.[2]

1/2	173	1/4	46	1/6	168	5/7	199
2/1	27	4/1	218	6/1	32	7/5	34
1/3	156	1/5	30	3/6	20	1/8	226
3/1	34	5/1	197	6/3	175	8/1	40
2/3	24	4/5	205	1/7	37	2/8	153
3/2	187	5/4	17	7/1	215	8/2	17

[2]Problem 12 is from Burton S. Guttman, *Biological Principles,* copyright © 1971, W. A. Benjamin, Inc., Menlo Park, California.

7 / Cytogenetics

(Or: How to become a chromosome mechanic.)

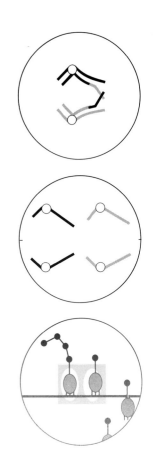

Now that we have derived the basic ground rules for linkage analysis, we can turn our attention to the actual physical structure in which the genes of eukaryotes are located, the *chromosome*. We can ask what its basic properties are and how the observed behavior of a chromosome corresponds to its genetic properties.

First, let's consider the purely descriptive aspect of chromosomes. From pictures you have seen, chromosomes may appear to be skinny, wormlike squiggles with no distinctive landmarks for identification other than size. However, closer inspection can and often does show considerable variation. Chromosomes may vary immensely in size from species to species, from less than one micron (written as $1\,\mu$ or 10^{-6} m) to several hundred microns in length. Even within a single organism, chromosomes may vary in size from tissue to tissue. Amphibian oocytes, for example, contain giant lampbrush chromosomes, which may be as much as $800\,\mu$ in length, whereas their somatic cell chromosomes are only a few microns. Techniques of electron microscopy and chemistry have revealed the chromosome as a highly condensed complex of nucleic acids and proteins.

At any given moment in cell division, segments of chromosomes may exhibit staining differences. Some regions (generally found near the centromere and at the tips, or *telomeres*) are darkly stained and have been designated *hetero-*

chromatin, whereas other areas are less densely stained and are termed *euchromatin.* The distribution of heterochromatin and euchromatin within chromosomes is usually constant from cell to cell and thus, in a sense, hereditary. Heterochromatin found in the same area of both homologs in a cell is called *constitutive.* There are other cases, such as the X chromosomes of mammalian females and the paternally derived chromosomes in some insects, in which one chromosome may be heterochromatic whereas its homolog is not. This is called *facultative* heterochromatin. Although the genetic difference between hetero- and eu- chromatin is not yet clear, there is good evidence that heterochromatic regions are genetically inactive in the cells in which they are seen.

Within a chromosome, there may be small areas of intense staining connected by lightly stained areas. These correspond to the chromomeres on which Belling built his crossover model. Recently, staining procedures have been developed that result in patterns of lightly and darkly stained regions, or bands. The banding patterns are highly specific to each chromosome pair and have allowed unequivocal identification of all 23 chromosome pairs in humans and are equally useful in identifying specific chromosomes in a wide range of other organisms. We have already encountered this banding in an earlier chapter (see Figure 5-26).

There are other landmarks for identification of specific chromosomes. The position of the centromere (alias spindle-fiber attachment or kinetochore) within a chromosome determines the length of the *arms* on each side and the chromosome's shape during anaphase. Thus, the terms telocentric, acrocentric, and metacentric describe chromosomes in which the centromere is located anywhere from either end to the middle, producing differences in shape that range from a simple rod through a J to a V (Figure 7-1).

The region of the centromere may appear pinched and is referred to as the *primary constriction.* One or more *secondary constrictions* indicate the position of the nucleolus organizer region. In some organisms, such as Lepidoptera, centromeres are "diffuse" so that spindle fibers attach all along the chromosome, in which case the chromosome's migration to one of the poles is parallel to the metaphase plate. When such a chromosome is broken, both parts can still migrate to the pole because they are both attached by spindle fibers. In contrast, a break in a chromosome that has a single centromere results in the inability of

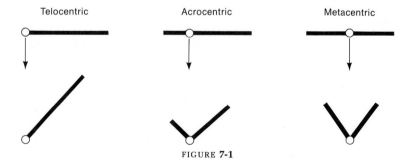

FIGURE 7-1

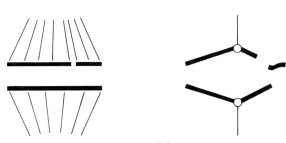

FIGURE **7-2**

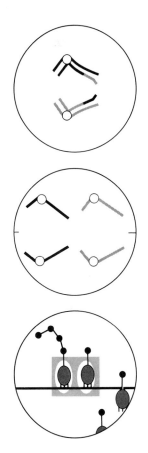

the *acentric* (without a centromere) fragment to move owing to the loss of the point of spindle-fiber attachment (Figure 7-2).

An important tool for the cytological study of chromosomes was supplied by the rediscovery of so-called giant chromosomes in certain organs of the Diptera. In 1881, E. G. Balbiani recorded structures in the nuclei of certain secretory cells of two-winged flies that were long and sausage-shaped and marked by swellings and cross striations. Unfortunately, he did not recognize them as chromosomes and the report remained buried in the literature. It was not until 1933 that Theophilus Painter, Ernst Heitz, and H. Bauer rediscovered them and realized they were chromosomes.

In secretory tissues, such as Malphigian tubules, rectum, gut, footpads, and salivary glands of Diptera, the chromosomes apparently replicate their genetic material many times without actual separation into distinct chromatids. Thus, as the chromosome increases in replicas, it elongates and thickens. *Drosophila* has an n number of 4, and only 4 chromosomes are seen in the cells of such tissues because, for some reason, homologs are tightly paired. Furthermore, they are joined at the *chromocenter,* which represents a coalescence of the heterochromatic areas around the centromeres of all four chromosome pairs. Salivary gland chromosomes are shown in Figure 7-3, in which L and R stand for arbitrarily assigned left and right arms.

Such giant chromosomes never divide and the process of their formation is called *endomitosis.* The multiple replicas of the chromatids are referred to as *polytene chromosomes.* The most commonly examined polytene chromosomes are in the salivary gland nuclei. Along the chromosome length, characteristic stripes called *bands,* which vary in width and morphology, can be observed and identified. In addition, there are regions that may appear swollen (puffs) or greatly distended (*Balbiani rings*) and are presumed to correspond to regions of genetic activity. Recently, it has been found in *Drosophila* that, in general, each band contains the genetic material of a single gene. (However, the significance of the bands in human chromosomes is not known.) The polytene chromosomes corresponding to linkage groups have been identified through the use of chromosomal aberrations. Such aberrations, as we shall see, have also been useful in specific localization of genes along the chromosomes.

So much for the cytological properties of chromosomes. The rest of this chapter is concerned with what has been termed "chromosome mechanics."

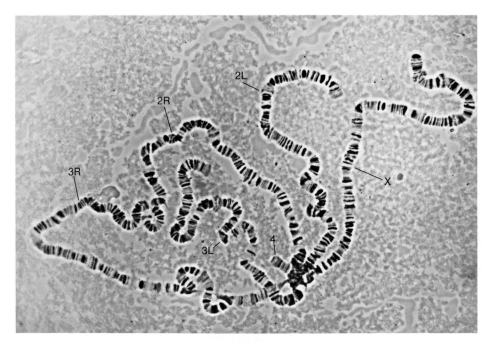

FIGURE **7-3**

Genetics, perhaps more than any other field of biology, makes extensive use of deviations from the norm and the study of chromosomes is no exception. Although this means of investigation may seem highly esoteric to some, the findings have proven to be very important in applied biology, especially in agriculture, animal husbandry, and medicine. Also many genetic tricks and devices that geneticists routinely use to build certain genotypes will be revealed. Finally, although we will not give this point the attention it deserves, many of the aberrations to be discussed have been important in building the theories of evolution and speciation. First, let's classify chromosomal aberrations:

1. Those concerning *chromosome structure.*
   a. Deletions.
   b. Duplications.
   c. Inversions.
   d. Translocations.
2. Those concerning *chromosome number.*
   a. Euploids. Variation of the number of sets of chromosomes; for example, n = monoploid, 2n = diploid, 3n = triploid, and 4n = tetraploid (multiples of n that are 3 or more are polyploids).
   b. Aneuploids. Addition or subtraction of chromosomes from entire sets; for example, 2n − 1 is monosomic, 2n + 1 is trisomic, 2n + 1 + 1 is double trisomic, and 2n − 2 is nullosomic (in which the two chromosomes lost are homologs).

It should be emphasized here that meiotic pairing is incredibly precise. Homologous regions generally pair, gene for gene, and chromosomes will go through remarkable contortions to fulfill the demands of homology. However, in the presence of more than two homologous regions, only two of them can pair at any one time or position. Bear this in mind throughout the following discussion.

Changes in Chromosome Structure

It is possible that a segment of a chromosome might be lost.

This type of change is called a *deletion* or a *deficiency*. The reciprocal of such a change would be its *duplication*.

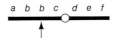

We can also conceive of a segment of a chromosome that has rotated 180° and rejoined the chromosome as an *inversion*.

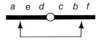

Finally, parts of two nonhomologous chromosomes might be exchanged to produce a *translocation*.

In fact, all of these types of chromosomal aberrations do occur and so we can examine their genetic and cytological properties.

DEFICIENCIES AND DUPLICATIONS

Obviously, the occurrence of chromosomal rearrangements results from a break or disruption in the linear continuity of a chromosome. Deficiencies and duplications can be produced by the same event if breaks occur simultaneously at different points in two homologs, which can be visualized as occurring when the homologs overlap (Figure 7-4). This does not mean that duplications and deficiencies are always reciprocal products of the same event.

In general, if a deficiency is made homozygous, it is lethal. This suggests that most regions of the chromosomes are vital for normal viability and that

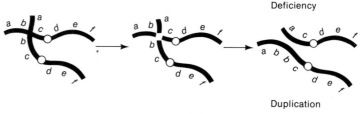

FIGURE 7-4

complete elimination of any segment from the genome is deleterious. Individuals heterozygous for deletions often do not survive because the genome has been finely "tuned" during evolution to require a specific balance or ratio of most genes; the presence of the deletion upsets this balance. Nevertheless, individuals with relatively large deficiencies can survive if the deficiencies are heterozygous with normal chromosomes. If meiotic chromosomes in such heterozygotes can be examined, the region of the deficiency can be detected by the failure of the corresponding segment on the normal chromosome to pair properly. You can see that deficiencies can be located on the chromosome by this technique (Figure 7-5).

Meiotic   Polytene

FIGURE 7-5

But what of the genetic properties of deficiencies? Cytological detection of a "deficiency loop" (e.g., in salivary gland chromosomes of *Drosophila*) confirms its presence, but there are genetic criteria for inferring the presence of a deficiency. One is the failure of the chromosome to survive as a homozygote, but that, of course, could also be produced by any lethal gene. Another is the suppression of crossing over in the region spanning the deficiency, but again this could occur with other aberrations and small deficiencies may have only minor effects on crossing over. The best criterion will make more sense after you have read Chapter 8. Let's just say that chromosomes with deletions can never revert to a normal condition. Another reliable criterion is the phenotypic expression of a recessive gene on a normal chromosome when the region in which it is located has been deleted from the homolog.

Such *pseudodominance* (i.e., the expression of a recessive gene when present in a single dose) also allows the cytological location of that gene when coupled with the chromosomal positioning of the deficiency loop. This technique permits a correlation between the genetic map (based on linkage analysis) and the cytological map (devised by marking the position of deficiency loops in specific cases of pseudodominance). By and large the maps correspond well—a satisfying cytological endorsement of a purely genetic creation.

MESSAGE

Deletion analysis proves that linkage maps are in fact reflections of chromosome maps.

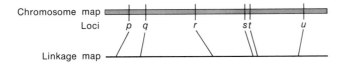

Deletions are sometimes identified and mapped *genetically* before cytological analysis by this unexpected appearance of a recessive allele in pseudodominance. To give you an example of how this works, suppose that you have a strain of flies having an X chromosome (which we'll designate as X?) that, in the cross X?/X♀ × X/Y♂, gives progeny in a ratio of 2 female:1 male. This suggests the presence of a sex-linked lethal factor that causes the death of X?/Y males. We can first ask, Where is the lethal factor genetically? We can find this out by crossing X?/X females with males carrying X chromosomes that have conveniently placed markers from one tip to the centromere (let's label them *a–b–c–d*). The cross is then X?/X♀ × *a b c d*/Y♂, which gives us X?/*a b c d*♀♀, which can be used to map the position of the lethal factor when crossed with wild males. Because we have a factor that is lethal in males, any crossover chromosome that carries the lethal factor will die in the male progeny of these females. Suppose that the defect is at the tip of the X chromosome between *a* and *b*. We will be able to map it immediately by the genotypes of the surviving males (Figure 7-6). You can see that one of each pair of reciprocal crossover products will always die in males. Nevertheless, map distances can be determined because all classes, even parentals, are halved by the lethal factor. If the lethal factor does not lie within a genetic region, then only one of the two reciprocal crossovers in that region will live. On the other hand, if the lethal factor does lie within a region, *both* reciprocal recombinants survive, but in fact they come from crossovers on either side of the lethal. The percentage of these two classes will give us the genetic position of the lethal factor within the region; that is, the percentage of + *b* and *a* + male recombinants are the genetic distances from *a* to lethal and from lethal to *b*, respectively.

If the total *a* to *b* recombinant frequency is much lower than the control or normal value, then we can suspect that the lethal factor is a deletion, because the

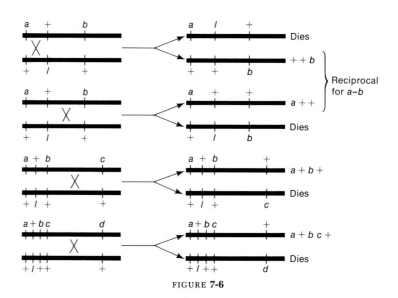

FIGURE 7-6

total physically paired region would be shorter. If our X? chromosome carries a lethal factor between *a* and *b* that reduces the genetic interval in this way, we can then cross X?-bearing females with males carrying different recessive genes known to reside in that interval. A map of genes in the tip region is:

y   dor   br   gt   swa w   rst   vt
├─ 0.3 ┼ 0.3 ┼ 0.3 ┼ 0.4 ┼0.2┼0.2┼ 0.6 ┼ ─ ─ ─ ─ ─

If we get all wild-type flies in crosses between X?/X females and males carrying *y*, *dor*, *br*, *gt*, *rst*, or *vt* but pseudodominance of *swa* and *w* with X? (that is, X?/*swa* is swa and X?/*w* is w), then we have good genetic evidence for a deletion of the chromosome including at least the *swa* and *w* loci but not *gt* or *rst*. As an exercise, work out the *genetic* results if a deletion spans *b*, if a deletion lies right at the end of the X chromosome, and if the X? chromosome carries two lethal factors, one between *a* and *b* and the other between *c* and *d*.

An interesting difference between animals and plants is revealed by deficiencies. In animals, a male that is heterozygous for a deficient chromosome and a normal one will produce functional sperm carrying each of the two chromosomes in approximately equal numbers. In other words, sperm seem to function to some extent regardless of their genetic content. In plants, on the other hand, the pollen produced by a deficient heterozygote is of two types: functional pollen carrying the normal chromosome and nonfunctional, or aborted, pollen carrying the deficient homolog. Thus, pollen cells seem to be sensitive to changes in *amount* of chromosome material and this might act to weed out any large changes in the genome.

Duplications are very important chromosomal changes from the standpoint of evolution because they supply additional genetic material potentially capable of assuming new functions. Adjacent duplicated segments may occur in *tandem sequence* with respect to each other (*a bc bc d*) or in *reverse order* (*a bc cb d*). The pairing patterns obtained in these two sequences are different and illustrate the high affinity of homologous regions for pairing. Thus, chromosomes in meiotic nuclei containing a normal chromosome and a homolog with a duplication are seen to pair in the configurations shown in Figure 7-7. Alternatively, duplicated segments may be nonadjacent, either in the same chromosome or in separate chromosomes.

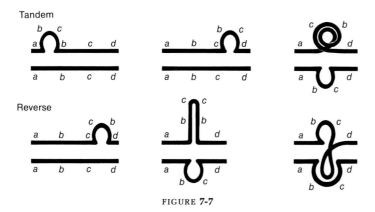

FIGURE 7-7

An organism that has evolved with a fixed number of genetic units would presumably be able to "spare" duplicated copies when they arise. Thus, loss or alteration of gene function due to changes in one of the duplicated genes would be covered by the duplicate copies. This provides an opportunity for divergence in function of genes, which could be potentially advantageous. Indeed, in situations in which different gene products with related functions such as the globins (which will be discussed in a later chapter) can be compared, there is good evidence that they indeed arose as duplicates of each other. Interestingly, once an adjacent duplication arises in a population, homozygosity for such a duplication can result in higher orders of duplication by crossing over when the chromosomes are *asymmetrically* paired (Figure 7-8).

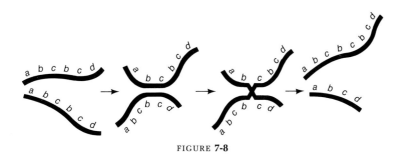

FIGURE 7-8

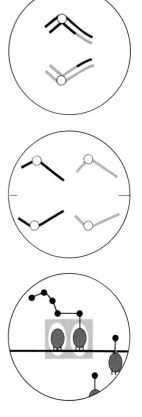

MESSAGE

Duplications are important chromosomal alterations that supply additional genetic material capable of evolving new functions.

Duplications of certain genetic regions may produce specific phenotypes and act like genes. For example, the dominant gene Bar in *Drosophila*, produces a slitlike eye instead of the normal oval one. Cytologically, in the polytene chromosomes Bar was found to be in fact a tandem duplication that probably resulted from an *unequal crossover* (Figure 7-9). Evidence for the asymmetrical

FIGURE **7-9**

pairing and crossing over in *Drosophila* comes from studying homozygous Bar females. Occasionally, such females produce offspring with extremely small eyes called "double bar." They are found to carry three doses of the Bar region in tandem (Figure 7-10).

FIGURE **7-10**

In general, duplications are hard to detect and are rare. However, they are useful tools and can be generated from other aberrations by tricks that we shall learn later.

Duplications and deficiencies are detected in human chromosomes. Deficiencies in particular are often associated with or responsible for hereditary defects. For example, a syndrome called cri du chat (which was so named because young infants with the defect cry endlessly in a mewing sound like that of cats) results from heterozygosity for a deletion of half of the short arm of chromosome 5. Figure 7-11 shows a pedigree involving cri du chat syndrome; examine it now, but don't worry about the details of generating the deletion until after we have dealt with translocations.

INVERSIONS

Homozygosity for a chromosome carrying an inverted gene sequence will result in a linkage map with a different gene order. In a heterozygote having a

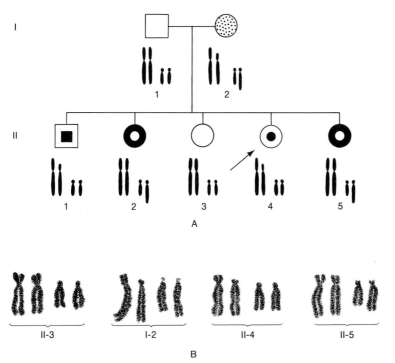

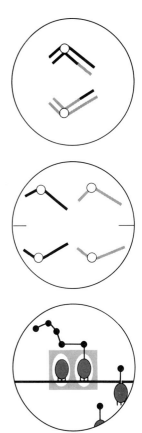

FIGURE **7-11**
  A. Pedigree with cases of cri du chat syndrome:
    I. Both parents are phenotypically normal.
    II. Of the offspring, II-1 and II-4 have cri du chat syndrome, II-2 and II-5 have the reciprocal chromosome abnormality, and II-3 is normal. Symbols represent different karyotypes, and the arrow points to the *proband* or affected individual with whom the study began.
  B. Chromosomes 5 and 13 of four persons in the pedigree.

chromosome that contains an inversion and one that is normal, there are important genetic and cytological effects. Because there is no net loss or gain of material, heterozygotes are usually perfectly viable. The location of the inverted segment can be recognized cytologically in the meiotic nuclei of such heterozygotes by the presence of an "inversion loop" in the paired homologs (Figure 7-12).

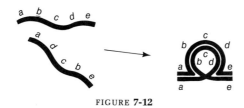

FIGURE **7-12**

The location of the centromere relative to the inverted segment determines the genetic behavior of the chromosomes. If the centromere is not included in the inversion, the inversion is *paracentric,* whereas inversions spanning the centromere are *pericentric* (Figure 7-13).

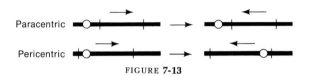

FIGURE 7-13

What do inversions do genetically? In a heterozygote for a paracentric inversion, crossing over within the inversion loop has the effect of connecting homologous centromeres in a *dicentric bridge,* as well as producing an acentric piece of chromosome, that is, without a centromere (Figure 7-14).

Thus, as the chromosomes separate during anaphase I, the disjoining centromeres will remain linked by means of the bridge. This orients the centromeres so that the noncrossover chromatids lie furthest apart. The acentric fragment cannot align itself or move and consequently it will be lost. Remarkably, in *Drosophila* and in plant eggs the dicentric bridge will remain intact long after the chromosomes have condensed after anaphase I and, as the second meiotic division begins, the noncrossover chromatids are directed to the outermost nuclei. Thus, the two inner nuclei will either be linked by the dicentric bridge or contain fragments of the bridge if it breaks, whereas the outer nuclei contain the noncrossover chromatids (Figure 7-15).

Fertilization of a nucleus carrying the broken bridge should produce defective zygotes that die because they have an unbalanced set of genes. Consequently, in a test cross the recombinant chromosomes would end up in dead zygotes and recombinant frequency would be lowered. However, in *Drosophila,* the presence of large inversions does not result in a large increase in zygotic mortality. Consequently, it has been inferred that the inner nuclei never participate in fertilization and that only one of the outer nuclei can be the egg nucleus. Thus, you can see that the chromatids participating in a crossover event will be selectively retained in the central nuclei, thereby allowing recovery of the noncrossover chromatids in the egg nuclei. This remarkable suggestion based on genetic results was later confirmed cytologically in *Drosophila.* It has also been shown in plants. (How would you test this suggestion in *Neurospora* using tetrad analysis?) Nevertheless the genetic consequence of inversion heterozygosity is the same; that is, the selective recovery of noncrossover chromatids from exchange tetrads. In addition, inversion heterozygotes often have pairing problems in the area of the inversion and this reduces crossing over and recombinant frequency in the vicinity.

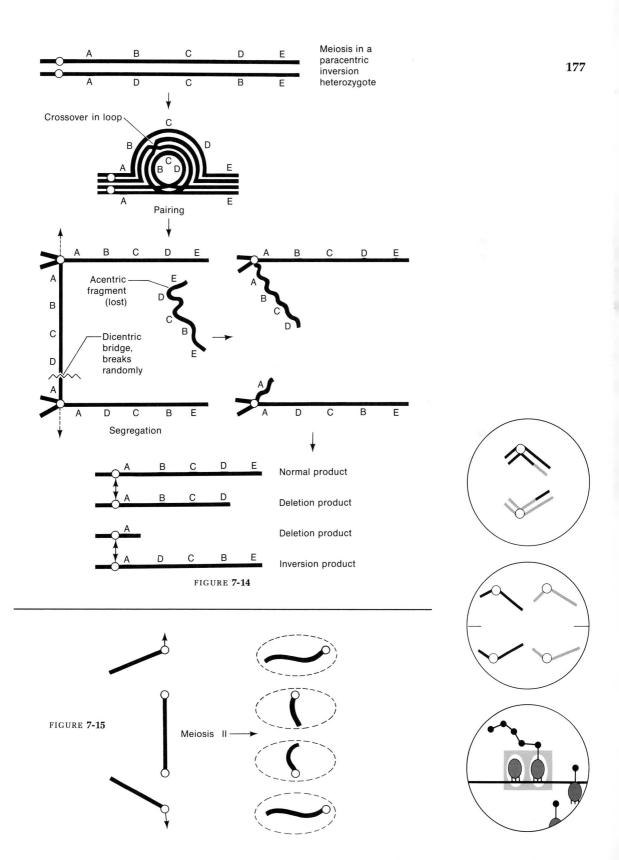

Meiosis in a paracentric inversion heterozygote

Crossover in loop

Pairing

Acentric fragment (lost)

Dicentric bridge, breaks randomly

Segregation

A B C D E — Normal product

A B C D — Deletion product

A — Deletion product

A D C B E — Inversion product

FIGURE **7-14**

FIGURE **7-15**

Meiosis II →

MESSAGE

Although inversion heterozygosity does reduce the number of recombinants recovered, it in fact does so by two mechanisms: one by inhibiting the process of chromosome pairing in the vicinity of the inversion and the other by selectively eliminating the products of crossovers in the inversion loop.

It is worthwhile noting that paracentric inversions have been useful in studying crossing over in organisms for which no genetic markers are known but which do have chromosomes that can be studied. Dicentric bridges are the consequence of crossovers within the inversion loop; and so their frequency is related to the amount of crossing over.

The net genetic effect of a pericentric inversion is the same as that of a paracentric one; that is, crossover products are not recovered, but for different reasons. In a pericentric inversion, because the centromeres are contained within the inverted region, disjunction of crossover chromosomes is normal. However, a crossover within the inversion produces chromatids that contain a duplication and a deficiency for different parts of the chromosome (Figure 7-16). In this case, fertilization of a nucleus carrying a crossover chromosome generally results in its elimination through zygotic mortality caused by an imbalance of genes.

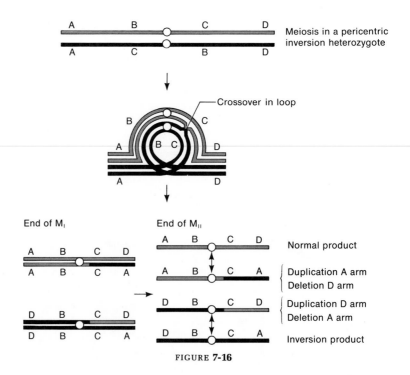

FIGURE **7-16**

Again, the result is the selective recovery of noncrossover chromosomes as viable progeny.

To generate a duplication "on purpose" as it were, it is possible to use a pericentric inversion having one breakpoint at the tip of the chromosome (Figure 7-17). A crossover in the loop produces a chromatid type in which the

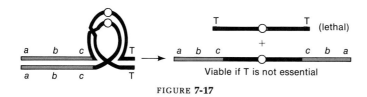

FIGURE **7-17**

entire left arm is duplicated, and, if the tip is nonessential, a duplication stock is generated for investigation. Another way to make a duplication (and a deficiency) uses two paracentric inversions whose breakpoints overlap (Figure 7-18). (These tricks are possible only in genetically well marked organisms.)

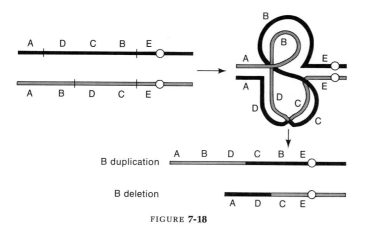

FIGURE **7-18**

So far, we have considered only single crossovers. You may ask, What happens in double-exchange tetrads? The consequences are predictable from a tetrad analysis of normal chromosomes (discussed in Chapter 5). We would expect two-, three-, and four- strand double exchanges within the inversion loop in a 1:2:1 ratio. Let's consider such tetrads in a paracentric inversion (Figure 7-19). In this case, viable crossover chromatids can be recovered but only if they are the products of both exchanges. Thus, comparison of the frequency of double crossovers within inversion loops of heterozygotes and in the same region of structurally normal chromosomes provides a measure of how much effect the inversion has on the occurrence of crossing over.

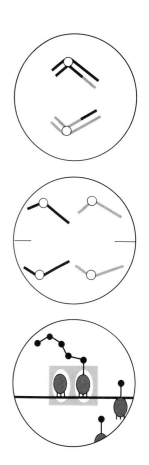

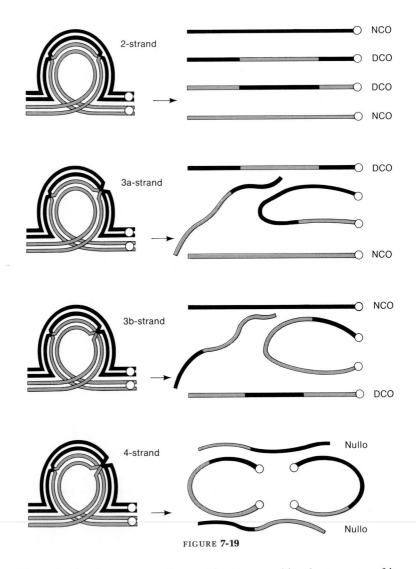

FIGURE 7-19

The reduction in recovery of recombinants caused by the presence of inversions is a useful property in laboratory experiments. Thus, if a particularly interesting combination of genes has been linked by crossing over but cannot be made homozygous (because some of the genes cause sterility, reduced viability, or lethality), an inversion of its homolog can be introduced to prevent disruption of that sequence by crossing over. The use of inversions in this way may, in fact, be an imitation of nature. In natural populations of organisms, inversions are commonly found. Thus, it has been inferred that inversions may be of value in reducing the recovery of crossovers in chromosomes that carry particularly advantageous combinations of genes that had arisen previously.

The exchange of chromosome parts between nonhomologs establishes new linkage relationships if the translocated chromosomes are homozygous. Furthermore, translocations may drastically alter the size of a chromosome, as well as the position of its centromere. For example,

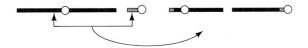

Here a large metacentric chromosome is shortened by half its length to an acrocentric one, whereas the small chromosome becomes a large one. Examples in natural populations are known in which chromosome numbers have actually been changed by translocation between acrocentric chromosomes and the subsequent loss of the resulting small chromosome elements (Figure 7-20).

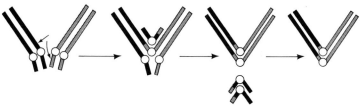

FIGURE **7-20**

In heterozygotes having translocated and normal chromosomes, the genetic and cytological effects are important. Again, the pairing affinities of homologous regions dictate a characteristic configuration when all chromosomes are synapsed in meiosis. In Figure 7-21, which illustrates meiosis in a reciprocally translocated heterozygote, the configuration is that of a cross.

Remember, this configuration lies on the metaphase plate with the spindle fibers perpendicular to the page. Thus the centromeres would actually migrate up out of the page or down into it. Homologous paired centromeres disjoin, translocation or not. Because Mendel's Second Law still applies to *different paired centromeres,* there are two common patterns of disjunction. The segregation of each of the structurally normal chromosomes with one of the translocated ones (T_1 with N_2 and T_2 with N_1) is called *adjacent-1 segregation.* Both meiotic products are duplicated and deficient for different regions and will produce unbalanced zygotes if fertilized by normal gametes. On the other hand, the two normal chromosomes may segregate together, as do the reciprocal parts of the translocated ones to produce $T_1 + T_2$ and $N_1 + N_2$ products. This is called *alternate segregation.* There is another event called adjacent-2 segregation

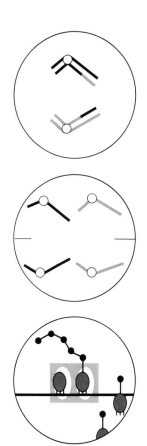

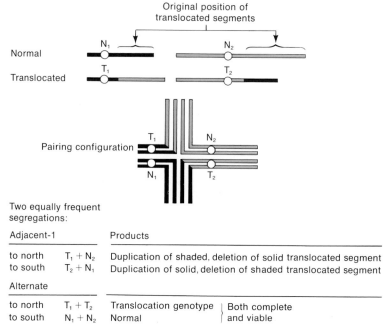

Original position of
translocated segments

FIGURE 7-21

in which homologous centromeres migrate to the same pole, but, in general, this is a rare occurrence.

Thus, adjacent-1 and alternate segregations occur with equal frequency and, in a cross between heterozygotes for the same translocation, you can see that a complete diploid set of the chromosomes is obtained only in certain gametic combinations (Figure 7-22). In this example, 6/16 of the zygotes carry a

		Male			
		T_1N_2	T_2N_1	T_1T_2	N_1N_2
Female	T_1N_2	−	+	−	−
	T_2N_1	+	−	−	−
	T_1T_2	−	−	+	+
	N_1N_2	−	−	+	+

FIGURE 7-22

complete diploid chromosome set. Of these six, one will be homozygous for the normal chromosomes and one will be homozygous for the translocation.

In a cross between translocation heterozygotes and structurally normal homozygotes, inviable and viable zygotes are produced in equal frequency.

	T_1N_2	T_2N_1	T_1T_2	N_1N_2
N_1N_2	−	−	+	+

Of the viable progeny, half will be homozygous for structurally normal chromosomes and the other half will remain heterozygous.

MESSAGE

Translocations reduce the production of viable offspring by generating zygotes that are unbalanced for some chromosome parts.

Genetically, markers on nonhomologous chromosomes will appear to be linked if these chromosome pairs are involved in a translocation (Figure 7-23).

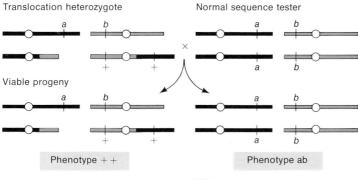

FIGURE **7-23**

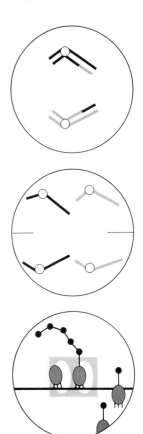

In fact, if all four arms of the meiotic pairing structure are genetically marked, a cross-shaped linkage group will result. Such interlinkage group linkage is often a giveaway for the presence of a translocation.

Translocations are economically important as well. In agriculture, the occurrence of translocations in certain crop strains can reduce yields considerably owing to the number of unbalanced zygotes formed. On the other hand, it has been proposed that the high incidence of inviable zygotes could be used to control insect pests by the introduction of translocations into the wild. Thus, 50% of the offspring of crosses between insects carrying the translocation and wild types would die and 50% of the survivors would carry the translocation. On the other hand, 2/3 of the surviving or fertile progeny between translocation-bearing insects would be heterozygous for the translocation.

Translocations occur in humans, usually in association with a normal chromosome set in a translocation heterozygote. Down's syndrome (commonly referred to as Mongolism) can arise in the progeny of an individual heterozygous for a translocation involving chromosome 21. The heterozygous person is phenotypically normal (and is called a "carrier"), but during meiosis an adjacent-1 segregation will produce gametes carrying duplicated parts of chromosome 21,

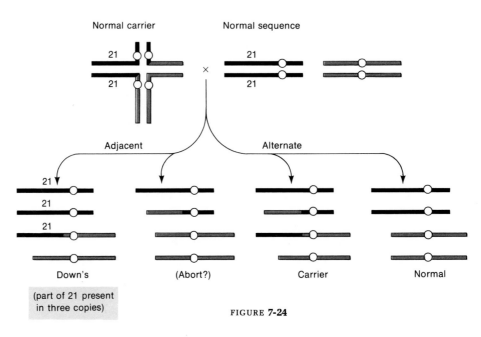

FIGURE **7-24**

and possibly a deficiency for some part of the other chromosome involved in the translocation. For unknown reasons, the extra chunk of chromosome 21 is the cause of Down's syndrome. Of the normal children of a carrier, half will themselves be carriers (Figure 7-24).

Later we will discuss another way in which Down's syndrome is generated. Bear in mind for now that under the present method there should be a high recurrence rate in the family pedigree involved: a translocation heterozygote or carrier can repeatedly produce Down's syndrome children, and of course the carriers can do the same in subsequent generations. The other method does *not* show this recurrence within a family. The factors responsible for numerous other hereditary disorders have been traced to translocation heterozygosity in the parents—as, for example, in the cri du chat pedigree in Figure 7-11.

Another method of producing duplications and deficiencies for specific chromosome regions makes use of translocations. Let's take *Drosophila* as an example. For reasons that are yet unclear, heterochromatin near the centromere, although very extensive physically, contains very few known genes. In fact, for a long time heterochromatin was considered useless, inert material. In any case, for our purposes, *Drosophila* can tolerate a loss or an excess of heterochromatin with little effect on viability or fertility.

Let's now select two different reciprocal translocations involving the same two chromosomes. Both of them have a breakpoint somewhere in heterochromatin and each has a euchromatic break on one side or the other of the region we want to be duplicated and deleted (Figure 7-25). You can see that, if

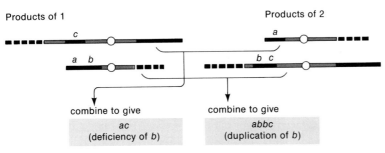

FIGURE **7-25**

we have a large collection of translocations having one heterochromatic break and euchromatic breaks at many different sites, duplications and deficiencies for many parts of the genome can be produced at will.

In fungi, a tetrad analysis can be very useful in detecting chromosome aberrations in general. In *Neurospora,* for example, any genome containing a deletion will produce an ascospore that does not ripen to the normal black color, and parents of crosses with high proportions of such "aborted" white ascospores usually contain rearrangements (duplications are generally recovered as black *viable* ascospores). Specific spore-abortion patterns sometimes identify specific rearrangements. The pattern resulting from a translocation is an equal number of asci having 8 black and 0 white spores (alternate-segregation meioses) and 0 black and 8 white spores (adjacent-segregation meioses) and some asci having 4 black and 4 white. The 4:4 asci are produced by crossing over between either centromere and the translocation breakpoint (Figure 7-26).

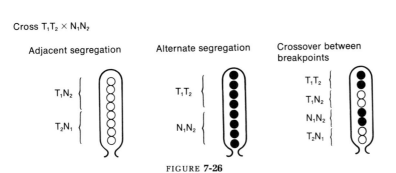

FIGURE **7-26**

Changes in Chromosome Number

EUPLOIDY

Monoploids. Monoploids contain a single chromosome set and are characteristically sterile. In germ cells, a proper meiosis does not occur because there are no pairing partners for the chromosomes. Of course, all the single chromosomes could go to one pole to produce a haploid gamete, but, if the chromosomes assort independently, you can see that this will happen only rarely. In fact, the probability of all of them going the same way is $(1/2)^{n-1}$ in which n is the number of chromosomes.

Triploids. Once into the realm of *polyploids* (more than two chromosome sets), we have to distinguish *autopolyploids*, formed by the duplication of one set of chromosomes (e.g., $n_1 + n_1$), from *allopolyploids*, formed by the fusion of separately derived sets (e.g., $n_1 + n_2$) that often come from different species. Artificially constructed triploids are always allopolyploids, formed from the fertilization of a diploid gamete by a haploid one ($2n + n = 3n$).

Once again there are problems at meiosis. In an autopolyploid, pairing can take place in several ways, but always between only two chromosomes at a time (Figure 7-27). The net result is the same, a $2n \leftrightarrow n$ segregation of three kinds:

$$\frac{1 + 2}{3} \qquad \frac{1 + 3}{2} \qquad \frac{2 + 3}{1}$$

This happens for every chromosome threesome and the probability of getting either a 2n or an n gamete will again be $(1/2)^{n-1}$. The rest will be unbalanced gametes, having two of one chromosome type, one of another, two of another, and so forth, and most will be nonfunctional. Even if functional, when unbalanced gametes are fertilized, inviable unbalanced genomes are formed. Seedless watermelons and other seedless produce are often triploids, formed by fertilizing n gametes (from a diploid) and 2n gametes (from a tetraploid). How are tetraploids formed? Read on.

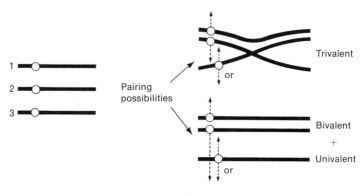

FIGURE 7-27

Tetraploids. Autotetraploids either occur naturally by the spontaneous doubling of a 2n chromosome set or artificially by induction with various spindle inhibitors like colchicine. Tetraploids are usually very stable and fertile. This is because pairing partners are available at meiosis, and diploid gametes are regularly produced (Figure 7-28). Both types of pairing patterns form 2n gametes.

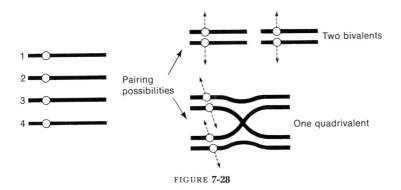

FIGURE **7-28**

A formal genetics can be developed for tetraploids. Let's use colchicine to double the chromosomes of an *Aa* plant into an *AA aa* autotetraploid. Assuming the locus is very close to the centromere we can examine the gamete types produced, if they pair as bivalents. Three pairing patterns are possible (Figure 7-29). All these events take place at random and occur equally frequently; so the ratio of *Aa* : *AA* : *aa* gametes will be 8 : 2 : 2 or 4 : 1 : 1. If such a plant is selfed, the probability of the aaaa phenotype appearing in the next generation is $1/6 \times 1/6 = 1/36$; in other words, a 35 : 1 ratio prevails. If the gene is very

Pairing	1 —O—— A 2 —O—— A 3 —O—— a 4 —O—— a			

	1 —O A 2 —O A	3 —O a 4 —O a	1 —O A 3 —O a	2 —O A 4 —O a	1 —O A 4 —O a	2 —O A 3 —O a
Gametes produced by random spindle attachment	$\frac{1+3}{2+4}$ Aa Aa		$\frac{1+2}{3+4}$ AA aa		$\frac{1+2}{3+4}$ AA aa	
	$\frac{1+4}{2+3}$ Aa Aa		$\frac{1+4}{2+3}$ Aa Aa		$\frac{1+3}{2+4}$ Aa Aa	

FIGURE **7-29**

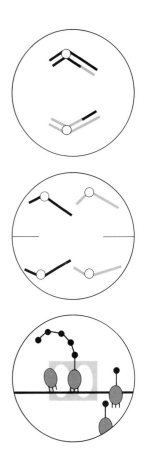

distant from the centromere, a 21:1 ratio ensues. (Can you prove it?) Usually a ratio that is in between the two ratios is obtained. In any case, as a rule a special kind of Mendelian analysis is possible with tetraploids and even-numbered polyploids.

Allopolyploids have probably been important in evolutionary speciation. The "classical" allopolyploid was an allotetraploid built by G. Karpechenko. He wanted to make a fertile hybrid between the cabbage and the radish that had the roots of a radish and the leaves of a cabbage (economically convenient—no waste!). Both species have 18 chromosomes and are closely enough related that one can be fertilized by the other. He crossed them and obtained an F_1 hybrid that also had 18 chromosomes but was sterile. It was sterile because the set of nine chromosomes from the cabbage parent was different enough from the radish set that homology was insufficient for normal synapsis and disjunction.

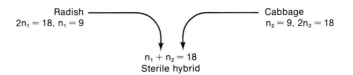

Radish
$2n_1 = 18, n_1 = 9$

Cabbage
$n_2 = 9, 2n_2 = 18$

$n_1 + n_2 = 18$
Sterile hybrid

However, one day a few seeds were produced by this (almost!) sterile hybrid. On planting, they were found to be allotetraploids of the type $2n_1 + 2n_2$ and had 36 *chromosomes*. Apparently a spontaneous accidental chromosome doubling had taken place in the plant, perhaps during a mitotic spindle failure. Thus, a patch of tissue (probably in a flower) was produced in which two sets of $n_1 + n_2$ were present and every chromosome had a pairing partner for meiosis. The allotetraploid was healthy and fertile, producing functional $n_1 + n_2$ gametes because every chromosome had a homologous pairing partner. This kind of tetraploid is often called an *amphidiploid*. (Unfortunately, for Karpechenko, his amphidiploid had the roots of a cabbage and the leaves of a radish.) Karpechenko's amphidiploid was sterile in crosses with either a radish or a cabbage plant. (Why?) Consequently, he thought he had created a new species of plant and named it *Raphanobrassica* (*Raphanus* for radish and *Brassica* for cabbage).

It is easy to make amphidiploids nowadays by applying colchicine to the sterile hybrid to induce chromosome doubling. Probably the production of a new species of grass, *Spartina townsendii* (n = 63), was due to amphidiploidy through the union of gametes of the American species *S. alterniflora* (n = 35) and those of the British *S. stricta* (n = 28). The union was made possible by commercial sea traffic between the two continents, which brought the two species into contact.

Polyploids are of great economic importance. They are generally much larger than their diploid ancestors. Thus, many commercially used grains (especially wheat), cotton, and tobacco are polyploids, and even in the counterculture, rumors of polyploid *Cannabis sativa* (marijuana) abound. Note that all of the examples of polyploidy have been plants. There are viable polyploids among animals but they're rare. This probably reflects the evolution of sex-determining mechanisms that depend on a balance between chromosome factors. In humans, as much as 15% of early spontaneously aborted fetuses are polyploid.

ANEUPLOIDY

Monosomics $(2n - 1)$. Monosomic chromosome complements are generally deleterious for two main reasons: First, the balance of chromosomes that is necessary for a finely tuned cellular homeostasis, carefully put together during evolution, is grossly disturbed. For example, if a genome consisting of two of each of chromosomes a, b, and c becomes monosomic for c (i.e., $2a + 2b + 1c$), the ratio of these chromosomes is changed from $1c : 1 (a + b)$ to $1c : 2 (a + b)$. Second, any deleterious recessive on the single remaining chromosome becomes hemizygous and may be directly expressed phenotypically. (Note that these are the same effects as those produced by deletions.)

Monosomics, trisomics $(2n + 1)$, and other chromosome aneuploids are probably produced by nondisjunction during mitosis or meiosis. In meiosis it can happen either at the first or second division (Figure 7-30). (You can ask yourself whether products of nondisjunction at these two times can be distin-

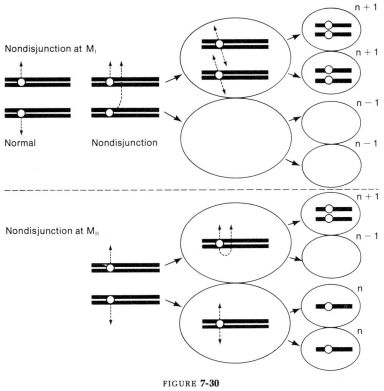

FIGURE **7-30**

guished genetically.) If an n − 1 gamete is fertilized by an n gamete, a monosomic (2n − 1) zygote is produced; an n + 1 and an n gamete gives a trisomic 2n + 1; and an n + 1 and an n + 1 gives a tetrasomic, if the same chromosome is involved, and so on. In *Neurospora* (a haploid) the n − 1 meiotic products abort and do not darken in the ascospore; so M_I and M_{II} nondisjunctions are detected as 4:4 and 6:2 ratios. (Diagram the chromosome content of the various spores to convince yourself of the relation of the spore pattern to nondisjunction.) For loci on the aneuploid chromosomes, what ascus genotypes are produced?

In humans, monosomics are well known, especially the sex chromosome monosomic (44 autosomes + 1 X), which produces a phenotype known as Turner's syndrome. People who have Turner's syndrome have a characteristic, easily recognizable phenotype: they are sterile females, are short in stature, and have webbed necks. Their intelligence varies from different degrees of mental retardation to normal. Monosomics for all autosomes except 13 and 18 die in utero.

If viable, monosomics are useful in locating newly found recessive genes on specific chromosomes in plants. In one such method, different monosomic strains lacking a different chromosome in each stock are obtained. Homozygotes for the new gene are crossed with each monosomic strain and the progeny of each cross are inspected for expression of the recessive phenotype. The cross in which the phenotype appears identifies its chromosomal location. (At least one-half of the gametes produced by most monosomic plants are nullisomic, or n − 1.)

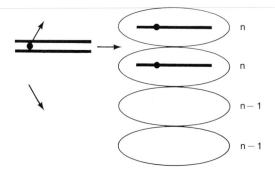

A similar trick can be used in humans. For example, two people whose vision is normal may produce a daughter who has Turner's syndrome and is also red-green color-blind. This shows that the allele for red-green color blindness is

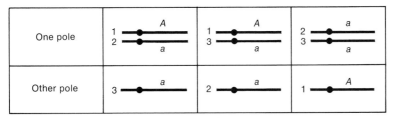

FIGURE **7-31**

recessive, that it was on the X chromosome of the mother, and that nondisjunction must have occurred in the father. (Can you see why?)

Trisomics $(2n + 1)$. For genes that are tightly linked to the centromere of a trisomic chromosome set, the random segregations can be represented as shown in Figure 7-31 in a trisomic Aaa. All types occur equally frequently and a gamete ratio of $1A:2Aa:2a:1aa$ is produced. Trisomics are sometimes recognized by these ratios, which are also useful in locating genes on chromosomes.

In humans there are several examples of trisomics: The combination XXY results in Klinefelter's syndrome, producing males that have lanky builds, are mentally retarded, and are often sterile. Another combination, XYY males, occurs in more than 1 in 3,000 births and a lot of excitement was aroused when an attempt was made to link the XYY condition with a predisposition toward violence. This is still hotly debated although it is now clear that an XYY condition in no way guarantees such behavior. Nevertheless, several enterprising lawyers have attempted to use the XYY genotype as grounds for acquittal or compassion in crimes of violence. The XYY males are usually fertile.

In a recent study of newborn babies in Toronto hospitals, 72,739 individuals were tested for the possession of extra X chromosomes. Forty XXY, sixteen XXX, one XO, one XXYY, and nine mosaic individuals were found. An additional 3,660 males babies were screened for the possession of an extra Y, yielding five XYY newborns.

We have already looked at the generation of Down's syndrome through adjacent segregation in translocation heterozygotes. Down's syndrome also occurs as a result of nondisjunction during meiosis and it is then called trisomy 21. In this form of Down's syndrome, there is generally no family history of the phenotype. However, the frequency of this form is dramatically higher among the children of older mothers (Figure 7-32). Trisomics for many other autosomes have been observed in humans, but they do not usually survive the nine-month gestation. It has been estimated that about five thousand out of

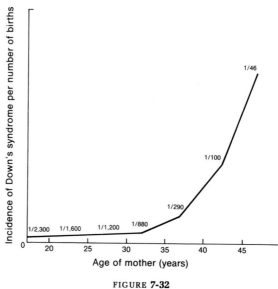

FIGURE 7-32

a million humans born live have serious genetic defects attributable to the nondisjunction of chromosomes. A high incidence of chromosome anomalies is found in examining tissue from aborted fetuses. The detection of abnormal chromosomes in humans has been greatly facilitated by the identification of specific banding sequences in each chromosome.

Problems

1. H. Sharat Chandra recovered triploids of mealybugs, *Planococcus* (n = 5). He found that, in the gonads, at the end of meiosis I all cells had 15 chromosomes and at the end of meiosis II cells had variable numbers of chromosomes ranging from 0 to 15. How do you interpret these results?

2. Given a telocentric chromosome and a ring chromosome that is homologous to the telocentric one, they will pair in the following fashion:

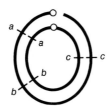

   a. Diagram the products of meiosis from a single crossover between *a* and *b* at the four-strand stage (remember, sister strands never crossover).
   b. Diagram the results of a double crossover occurring between *a* and *b* and between *b* and *c* when it is
      i. a two-strand double crossover
      ii. two kinds of three-strand double crossovers
      iii. a four-strand double crossover
      (Show the results at the end of meiosis I and II.)

3. A corn plant *pr/pr* that has standard chromosomes was crossed with a plant homozygous for a reciprocal translocation between chromosomes 2 and 5 and for the *Pr* allele. The F_1 was semisterile and phenotypically Pr (a seed color). A backcross to the parent with standard chromosomes gave: 764 stemisterile Pr; 145 semisterile pr; 186 normal Pr; and 727 normal pr. What is the map distance of the *Pr/pr* locus from the translocation point?[1]

4. Suppose that you discover a *Drosophila* male that is heterozygous for a reciprocal translocation between the second and third chromosomes, each break having occurred near the centromere (which for these chromosomes is near the center).
   a. Draw a diagram showing how these chromosomes would synapse at meiosis.

[1] From *General Genetics*, 2d ed., by Adrian M. Srb, Ray D. Owen, and Robert S. Edgar. W. H. Freeman and Company, Copyright © 1965.

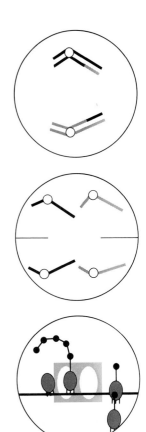

b. You find that this fly has the recessive genes *bw* (brown eye) and *e* (ebony body) on the nontranslocated second and third chromosomes, respectively, and wild-type alleles on the translocated ones. It is crossed with a female having normal chromosomes that is heterozygous for *bw* and *e*. What type of offspring would be expected and in what ratio? (Zygotes with an extra chromosome arm or deficient for one do not survive. There is no crossing over in *Drosophila* males.)

5. An *insertional* translocation consists of the insertion of a piece from the center of one chromosome into the middle of another (nonhomologous) chromosome. Thus

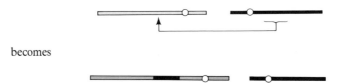

becomes

How will genomes heterozygous for such translocations pair at meiosis? In *Neurospora*, what spore abortion patterns will be produced and in what relative proportions in such translocation heterozygotes? (Remember, duplications survive and have dark spores, but deficiencies are light-spored.)

6. Suppose that you are studying the cytogenetics of five closely related species of *Drosophila*. You find the following gene order (the letters indicate genes identical in all five species) and chromosome pairs in each species:

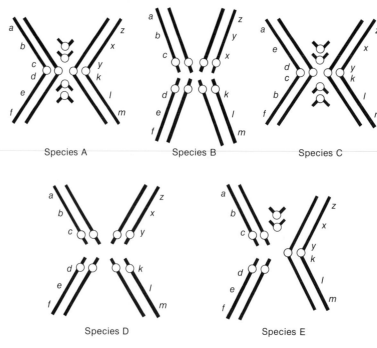

Show how these species probably evolved from each other, describing the changes that occurred at each step. *Note:* Be sure to compare gene order carefully.

7. An autotetraploid is heterozygous for two gene loci, *FFff* and *GGgg*, each locus affecting a different character and located on a different set of homologous chromosomes very close to their respective centromeres.

   a. What gametic genotypes will be produced by this individual, and in what proportions?
   b. If the individual is self-fertilized, what proportion of the progeny will have the genotype *FFFf GGgg*? the genotype *ffff gggg*?

8. In *Neurospora*, a cross is heterozygous for a paracentric inversion. The break points of the inversion are known to be very close to two loci that recombine with an RF of 10%.

   a. Using the mapping function $RF = 1/2(1 - e^{-m})$, calculate the *mean* number of exchanges expected in the inversion loop per meiosis.
   b. Use this mean frequency to calculate the frequency of meioses with
      i. no exchanges in the loop
      ii. one exchange in the loop
      iii. two exchanges in the loop

   The Poisson formula is $e^{-m}\left(\dfrac{1}{0!} + \dfrac{m}{1!} + \dfrac{m^2}{2!} + \cdots\right)$.

   c. Remembering that ascospores bearing deficient chromosome complements do not darken in *Neurospora*, predict how many light and dark ascospores you would find in 8-spored asci resulting from meioses in which there had been
      i. no crossovers in the loop
      ii. one crossover in the loop
      iii. two crossovers in the loop
   (Remember that there are three kinds of double crossovers and so the progeny population of asci from part iii could be heterogeneous.)
   d. Using your predicted frequencies of 0, 1, and 2 crossover meioses, what *overall* frequencies of
      i. 8 dark:0 light
      ii. 0 dark:8 light
      iii. 4 dark:4 light
   asci would you find from this cross? (*Note:* Your total from part b should be less than 100% because we ignored triple and\higher exchanges, but it should be close to 100% because these events are rare.) Simply make your total for this part equal your total for part b.

9. The New World cotton species *Gossypium hirsutum* has a 2n chromosome number of 52. The Old World species *G. thurberi* and *G. herbaceum* each have a 2n number of 26. Hybrids between these species show the following chromosome pairing arrangements at meiosis:

Hybrid	
G. hirsutum × *G. thurberi*	13 small pairs (bivalents) + 13 large univalents
G. hirsutum × *G. herbaceum*	13 large bivalents + 13 small univalents
G. thurberi × *G. herbaceum*	13 large univalents + 13 small univalents

   Draw diagrams to interpret these observations phylogenetically, indicating clearly the relationships between the species. How would you go about proving that your interpretation is correct?

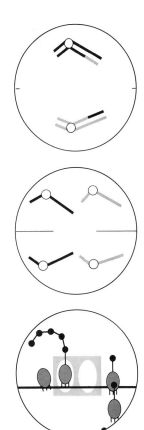

10. In *Neurospora,* a cross between the multiply marked chromosomes $a + c + e$ and $+ b + d +$ gave one product of meiosis that grew on minimal medium (assume a, b, c, d, and e are nutritional markers). The rare colony was grown up and some *asexual* spores were $a + c + e$ in genotype, some were $+ b + d +$, and the rest grew on minimal medium. Explain the origin of the rare product of meiosis.

11. In *Neurospora,* the genes a and b are on separate chromosomes. In a cross of a standard $a\,b$ strain with a wild type obtained from nature, the progeny are as follows: $a\,b$ 45%, $+ +$ 45%, $a +$ 5%, $+ b$ 5%. Interpret these results and explain the origin of all the progeny types under your hypothesis.

12. Assume that, in a study of hybrid cells, three genes in humans have been assigned to chromosome 17. These genes are a, b, and c and are concerned with making the compounds a, b, and c—all of which are essential for growth. If $a^- b^- c^-$ mouse cells are fused with $a^+ b^+ c^+$ human cells, assume that you find a hybrid in which the only human component is the right arm of chromosome 17 (17R), translocated by some unknown mechanism to a mouse chromosome. The hybrid can make the compounds a, b, and c. Treatment of cells with adenovirus causes chromosome breaks. Let's assume that you can isolate 200 lines in which bits of the translocated 17R have been clipped off. These lines are tested for ability to make a, b, and c and the results are below:

Number of lines	Can make
0	a only
0	b only
12	c only
0	a and b only
80	b and c only
0	a and c only
60	a, b, and c
48	nothing

a. How would these different types arise?
b. Are a, b, and c all on the right arm of 17?
c. If so, could you draw an approximate map indicating relative positions?
d. How would quinacrine dyes help you in this? (*Note:* This *kind* of approach has been actually used, although the details of this particular question are largely hypothetical.)

13. In humans, only autosomal trisomics for chromosomes 13, 18, or 21 survive until birth. All three types are severely deformed. If you were a medical geneticist, how would you go about studying aneuploidy? Do you think aneuploids for the other chromosomes never occur? or are they very rare?

14. The incidence of trisomy 21 or Down's syndrome increases some 40-fold among mothers over the age of 35. What kinds of explanations for this increase in nondisjunction can you think of? (We don't know the answer yet; so if you come up with a super idea, who knows?)

15. Several kinds of sexual mosaics are well documented in humans, some examples of which are given below. Suggest how each may have arisen.

> XX/XO (i.e., there are two cell types in the body, XX and XO)
> XX/XXYY
> XO/XXX
> XX/XY
> XO/XX/XXX

16. A *Neurospora* heterokaryon is established between nuclei of the following genotypes in a common cytoplasm

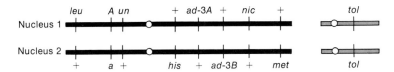

in which *leu, his, ad, nic,* and *met* are all recessive alleles causing specific nutritional requirements for growth. *A* and *a* are the mating-type alleles (for a cross to occur one parent has to be *A* and the other *a*). Usually (*A* plus *a*) heterokaryons are incompatible, but the recessive mutant *tol* suppresses this incompatibility and permits heterokaryotic growth on vegetative medium. The allele *un* is recessive, prevents the fungus from growing at 37°C (it is a temperature-sensitive allele), and cannot be corrected nutritionally. This heterokaryon grows well on minimal medium, as do most of the cells derived mitotically from it. Some rare cells are found, however, that show the following traits:

a. They will not grow on minimal medium unless it is supplemented with leucine.
b. When transferred to a crossing medium, a cross does not occur (i.e., they will not self).
c. They will not grow when moved into a 37°C temperature, *even* if supplied with leucine.
d. When haploid wild-type *a* cells are added to these aberrant cells, a cross occurs, but the addition of *A* does not cause a cross.
e. From the cross with wild-type *a*, progeny with the genotype of nucleus 1 are recovered, but no alleles from nucleus 2 ever emerge from the cross.

Formulate an explanation for the origin of these strange cells in the original heterokaryon, and account for observations a–e concerning them.

17. In *Neurospora* a reciprocal translocation of the following type is obtained

and the following cross is made.

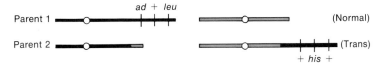

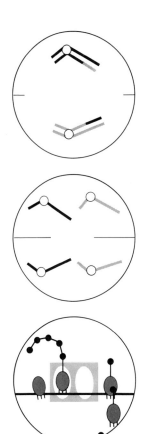

Assuming that the small, dotted piece of the chromosome involved in the trans- location does not carry any essential genes, how would you select products of meiosis that are duplicated for the translocated part of the solid chromosome?

18. Assume that gene x is a new mutant in corn. An xx plant is crossed with a triplo-10 individual (trisomic for chromosome 10) carrying only dominant alleles at the x locus. Trisomic progeny are recovered (how?) and crossed back to xx ♀ ♀. What ratio of dominant to recessive phenotypes can be expected if the x locus is *not* on chromosome 10? If it *is* on chromosome 10? What strains do you need to definitely locate *any* new mutation on a specific chromosome?[2]

19. A *Drosophila* geneticist has a strain of fruit flies that is true breeding and wild type. If she crosses this strain with a multiply marked X chromosome strain carrying the recessive genes y—yellow, cv—crossveinless, v—vermilion, f—forked, and car —carnation (which are equally distributed along the X chromosome from one end to the other), collects the heterozygous F_1 female offspring, and crosses them with $y\ cv\ v\ f\ B\ car$ males, she gets the following classes among the male offspring.

1. $y\ cv\ v\ f\ car$
2. $+\ +\ +\ +\ +$
3. $y\ +\ +\ +\ car$
4. $+\ cv\ v\ f\ +$
5. $y\ cv\ +\ f\ car$
6. $+\ +\ v\ +\ +$
7. $y\ cv\ +\ +\ car$
8. $+\ +\ v\ f\ +$
9. $y\ +\ +\ f\ car$
10. $+\ cv\ v\ +\ +$
11. $y\ cv\ v\ f\ B\ car$

a. How would you account for the results in classes 1 through 10?
b. How can you account for class 11? You should be able to give two ways.
c. How would you test your hypotheses?[3]

---

[2] From *General Genetics*, 2d ed., by Adrian M. Srb, Ray D. Owen, and Robert S. Edgar. W. H. Freeman and Company. Copyright © 1965.
[3] Problem 19 courtesy of Tom Kaufmann.

8 / Genetic Change

(Or: The ultimate source of all biological diversity.)

The science of genetics depends on the existence of inherited differences; that is, at least two allelic states of a gene. The process of change from one hereditary state to another is called *mutation*. To confuse it a bit, a new allele of a normal gene is also called a mutation. An individual carrying a mutation and expressing it phenotypically is called a "mutant." We have already considered heritable changes in chromosome structure or number, which we can call *chromosomal mutations*. In this chapter, we will focus on mutations of single genes from one allele to another, which we can call *point mutations*. Because mutation can occur in somatic as well as germ cells, we must bear in mind that mutations are "potentially" but not necessarily heritable from generation to generation.

The "direction" of mutation can be arbitrarily defined once we establish the allelic state that will be regarded as standard or wild type. Any change from the wild-type allele, then, is called a "forward" mutation, whereas a change from mutant to wild type is called a "back" or "reverse" mutation (or a *reversion*). Bear in mind how arbitrary this is: the wild type of today may have been a mutation in our evolutionary past.

The phenotypic consequences of a mutation may be so subtle as to require refined biochemical techniques to detect a difference from wild type or so severe as to produce gross morphological defects or death. The tremendous stability and constancy of form from generation to generation suggests that mutation

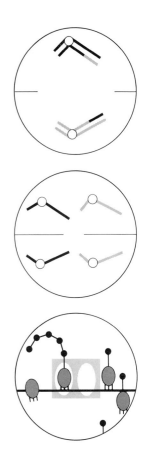

must occur very rarely or, if it is frequent, the change must be very small.

If we want to study the process of mutation itself, we must immediately ask, How can we quantify the process? One of the simplest methods is to look for the change from a dominant allele to a recessive. For example, in corn, the rate of change from C (colored kernel) to c (white kernel) can be easily measured in the cross $CC \times cc$ by simply looking for the white kernels that arise from mutation in the CC parent. Lewis Stadler used such an approach to measure mutation rates of a number of different loci. Some of his results are given in Table 8-1.

TABLE **8-1**

Gene	Number of gametes tested	Number of mutations	Average per million gametes
$R \to r$	554,786	273	492.0
$I \to i$	265,391	28	106.0
$Pr \to pr$	647,102	7	11.0
$Su \to su$	1,678,736	4	2.4
$Y \to y$	1,745,280	4	2.2
$Sh \to sh$	2,469,285	3	1.2
$Wx \to wx$	1,503,744	0	0.0

Several points come out of these results right away.

1. These studies are a lot of work. Counting a million of *anything* is no small task.
2. Mutation is a relatively rare event.
3. Mutation rates vary considerably from one gene to another.
4. Some genes never appear to mutate, at least at a frequency detectable by experimentation.

Very few people have the patience or the resources to score millions of progeny in crosses of plants or animals. Furthermore, it is possible that some genes mutate at a frequency of less than one in a million or ten million, in which case they would be impossible to detect.

Microorganisms have been a key to mutation studies because millions can be grown in a single test tube. In Chapter 6 we saw how Lederberg and Tatum were able to select rare wild-type recombinants in *E. coli* by observing their growth on minimal medium on which the parental classes did not grow. This kind of selective procedure can also be applied to mutation. The detection of a rare mutant depends on how cleverly the investigator can design a selective screen. Geneticists are basically lazy people who have devised many ingenious ways of letting mutants themselves announce their presence. Because mutants often occur at frequencies as low as 10^{-6} or lower, such selective methods are often essential. For example, adenine-3 mutant colonies in *Neurospora* accumulate a purple pigment and the morphology is easily identifiable against a background of millions of white adenine-3[+] colonies. Forward mutation systems like this, which specifically select for $+ \rightarrow m$ changes at a specific locus, are rare.

More commonly, one can select general classes of mutants in a forward direction. Taking *Neurospora* as an example again, a population of spores can be allowed to germinate in minimal medium in which any mutant with a nutritional requirement cannot grow. These mutants can be easily recovered by simply filtering off the wild-type individuals through a glass-wool filter. The nongerminated mutant spores will pass through the filter, whereas the wild types will have formed fibrous colonies, which get hung up on the filter. The mutants coming through can be grown on complete medium, then tested to see which specific compound they need for growth. This process is called *filtration enrichment*. It can be made more specific by growing the filtered mixture on minimal medium that contains everything the cells might possibly need except adenine, for example. Of course there could be many different loci in which a mutation might lead to an adenine requirement, and these would have to be distinguished later. Mutants that have a nutritional requirement are called *auxotrophs*, whereas the wild types are called *prototrophs* (Figure 8-1).

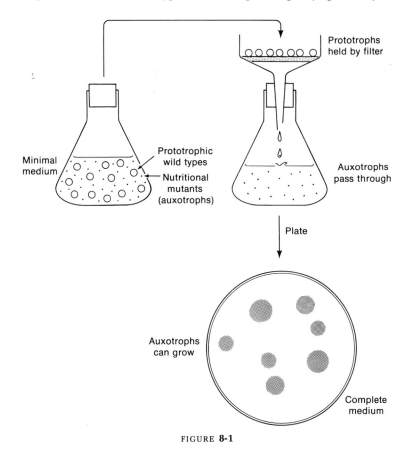

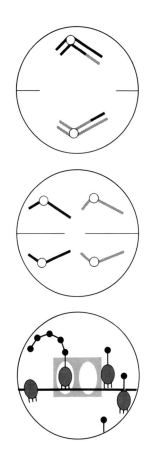

FIGURE **8-1**

Maximum enrichment for mutants can be obtained in which only the mutant is recovered and all other genotypes are eliminated. Thus, for example, reversion of auxotrophs can be measured because only the wild-type mutants can grow on minimal medium. In this way, it can be shown that mutation is indeed a rare event in any locus and that mutation frequencies are constant for any specific allele but may vary from one allele to another. Where forward and reverse mutations of a locus can be studied (as in adenine-3 mutants in *Neurospora*), it is found that the rates of change forward and back are not the same, with reversion usually being lower. (After reading the next two chapters, you may be able to suggest reasons for this.)

At this point, we would like to stress that, although we now know that bacteria have highly developed genetic systems, this was not always commonly accepted. When bacteria are subjected to an agent that kills them (such as a phage or a chemical), most of them are destroyed. However, within a very short time the culture is repopulated with cells, all of which are resistant to the agent. Has this come about from selection of a preexisting rare mutant among a large cell population or do all cells have a similar but low probability of becoming resistant *in response to* the selective agent? We hope that you can see the distinction between these two possibilities: the former is an explanation in standard genetic terms, whereas the latter suggests a potential for some kind of physiological adaptation to stress, which, if it occurs, can then be inherited.

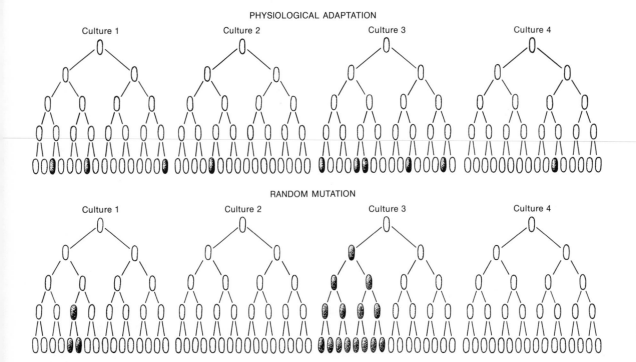

FIGURE 8-2

How can we distinguish between these two alternatives? Let's look at changes to resistance to T1 phage in *E. coli* cells and assume that the resistant cells result from a mutation that can occur at any time during growth. If we initiate a bacterial culture with a small number of cells and then the population increases, random mutation to resistance to T1 phage can occur in any cell at any time. If we take a large number of small populations of cells and let each one expand into a large population, which is then exposed to T1 phage, the number of mutations in each population will vary considerably depending on *when* the mutation occurred. On the other hand, if each cell has the same probability of being resistant by physiological adaptation, then in each population of cells, there should be the same general frequency of survivors and little variation of that value (Figure 8-2).

Salvadore Luria and Max Delbrück carried out such a "fluctuation" test. Into each of twenty cultures containing 0.2 ml of medium and into one containing 10 ml they introduced 10^3 *E. coli* cells per milliliter and incubated them until they obtained about 10^8 cells per milliliter. Each of the twenty 0.2-ml cultures were spread on plates that had a dense layer of T1 phages. From the 10-ml culture ten 0.2-ml volumes were withdrawn and plated. Many colonies were T1-resistant, as shown in Table 8-2.

TABLE **8-2**

Individual cultures		Bulk culture	
Number of culture	T1-resistant	Number of culture	T1-resistant
1	1	1	14
2	0		
3	3	2	15
4	0		
5	0	3	13
6	5		
7	0	4	21
8	5		
9	0	5	15
10	6		
11	107	6	14
12	0		
13	0	7	26
14	0		
15	1	8	16
16	0		
17	0	9	20
18	64		
19	0	10	13
20	35		
Mean	11.3	Mean	16.7

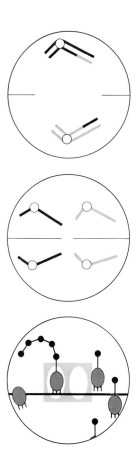

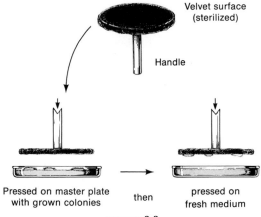

Pressed on master plate
with grown colonies   then   pressed on
fresh medium

FIGURE 8-3

The tremendous amount of variation from plate to plate in the individual 0.2-ml cultures cannot be explained by physiological adaptation because all the samples spread had the same number of cells. The simplest explanation is random mutation. This elegant analysis suggests that the resistant cells are *selected* by the agent rather than produced by it. Can the existence of mutants in a population *before* selection be determined? This was done by means of a *replica plating* technique developed by Joshua Lederberg and Esther Lederberg. When cells are grown on medium in a petri plate, those that are clonally related form a colony, which protrudes from the surface of the medium. A sterile piece of velvet placed lightly on the surface of the petri plate will pick up cells wherever there is a colony (Figure 8-3). (Under a microscope, velvet looks like a series of needles, which explains why lint adheres to it so well and why it picks up colonies of cells.) On touching the velvet to another sterile plate, some of the cells clinging to the velvet will be inoculated onto the plate in the same relative positions as the colonies on the original "master" plate. This simple technique allows us to grow cells on a nonselective medium (either complete nutrients or free of antibiotics or phages) and then to transfer a copy of the colonies to a selective medium (either minimal or plus antibiotics or phages), as shown in Figure 8-4. The cells from some of the colonies on the master plate form colonies when transferred to the selective plate, in which case cells from those colonies on the master plate can be retested. If they are found to be mutant (in this example, resistant), we have proof that the mutation occurred *before* any selection was applied for the mutant. (It should be noted that there are some situations in which physiological adaptation can occur. The point for now is that mutations do not arise by physiological adaptation.)

MESSAGE

Mutation is a random process that can occur in any cell at any time.

Master plate containing
10^7 colonies of Ton^s *E. Coli* (T1-sensitive)

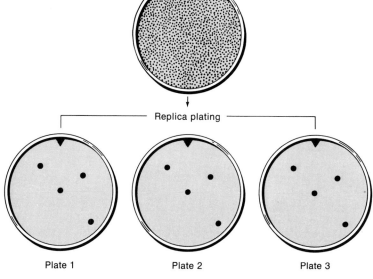

Replica plating

| Plate 1 | Plate 2 | Plate 3 |

Series of replica plates containing high concentration
of T1 phage and four Ton^r colonies

FIGURE **8-4**

Compared with finding specific gene changes in microorganisms, the task of finding such changes in multicellular organisms is tremendously complex. Hermann J. Müller devised a method for searching for a mutation in any gene on the X chromosome in *Drosophila* that causes death. He was able to do this by constructing a chromosome called "ClB," which carries an inversion (labelled *C* for crossover suppressor), a lethal (*l*), and the dominant eye marker Bar. *ClB*/Y males die because of hemizygosity for the lethal gene, but the chromosome can be maintained in heterozygous *ClB*/ + females. By mating wild-type males with *ClB*/ + females, one can test for lethal mutations anywhere on the X chromosomes in samples of the male gametes by mating single *ClB*/ + F$_1$ females with other wild-type males (Figure 8-5). You can see that, if a mutation has occurred on an X chromosome in one of the original male gametes sampled, then the F$_1$ female carrying that chromosome will not produce any viable male progeny. Note that only rarely will the new lethal mutation on the X chromosome be an allele of the lethal gene on the ClB chromosome (of course, when it is, the *ClB*/*l* female will die). The absence of males in a vial is very easily seen under low-power magnification and readily allows one to score for the presence of lethal genes on the X chromosome. Müller found their frequency to be about 1.5×10^{-3}, still a relatively low value for an entire chromosome with its many different genes.

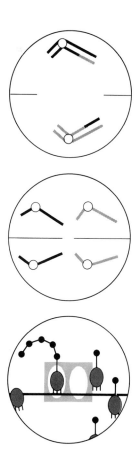

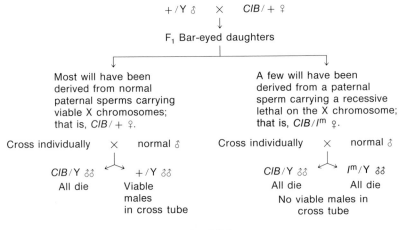

FIGURE **8-5**

Müller's experiments showed that mutation is a rare event that occurs spontaneously at most loci. He then asked whether there were any agents that would increase the rate of mutation. Using the ClB test, he scored for sex-linked lethal frequencies after irradiating males with X rays and discovered a striking increase. This was the first indication of a *mutagen* (an agent that induces mutations), which has been an invaluable tool not only for studying the process of mutation itself but for increasing the yield of mutants for other genetic studies. It is now known that any kind of radiation will increase mutations (Table 8-3).

The harnessing of nuclear energy for weapons and fuel has become a social issue because of the mutagenic effect of radiation. We will leave it to you to consider the pros and cons of the use of nuclear energy, but let's address one point that is often poorly understood. For any organism, the vast majority of newly formed mutations is deleterious. But you may ask, If organisms evolved through an advantage conferred by a mutant condition, why aren't many mutations an improvement? To answer this question, let's use an analogy often cited in this connection, in which a cell is compared to a highly developed Swiss

TABLE **8-3**

Type of radiation	Sex-linked recessive lethals per 1,000 r*	Percentage of irradiated male X chromosomes
Visible light (spontaneous)	0.0015	0.15
X rays (25 Mev)	0.0170	1.70
β rays, γ rays, hard X rays	0.0290	2.90
Soft X rays	0.0250	2.50
Neutrons	0.0190	1.90
α rays	0.0084	0.84

*r stands for roentgen, a unit of radiation energy.

watch that has many jewelled parts. This watch evolved through generations of minor changes in the workings of the machine. If we expose the cogs and wheels by removing the back casing, close our eyes, and plunge a thick needle (analogous to a mutagen) into the workings, there is a remote possibility that this random hit will improve the efficiency of the watch, but the chances are overwhelmingly high that any change inflicted in this way will damage it. So it is with a cell or an organism.

MESSAGE

Most mutations are deleterious. Evolution uses those very few that are not.

We will consider good molecular reasons for the detrimental nature of mutations in Chapter 12.

Many people have studied and are still studying the genetic effects of radiation. It is clear that, within a certain range of radiation dosage, the frequency of mutation induction is additive; that is, if we double or halve the radiation level, the number of mutants produced will vary accordingly. From the kind of graph shown in Figure 8-6 it is possible to extrapolate to very low radiation levels and to infer very low frequencies of mutation induction. Because

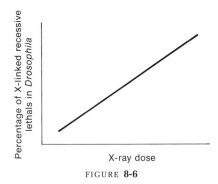

FIGURE **8-6**

exposure to radiation from X-ray machines, to radioactive fallout from bomb testing, and to contamination from nuclear plants is very low, it is possible to conclude that the effects are negligible. Yet every year 200 million new gametes form 100 million new babies in the world. In this very large annual "mutation experiment" even low mutation frequencies are potentially translatable into large numbers of mutant babies.

Radiation doses are cumulative. If a population of organisms is repeatedly exposed to radiation, the frequency of mutations induced will be in direct proportion to the *total amount* of radiation absorbed. However, there are exceptions to both additivity and cumulativeness. For example, if mice given *x* rads (a biological measure of a dose of radiation) in one short burst (called an

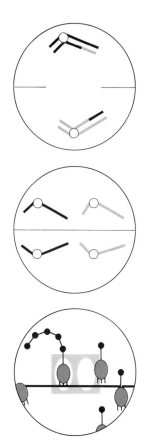

"acute" dose) are compared with those given the same dose for a protracted period of weeks or months (a "chronic" dose), significantly fewer mutations will be found in the chronically exposed group. This has been interpreted to mean that there is some form of repair of radiation-induced genetic damage with time.

We have not considered *how* ionizing radiation (the type discussed so far) causes mutation. Perhaps the simplest way to explain it is to compare a particle of radiation to a bullet; when the particle enters living cells, it strikes and breaks the genetic material directly. Or, it breaks molecules, such as water, into entities, which themselves become highly reactive chemically and interact secondarily with the genetic material. We will consider ultraviolet radiation, a nonionizing type, in detail in Chapter 12.

Of even greater importance for geneticists was the discovery that certain chemicals may also be mutagenic. The first demonstration of chemical mutagenesis was made by Charlotte Auerbach and J. M. Robson, who conducted experiments on mustard gases that were used in gas warfare. (You might be interested to know that their discovery was kept from publication by the British military for several years.) This initial study opened a floodgate to research into the mutagenic effects of a wide variety of chemicals.

The mutagenicity of chemicals is an important phenomenon because, in many cases, the chemical reactions responsible for the mutagenic action of a compound can be determined, thereby providing a clue to the molecular basis of the mutation. Furthermore, many chemicals are much less toxic to an organism than radiation and yet give much higher frequencies of mutation. So, as a tool, chemical mutagens have been used to induce much of the wide array of mutants now available for genetic studies. Finally, great controversy now exists over the potential mutagenic effects of a host of molecules in the human environment, ranging from caffeine to pollutants, pesticides, and LSD. The molecular basis of mutation will be considered in Chapter 12.

The Mutant Hunt

Mutations are very useful. In the same way that we can learn how the engine of a car works by tinkering with its parts one at a time to see what effect it has, we can see how a cell works by altering its parts one at a time by means of induced mutations. The *mutation analysis* is a further aspect of the process we have called genetic dissection of living systems.

MESSAGE

The two main tools of genetic dissection are recombination analysis and mutation analysis.

But first we have to find out how to go about obtaining some mutants.

If you are going to induce mutations, the first thing needed is a "survival curve" in order to judge which dose to use for the best yield. Some standard survival curves after irradiation are shown in Figure 8-7. As you can see, some have a

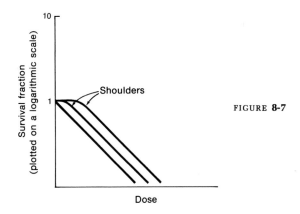

FIGURE **8-7**

shoulder. The meaning of this shoulder has been the topic of much debate, but a formal explanation is found in the target theory, which is interesting in itself anyway.

In the target theory, the mutagen is compared to a bullet fired from a gun at targets. The "hits" are mutations that inactivate the targets and cause a decrease in survival percentage. What are the targets? Genes? Chromosomes? Nuclei? Cells? This also has been the subject of great debate. The fact is that it is difficult to state precisely what the targets are, but let's develop the theory anyway and worry about the targets later.

We will consider a situation in which each cell has two targets. We will define dose (d) very simply as the average number of hits per target. The Poisson distribution will describe the actual distribution of hits, and the fraction of targets with *no* hits will be e^{-d}. Say that we have two targets, both of which have to be hit at least once to cause death.

$$p(\text{first target hit}) \quad = 1 - e^{-d}$$
$$p(\text{second target hit}) \quad = 1 - e^{-d}$$
$$p(\text{both targets hit}) \quad = (1 - e^{-d})^2$$
$$\text{Survival fraction therefore} = 1 - (1 - e^{-d})^2$$

This is quite a nasty expression to simplify, but it can be done as follows:

$$\text{Survival fraction} \left(\frac{N}{N_0}\right) = 1 - (1 - e^{-d})^2$$
$$= 1 - [1 - 2e^{-d} + (e^{-d})^2]$$

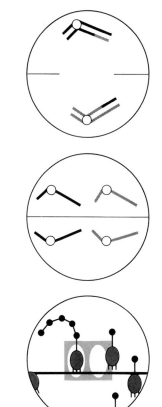

For high doses the last term is negligible, giving us $\frac{N}{N_0} = +2e^{-d}$.

Taking logarithms of both sides, $\ln \frac{N}{N_0} = -d + \ln 2$. Applying this to the general graph equation $y = mx + c$, we see that, if we plot the logarithm of the survival fraction against dose, the slope in this case will be -1 (not of interest), but the intercept on the $\ln \frac{N}{N_0}$ axis will be the target number. This intercept is obtained by extrapolating the straight-line part of the curve back, and so it is called the *extrapolation number* and is equal to the number of targets (in our example it is 2, as shown in Figure 8-8).

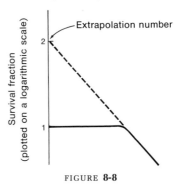

FIGURE **8-8**

Thus we see that the shoulder represents a situation in which survival does not drop off immediately because there is more than one target. The presence of the shoulder gives us an extrapolation number that tells us the number of targets.

Well, what *are* targets in a genetic system? We have reached a level of understanding of the genome that allows us to think about this. In a haploid cell, one can see that a mutation in virtually any gene will inactivate (kill) the cell, and so it seems that the genome is the target. In a diploid, however, the situation is more complex because an inactivated gene on one chromosome will nearly always be complemented by an untouched gene on its homolog. Only simultaneous hits in the *same locus* of both homologs or a single hit that produces a dominant effect will cause inactivation of diploids. So, is the *gene* the target in a diploid cell? This is obviously closer to the truth, but it is still not fair to state that genes are the targets because *any* two hits in homologous loci will cause inactivation. It is like rolling two dice and saying *any pair* of numbers will cause inactivation. Thus, in diploids, the concept of the target is difficult to translate into real entities. Suffice it to say that in haploids or prokaryotes an extrapolation number of 1 is common, whereas in diploids a shoulder is nearly always seen, often with an extrapolation number of 2. The situation is further complicated by repair systems that can fix up the targets after a hit before death occurs. Also, the type of mutagen and the genetic ancestry of the strains often greatly affect the shape of the survival curve. So, target theory may be difficult to translate literally, but it does provide one of the few interesting and useful models for explaining survival curves.

Once a survival curve is obtained, a dose is chosen that will not give too low a survival. The best dose varies with mutagenic agent and organism. At low survival levels there is always the possibility of picking up multiple mutations and these are to be avoided. A *double* mutant is not always immediately recognized and if the stock is treated as though it carried a single mutation, all sorts of strange results can be obtained. For example, what would happen if you tried to map a sex-linked lethal in *Drosophila* that really is two lethals?

KINDS OF MUTANTS

On a mutant hunt the mutants that are picked up are restricted to those that are recognizably different from the original (normal or wild type). There are four main classes of recognizable mutants that are named simply after the way in which we recognize them.

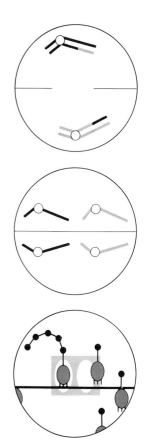

Morphological. "Morph" means form, but included in this class is any kind of altered appearance, whether it is shape, color, or size. Albino spores in *Neurospora,* curly wings in *Drosophila,* dwarf peas, rapid lysis plaques of T4 are classified as morphological mutants.

Lethal. In this class the new allele is recognized by its mortal effect on the organism. Death caused by mutagen treatment is not always due to a mutational event. Nevertheless, the heritability of lethality is quite well known. There are also mutations that reduce the probability of survival, but not to zero. These may be referred to as semilethals or subvitals.

Conditional. In this class, alleles express a mutant phenotype under certain conditions (called *restrictive*) but are normal under others (called *permissive*). Conditional mutants can be produced by very minor changes at the molecular level, as we will see later. Temperature, radiation, chemicals, drugs, other genes, or parasites can be used to alter the expression of conditional mutants, which are then designated as resistant or sensitive alleles.

Many mutants are less vigorous than normal organisms; so conditional mutants are coveted by geneticists because they can be grown under the permissive condition and shifted to the restrictive condition at the time of experiment.

Biochemical. This class is identified by the loss of a biochemical function of the cell, which can often be restored by supplementing the cell's particular medium with the appropriate nutrients. For example, adenine auxotrophs can be grown only if adenine is supplied, whereas wild types do not need the adenine supplement. So far, we have treated such mutants simply as convenient genetic markers, but the next chapter will reveal their importance in elucidating the nature of gene function.

Clearly, these four classes are not mutually exclusive. In the final analysis, all mutants have some "biochemical" lesion. Some known biochemically defective mutants such as auxotrophs are conditional, and so on. The classes are simply convenient descriptions of a wide variety of inherited defects.

In mutagen treatments the experimenter is usually concentrating on single cells, which when mutated will generate whole mutant individuals (e.g., gametes, bacterial cells, and fungal spores). However, some mutations occur in cells that are part of nongerminal (or somatic) tissue in a multicellular organism. Such mutations are *not* transmitted to subsequent generations (consider, for example, a cell in your hand that undergoes mutation, or a mutation in a plant root cell); however, they may be responsible for some forms of cancer. In plants somatic mutations can be transmitted by cuttings: many important commercial crops are derived from spontaneous mutations that occurred on the branch of a single plant—for example, the golden delicious apple.

The utility of mutations goes far beyond confirming Mendel's first and second laws. The selection of defects in specific biochemical processes has allowed many of the advances in molecular biology that we will soon encounter.

Mutationlike Phenomena

There are some events that seem at first to be mutations but, on further analysis, can be shown to have a different basis. We will consider a couple of them at this point to illustrate how such phenomena are investigated genetically.

POSITION-EFFECT VARIEGATION

Most of you are familiar with *variegated* phenotypes; an example is a plant whose leaves are striped or whose petals have patches of different color. You may have wondered whether the cells in the areas that are different in color are a clone of somatically mutated cells. Somatic mutability may indeed occur in some cases (either in chromosomal or in nonchromosomal genes—see Chapter 13). However, this discussion will be restricted to one class of variegation that is different.

The locus for white eye color in *Drosophila* is near the tip of the X chromosome. If the tip of a chromosome carrying w^+ is translocated to a heterochromatic region, say of chromosome 4, then in a heterozygote for the translocation and a normal chromosome 4, and with the normal X chromosome carrying w, the eye color is a mosaic of white and red cells. Because the translocation carries the dominant w^+ allele, we would expect the eye to be completely red. How have the white areas come about? We could suppose that,

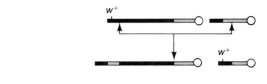

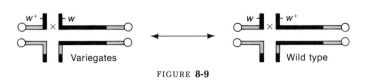

Variegates   Wild type

FIGURE **8-9**

when the translocation was formed, the w^+ gene itself was somehow changed to a state that made it more mutable in somatic cells; so the white eye tissue reflects cells in which w^+ has mutated to w.

Burke Judd tested this by crossing the w^+ out of the translocation and onto a normal X chromosome and by crossing a w gene from the normal X onto the translocation (Figure 8-9). He found that when the w^+ on the translocation was crossed onto a normal X chromosome and w was inserted into the translocation, the eye color was red; so obviously the w^+ was not defective. A w^+ gene that is crossed back onto the translocation again variegates. We can conclude, then, that for some reason, the w^+ gene in the translocation is not expressed in some cells, thereby allowing expression of w. This kind of variegation is called *position-effect variegation* because the unstable expression of a gene is a reflection of its position in a rearrangement.

MESSAGE

The expression of a gene can be negated by its position in the genome.

TRANSFORMATION

A puzzling observation was made by Frederick Griffith in the course of doing experiments on a pneumococcus bacterium in 1928. This bacterium causes pneumonia in humans and is extremely pathogenic in mice. Griffith had two different pneumococcal strains: one was "virulent" and caused mice to die soon after it was injected into them; Another was "nonvirulent" and would grow in mice but was not lethal. Griffith killed the virulent cells with heat and injected the heat-killed cells into mice. The mice survived quite well, thus showing that the carcasses of the cells would not cause death. However, if he injected the mice with a mixture of the debris from the heat-killed virulent strain and live nonvirulent cells, the mice died! Furthermore, he could then recover live virulent cells from the dead animals. Somehow the dead cell debris was able to transform the nonvirulent strain to the virulent type. This was called *trans-*

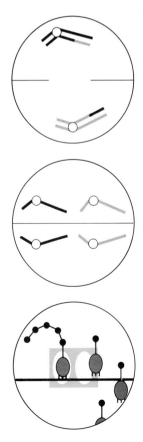

formation and resembled a kind of mutation induced in the live cells by the remnants of the dead cells.

This basic technique was utilized to determine the actual molecule responsible for transformation. In 1944, Oswald Avery, C. M. MacLeod, and M. McCarty examined morphological phenotypes of the two strains of pneumococcus. Cells of the virulent strain are encased in a polysaccharide shell, which gives a colony a smooth, shiny texture. Hence, the virulent strain is labelled S, for smooth. The nonvirulent strain, on the other hand, lacks the polysaccharide capsule; therefore, the colonies produced have a rough, irregular shape. This strain is labelled R, for rough. They found that adding dead cells of the S strain to live cells of the R strain did indeed result in the "transformation" of some cells to the S phenotype. They then set out to find out which class of molecules was responsible for transformation. They separated different classes of molecules and tested them for transforming ability one at a time. They showed that polysaccharides themselves did *not* transform the rough cells and therefore must be the phenotypic expression of the "transforming principle." Screening the different groups, they found that only deoxyribonucleic acid (DNA) induced transformants (Figure 8-10). This suggests that DNA is the agent that deter-

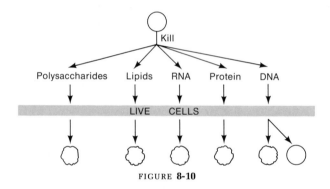

FIGURE **8-10**

mines the change to the polysaccharide capsular phenotype of the virulent, smoothly textured cells. It seemed, in fact, that providing R cells with the DNA was tantamount to providing them with S genes.

MESSAGE

The demonstration of transformation by DNA was the first indication that genes are composed of DNA.

It seems that the transforming principle, or DNA, is actually physically incorporated into the bacterial chromosome by a physical breakage and insertion process quite similar to crossing over. Thus, if ^{32}P-labelled DNA of donor arg^+ bacteria is isolated and added to unlabelled recipient arg^- cells, which are then plated on minimal medium, the arg^+ transformants can be shown to contain

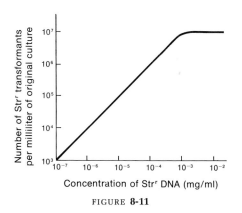

FIGURE **8-11**

some of the ^{32}P. We will have more to say about DNA later. For now, let's consider only the genetics of transformation.

Transformation provides another means of mapping bacterial chromosomes. Transformation of a single marker is relatively rare, ranging from one transformant per 10^6 cells to one per 10^3, depending on how much DNA is added (Figure 8-11). Generally, DNA extracted from $a^+ b^+$ cells and added to $a^- b^-$ cells yields double transformants at a frequency that is the product of the frequencies of a^+ and b^+ single transformations. Genetically, this means that the two markers are unlinked in the transforming event. This could be interpreted to imply that the transforming principle comprises pieces of the genome that are absorbed by the recipient cells. When a piece is inside the recipient cell, a merozygotic situation is once again set up and the incorporation of any genes into the recipient endogenote must be done by a double exchange event. Double ($a^+ b^+$) transformants *can* occur at a frequency higher than the product of their separate frequencies, for specific genes. This suggests that some loci are sufficiently close on the donor chromosome to be borne on the same piece of transforming DNA. In this case, linkage can be studied.

Using an $arg^+ ser^+$ donor as the source of DNA and $arg^- ser^-$ cells as the recipients, we can plate the transformed cells on different media.

1. Contains arginine alone: $arg^+ ser^+$ and $arg^- ser^+$ will grow.
2. Contains serine alone: $arg^+ ser^+$ and $arg^+ ser^-$ will grow.
3. Is minimal: $arg^+ ser^+$ will grow.

Medium 1 minus minimal (3) gives us total $arg^- ser^+$; medium 2 minus minimal gives us total $arg^+ ser^-$; and minimal gives us total $arg^+ ser^+$. The ratio of single transformants to the total is an index of the map distance between the loci; in this experiment the ratio is

$$\frac{\text{Number of } arg^- ser^+ + \text{number of } arg^+ ser^-}{\text{Number of } arg^- ser^+ + \text{number of } arg^+ ser^- + \text{number of } arg^+ ser^+}$$

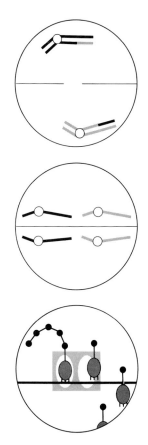

Observe that, as in all merozygote genetics, what is basically done is to examine one gene incorporated from the exogenote and then to see what other genes from the donor came with it.

You can see by looking at the ratio that the closer two loci are, the smaller the number of single transformants there will be and therefore the lower the ratio will be. This method depends on the constancy of the size of the transforming fragments (exogenotes), which, unfortunately, is not always controlled from experiment to experiment.

Summary

We have tried to show that mutation is defined operationally by the appearance of a new heritable condition. If a great deal of genetic information is known about an organism, a new heritable condition can usually be fairly readily assigned as either a chromosomal mutation or a gene mutation. On the other hand, what may seem initially to be mutationlike phenomena may, on further study, be found to have a different basis. Thus, transduction and even conjugation in bacteria could initially have been thought of as special cases of mutation, as could transformation and position-effect variegation.

We have included mutationlike phenomena to illustrate the point that, when a totally unexpected change is observed in the phenotype of a stock, a geneticist is apt to run through a mental catalog of known generators of genetic change to explain the observation. He must be prepared to consider possibilities other than mutation because phenomena resembling transformation and transduction have been reported in studies on eukaryotic multicellular organisms. One of the first interpretations for all these phenomena was mutation, which was not ruled out in some cases until much later.

MESSAGE

Nature has invented many vehicles for genetic change.* Some or all of them could be operating in the system you happen to be examining.

---

*Mutation, transduction, transformation, episome transfer, meiotic and mitotic recombination, and possibly more as yet undiscovered.

Problems

1. Suppose that you cross a single male mouse from a homozygous wild-type stock with several homozygous wild-type, virgin females and find in the F_1: 38 wild-type females and males and 5 black males and females. How could you explain this?

2. Suppose that you want to determine whether caffeine induces mutations in higher organisms. Describe how you might do this (include control tests).

3. Suppose that you are given a *Drosophila* stock from which you can get males or virgin females at any time. The stock is homozygous for a second chromosome, which has an inversion to prevent crossing over, a dominant gene (*Cu*) for curled wings, and a recessive gene (*pr*, purple) for dark eyes. The chromosome can be drawn as:

You have irradiated sperm in a wild-type male and wish to determine whether recessive lethal mutations have been induced in chromosome 2. How would you determine this? (*Hint:* Remember that each sperm carries a *different* irradiated second chromosome.) Indicate the kinds and number of flies used in each cross.

4. Some mealybugs called coccids have a diploid number of 10. In cells of males, 5 of the chromosomes are always seen to be heterochromatic and 5 are euchromatic. In female cells, all 10 are always euchromatic. Spencer Brown and Walter Nelson-Rees gave large doses of X radiation to males and females and obtained the following results:

Female (X-rayed) × Male (non-X-rayed)
↓
No surviving progeny

Female (non-X-rayed) × Male (X-rayed)
↓
Lots of male progeny
but no female progeny

Interpret these results.

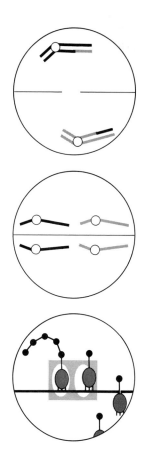

5. Joe Smith accidentally receives a heavy dose of radiation in the gonadal region. Nine months later his girlfriend has a daughter, Mary. Mary appears perfectly normal and marries a homozygous normal man. This is their pedigree:

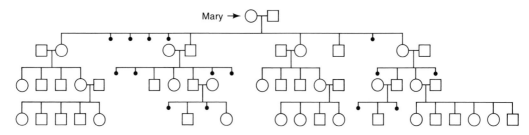

• Early miscarriages, sex undetermined

    a. What possible genetic mechanisms could explain these results?

    b. How would you prove which explanation is correct?

6. Suppose that you have been supplied with the following stocks of *Drosophila*.

Stock 1. Both males and females are *CxD/Sb*. Both *CxD* and *Sb* are associated with inversions on chromosome 3 that prevent crossing over and are recessive lethals; *CxD* has a dominant phenotype of wings held out from the body and *Sb* has a dominant phenotype of stubby bristles.

Stock 2. Females carry a Y chromosome and an attached-X chromosome homozygous for *y* and *w*, which produce yellow body and white eyes, respectively. Males have normal wild-type X and Y chromosomes.

Stock 3. Females are homozygous for *y* on normal X chromosomes. Males carry *y* on the X chromosome and a Y chromosome into which *y*$^+$ has been inserted. Males, therefore, are wild-type in body color.

Stock 4. Homozygous wild type.

Stock 5. Flies are homozygous for dominant genes *M*, *N*, *O*, and *P* on chromosome 3: *M* is at the left end, *P* is at the right end, and *N* and *O* are equidistant from each other and their respective ends.

Stock 6. Males and females have X chromosomes that carry *y* and *B*.

Stock 7. Males and females have X chromosomes that carry *w*.

Stock 8. Both males and females are *Cy/Pm*. *Cy* and *Pm* are associated with inversions on chromosome 2 that prevent crossing over and are recessive lethals; *Cy* has a dominant phenotype of curled wings and *Pm* has a dominant phenotype of dark brown eyes.

Note: Be sure that you understand the symbols of each stock and how the genes and chromosomes are inherited. Now with these eight stocks, you are going to be a geneticist. Decide which stocks to use and diagram the crosses you would make and the offspring you would score for the following experiments. (Specify all details such as virginity, genotype, and number of parents per culture vial where it's important, so that someone could reproduce your experiments.)

a. Determine the frequency of nondisjunction of the sex chromosomes in males and females in one cross.

b. Determine whether a male you've found in the wild carries a translocation between chromosomes 2 and 3 (do it genetically, not cytologically).

c. After exposing Stock 4 males to X radiation, you determine the frequency of lethals induced on chromosome 2.

d. You have detected a lethal mutation on chromosome 3. How would you determine its genetic position?

7. The government wants to build a nuclear reactor near the town of Poadnuck. The townspeople are very upset about the possibility of an accidental release of radioactivity and are putting up a stiff fight to keep it out. They call on you, a geneticist, to inform them about the biological hazards of an accident. What would you say in your speech?

8. The nuclear plant has been built and you are now a professor at Poadnuck State College. A radical group manages to infiltrate the plant and blow it up, releasing a large amount of radioactivity in the area. Fortunately, a thunderstorm washes most of the radioisotopes to the ground, preventing widespread contamination. You set out immediately to determine the genetic effects of the radiation. How do you go about it? (Remember, you must have controls.)

9. One of the first genetic studies done on the survivors of the atomic explosions over Hiroshima and Nagasaki was to measure the sex ratio among the offspring of exposed people. Why?

10. More than 10,000 new molecules are synthesized every year. Many of them could be mutagenic. How would you readily screen large numbers of compounds for their mutagenicity?

11. A man employed for several years in a nuclear power plant becomes the father of a hemophiliac boy, the first in the extensive family pedigrees of both his own and his wife's ancestry. Another man, also employed for several years in the same plant, has an achondroplastic dwarf child, the first occurrence in his ancestry and in that of his wife. Both men sue their employer for damages. What would your testimony as a geneticist be before the court?[1] (*Note:* Hemophilia is an X-linked recessive and achondroplasia is an autosomal dominant.)

12. Using the filtration-enrichment technique, you do all your filtering with minimal medium and your final plating on complete medium that contains every known nutritional compound. How would you go about finding out what *specific* nutrient was required? After replica plating onto every kind of medium supplement known to man, you still can't identify the nutritional requirement of your new yeast mutant. What could be the reason(s)?

13. In a large maternity hospital in Copenhagen, there were 94,075 births. Ten of the infants were achondroplastic dwarfs. (Achondroplasia is an autosomal dominant showing virtually full penetrance.) Only two of the dwarfs had a dwarf parent. What is the mutation rate to achondroplasia? Do you have to worry about reversion rates in this question? Explain.

---

[1] From *Principles of Human Genetics*, 3d ed., by Curt Stern. W. H. Freeman and Company. Copyright © 1973.

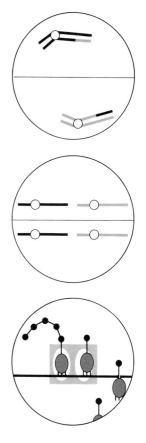

14. A new high-yielding strain of wheat suffers from the disadvantage that it tends to "lodge" or fall over during storms. You have an X-ray source, and a couple of years, to correct this defect. How would you proceed? State which part of the plant, or which stage of its life cycle, you would treat, what would you look for, which generation you might expect your results in, and so forth.

15. Transformation in bacteria occurs by the uptake of DNA by a recipient cell. This exogenotic material must then be integrated into the endogenote. Design an experiment to determine whether integration occurs by breakage-and-reunion or by copy choice.

16. A transformation experiment was performed with a donor strain that is resistant to four drugs: A, B, C, and D. The recipient is sensitive to all four drugs. The treated recipient cell population was divided up and plated on medium containing various combinations of the drugs. The results were as follows:

Drugs added	Number of colonies
None	10,000
A	1,156
B	1,148
C	1,161
D	1,139
AB	46
AC	640
AD	942
BC	51
BD	49
CD	786
ABC	30
ABD	42
ACD	630
BCD	36
ABCD	30

a. One of the genes is obviously quite distant from the other three, which appear to be tightly linked. Which *is* the distant gene?
b. What is the probable order of the three tightly linked genes?[2]

17. *Bacillus subtilis* forms spores in which the genetic material is inert and present in a single dose. In a growing culture of *B. subtilis,* not all genes duplicate at the same time. Can you suggest ways of testing genetically for the number of copies of different genes at different times in the cell cycle? (This is a tough problem, so don't worry if you can't solve it. The answer will come later in the book.)

---

[2] Problem 16 is from *The Mechanics of Inheritance,* 2d ed., by Franklin Stahl, © 1969, p. 166. Reprinted by permission of Prentice-Hall, Inc., Englewood Cliffs, New Jersey.

9 / The Nature of the Gene

(Or: The death of the bead theory.)

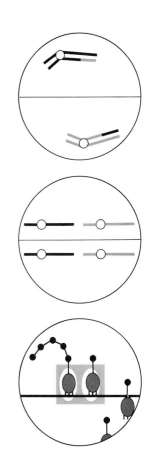

So far, our analyses, both genetic and cytological, have led us to regard the chromosome as a linear (one-dimensional) array of genes, strung rather like beads on an unfastened necklace. Indeed, this view is sometimes called the bead theory. There are several points about the bead theory worth emphasizing.

1. The gene is viewed as a fundamental unit of structure, indivisible by crossing over. Crossing over occurs between genes (i.e., beads) but not within them.

2. The gene is viewed as a fundamental unit of change, or mutation. It changes from one allelic form to another, but there are no smaller components within it that can change.

3. The gene is viewed as the basic unit of function. Parts of a gene, if they exist, can't function (although the precise function of the gene is uncertain).

4. The chromosome is viewed merely as a vector or transporter of genes and exists simply to permit their orderly segregation and to shuffle them in recombination.

In this chapter we will see the demise of the bead theory on all four points. Although there will be more to say about point 4 in a later chapter, it is a convenient starting point in our attack on the bead.

One of the first studies of this point was carried out on the sex-linked dominant mutation Bar (*B*) in *Drosophila*. We have already considered the nature of Bar eyes in Chapter 7; if you remember, the phenotype is caused by a

tandem duplication (Figure 9-1). Sturtevant was able to obtain several genotypes in which there were various combinations of Bar and wild type. In each of
these he counted the number of "facets," or components, of the compound eye
in order to measure eye area and obtained the following results.

$$+/+ = 779 \qquad +/B = 358$$
$$B/B = 68 \qquad BB/B = 36$$
$$BB/+ = 45 \qquad BB/BB = 25$$

(The symbol *BB* stands for double bar, which was shown to be a *triplication* of
the wild-type region.) Sturtevant pointed out that the difference in the position
of *B* (for example, *B/B* versus *BB/+*, both of which contain four total regions)
has a striking effect on the phenotype. So the arrangement of the region of the
chromosomes affects the end phenotype. This is called a *position effect* and
suggests that a gene's immediate neighbors are important and that chromosomal
location could have *functional significance.* (Position-effect variegation, which
was discussed in Chapter 8, concerns an analogous phenomenon.)

One of the first setbacks to the concept of the indivisibility of the gene also
came from *Drosophila* work. Edward B. Lewis studied a locus near the left end
of chromosome 2 in *Drosophila.* He had two mutants: one was the dominant Star
(or *S*), which produces a slightly smaller eye than wild type in *S/+* flies; the
other mutant, mapping in the same region, was a recessive, asteroid (*ast*), which
gives a much smaller eye in *ast/ast* homozygotes. *S/ast* heterozygotes have even
smaller eyes, thus suggesting that *S* and *ast* must be functionally related and
therefore alleles. The order of eye size in the genotypes is: $+/+ > S/+ >$
$ast/ast > S/ast.$ Lewis marked the S chromosome with closely linked flanking markers, *al* and *ho*, and made the following cross: *al S ho/+ ast + ♀* ×
al ast ho/al ast ho ♂.

In 57,000 progeny of this cross he recovered 16 flies with wild-type eyes! All
were of the genotype *+ + ho,* which is clearly derived from a crossover event
somewhere between *al* and *ho.* Presumably that crossover also generated a
wild-type allele for eye shape. This is understandable if we assume that the S
and ast phenotypic effects result from changes at *different positions within the*
S-ast *gene.* Thus, a crossover between *S* and *ast* could combine the wild-type
parts of the gene. It was decided that, because there was recombination between
S and *ast,* they could not be true alleles, and therefore they were called *pseudoalleles.* To generate the *+ + ho* recombinants, the left part of the ast chromosome and the right part of the S are joined. We can explain these results if we
construct a linear linkage map in which S is to the left of *ast:*

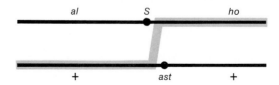

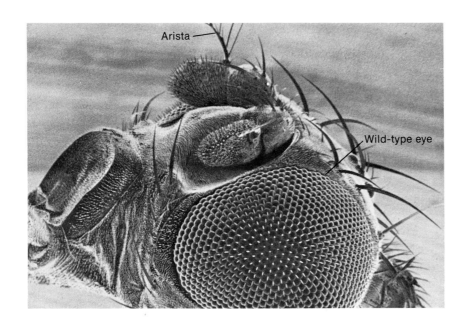

FIGURE 9-1
Scanning electron micrograph of *Drosophila* wild-type eye (above) and bar eye (below).

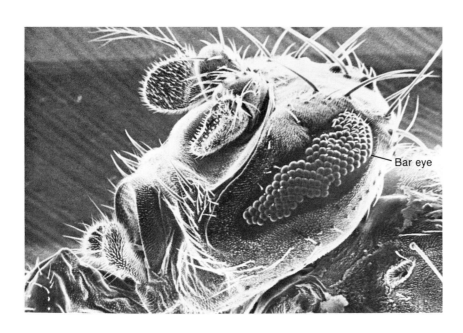

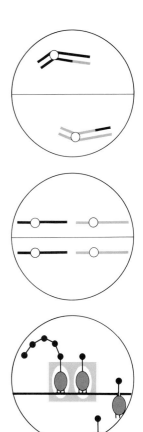

The reciprocal recombinant was also recovered but it is much harder to detect because it carries both S and ast and is not easily distinguishable from the single allele phenotypes. (How would you distinguish S ast from S and ast genetically?) Again, we see a striking position effect in that the phenotypes of $S + / + ast$ (very small eyes) and S $ast/ + +$ (eyes only slightly smaller than wild type) are quite different even though their total genetic content is similar.

The pseudoallele situation was verified at another locus, lozenge (lz), in *Drosophila* by Mel Green and Kathleen Green. Here mutant alleles produce eyes with a glossy, smooth surface. They had several independently isolated alleles of lozenge, all mapping at one chromosomal position. All heterozygotes carrying two different mutants were lozenge in phenotype, thereby showing that they were indeed functionally allelic. They intercrossed different lz mutants and testcrossed heterozygotes with different lz alleles (very much like S/ast) and also found wild-type recombinants! Were these pseudoalleles too, each having a slightly different chromosomal location? It began to look that way when they further showed that wild-type recombinant frequency could be used to draw a linear genetic map of the pseudoallelic sites!

But are these pseudoalleles really any different from "true" alleles? Is the real difference between alleles and pseudoalleles simply that in the case of alleles no recombinants have been detected, or *looked for* hard enough? After all, recombination between pseudoalleles is very rare. In the coat-color allelic series in rabbits (between chinchilla and Himalayan, for example), is it possible that, if we studied tens of thousands of rabbit progeny, we could detect recombinants (denoted by c^+, or C)?

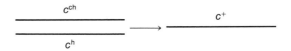

Pseudoalleles, of course, segregate almost perfectly $1:1$ in a true Mendelian fashion because crossing over between them is so rare; so this hypothesis seems tenable. Rearing and scoring 100,000 flies is quite a job, let alone that number of rabbits.

A great deal of light was shed on the nature of alleles by some beautiful experiments performed by Seymour Benzer. Benzer capitalized on the fantastic *resolving power* made possible by using selective systems for rare events in phages. He worked with the virulent *E. coli* phage T4 and a class of mutants that produces a phenotype called rapid lysis (r). Rapid lysis strains produce large circular plaques, whereas the plaques of normal phages are small and irregular. Several different loci can mutate to cause rapid lysis. Benzer worked on one group of rapid lysis mutants called *rII* for one important reason, which was that all *rII* alleles are *conditional* lethals; that is, *rII* mutants will grow and are recognized as large plaque formers on *E. coli* strain B, but will not grow at all on strains of *E. coli* that are lysogenic for λ, called strain K(λ). (We don't have to worry about what causes this difference between B and K(λ) right now.) In contrast, *rII*⁺ phages will grow and form small ragged plaques on both strains. Thus, the relations are:

		E. coli strain	
		B	K(λ)
T4 phage strain	*rII*	+ (smooth)	−
	rII⁺	+ (fuzzy)	+ (fuzzy)

where a plus indicates growth and a minus none.

This conditional growth of *rII* strains proved essential to the analysis because it allowed the use of selective techniques to identify recombinants. Benzer started with an initial sample of eight independently derived *rII* mutant strains and set about crossing them in all possible combinations of pairs the (simply by the double infection of *E. coli* B). He took the lysate (progeny phages) and plated them onto a lawn of *E. coli* K(λ), on which the *rII* genotypes will not grow. But plaques *were* seen on K(λ), indicating that recombination had given rise to *rII*⁺ genotypes (Figure 9-2). (The plaques were found at frequencies too high for

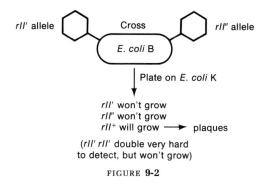

FIGURE **9-2**

them to be back mutations.) Also, as with the lozenge locus, the alleles, or pseudoalleles (or whatever), could be mapped unambiguously to the right or left of each other to give what we now call a gene map.

*rII*¹		*rII*⁷	*rII*⁴	*rII*⁵		*rII*²		*rII*⁸	*rII*³		*rII*⁶

At this point, it seems just as well to drop the notion of pseudoalleles. It was apparent, and became even more apparent when allele maps were made of many loci in microorganisms, that there was nothing pseudo about pseudoalleles. In other words, a mutant allele can be thought of as a length of genetic material

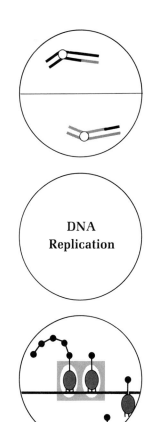

DNA Replication

(the gene) that has a damaged or nonwild part (a mutant site) somewhere that results in the nonwild phenotype. Thus, an allele a^1 can be represented as

$$+ + + + + + + + + +^* +$$

where the asterisk represents the nonwild area within an otherwise normal gene delineated by the pluses. A cross between a^1 and a^2, in general then, is:

and it is easy to see how a

$$+ +$$

is generated by a simple crossover between the two asterisks.

Benzer extended his *rII* analysis to use many hundreds of alleles. He found that he *never* obtained rII^+ recombinants in a heteroallelic (different alleles) cross at frequencies of less than 0.01%, although his analytical system was sufficiently fine to have allowed detection of values as low as 0.0001% had they occurred. He concluded that there was a minimal recombination distance within a gene. This smallest genetic interval separates *recons,* which are defined as the regions within a gene *between* which recombination can occur. Once again, we have a hypothetical entity, based on genetic analysis, that was soon to assume physical reality.

RECON

A gene is composed of units that are not divisible by recombination called recons.

The map of a gene is a completely linear sequence of recons.
Benzer also coined the term *muton.*

MUTON

The smallest element within a gene that, if altered, can give rise to a nonwild or mutant phenotype.

The concept of the muton is suggested by the fact that *part* of a gene can change, as in $+ + +^* + + + + + +$. Therefore, the fundamental principle of the bead theory, which would have the *entire* gene as the muton, is knocked down. Of course, a muton is not necessarily the same as a recon, as is apparent in Figure 9-3, which illustrates a *hypothetical* gene composed of different recons and mutons. (We shall see later, however, that a muton *is* the same as a recon.)

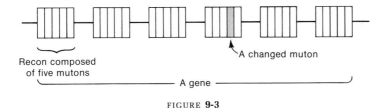

Recon composed
of five mutons

A changed muton

A gene

FIGURE **9-3**

Deletion Mapping

We digress a bit to describe the way in which Benzer was able to rapidly locate new mutant sites in his allele map of *rII*. He found some *rII* mutants that would not give recombinants when crossed with any of several other mutants, which had been shown to be different recons, but would when crossed to still others. He realized that such mutants behaved as short deletions within the *rII* region and that they could be used to rapidly locate mutant sites in newly obtained mutant alleles. For example, the following gene map contains twelve separable sites.

Mutant D_1 will not give *rII*⁺ recombinants when crossed with *1, 2, 3, 4, 5, 6, 7,* or *8*; therefore D_1 behaves as a deletion of sites *1* through *8*.

Mutant D_2 will not give *rII*⁺ recombinants when crossed with *5, 6, 7, 8, 9, 10, 11,* or *12*; therefore D_2 behaves as a deletion of sites *5* through *12*.

These overlapping deletions now define three areas of the gene—call them i, ii, and iii.

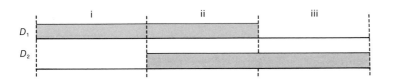

New mutants that give *rII*⁺ recombinants when crossed with D_1 only must be in area iii, those that give *rII*⁺ recombinants with D_2 only must be in area i, and those that give *rII*⁺ recombinants with neither must be in area ii. To illustrate

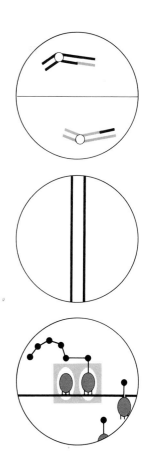

this, let's consider mutants in area iii crossed with D_1:

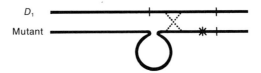

The more deletions there are in the tester set, the more areas will be covered, and new mutant sites can be located more quickly. Once assigned to a region, a mutant can be crossed with other alleles in the same region to provide an accurate position.

Deletions themselves can be intercrossed and mapped just like point mutations. The deleted region is represented by a bar, and, if no wild-type recombinants are produced in a cross between different deletions, the bars are shown as overlapping. A typical deletion map might be:

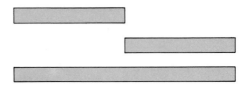

Such deletion maps are useful in delineating regions of the gene to which new point mutants can be assigned.

In another part of his studies, Benzer found that, although *rII* mutants could not infect *E. coli* K(λ) cells individually, some specific pairs of mutants could "help each other out" or *complement* each other to produce the infection of K(λ) that led to lysis of the cells. In fact, all *rII* mutants tested could be divided into two groups, A and B. Any member of group A could complement any member of group B, but within A or B no complementation between alleles was possible. Thus, mutants in the A group are apparently defective in a function that mutants in the B group can supply and vice versa. Very interestingly, all of the A group alleles mapped to one half of the *rII* locus, and those in the B group all mapped in the other half.

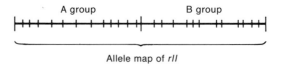

Allele map of *rII*

He called the A and B groups different *cistrons*, which he defined as follows:

CISTRON

A genetic region within which there is no complementation between mutations.

---

	cis test	*trans* test
Mutant sites in same cistron	A B	A B

+	−
+	+

| Mutant sites in different cistrons | A   B | A   B |

FIGURE **9-4**

The cistron gets its name from the test that is performed to see if two mutant sites are within the same cistron or are in different ones. It is called the *cis-trans* test and consists of setting up the complementation test with the mutant sites on the same chromosome (*cis*) or on opposite chromosomes (*trans*), as shown in Figure 9-4, in which plus and minus refer to successful or unsuccessful complementation. The *trans* test is the critical one, with the *cis* test serving mainly as a control. A complementation in a *trans* test means that the mutant sites are in different cistrons; a failure to complement means that they are in the same cistron. (Think back on the *S-ast* combinations.)

This has very fatal consequences for the bead-theory tenet that states that the gene is the basic unit of function. Obviously, in the *rII* locus, the *cistron* is the unit of function! Both the *rIIA* and the *rIIB* have to be undamaged or unchanged for the locus to function normally (i.e., to produce normal slow lysis). It is now known that some genes consist of only one cistron; some consist of two, three, or even more. Thus, the mutants miniature (*m*) and dusky (*dy*) both decrease wing size and map in the same part of the X chromosome of *Drosophila*. But when brought together in a *dy* + / + *m trans* heterozygote (the equivalent of doing a *trans* test), the phenotype is normal, an indication that the locus concerned with wing size is composed of at least two cistrons. The word "gene" therefore has become more or less synonymous with "cistron," whereas we usually call the region of the A and B cistrons the *rII locus*. It is not necessary to use hard and fast rules because, as we shall see later, even the concept of the cistron breaks down when pushed a little harder. Some geneticists use the word allele in reference to a mutant within a locus; so an allele may or may not complement other alleles of the locus. Others use allele for a mutant in a cistron, in which case, by definition, there is never complementation between alleles. We shall use the latter: the *cis-trans* test is a test for true allelism. It is important to note that each term—recon, muton, and cistron—has an operational definition; that is, by recombination, mutation, and function.

The problem of distinguishing or defining gene, cistron, and locus centers on the fact that functionally related cistrons often occur together at one chromosomal position. There are many examples of this and we will examine some of the reasons why in Chapter 13. For now let it suffice to say that functionally related genes often act simultaneously and one of the mechanisms for this involves their being immediately adjacent to one another.

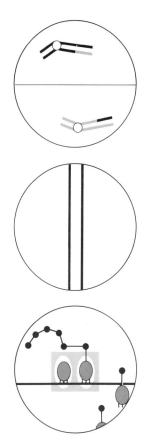

We have said that the cistron is a unit of function. What is a cistron in real terms, and how does it function? The first clues to the nature of primary gene function came from studies of humans, unlikely subjects for doing genetic work. Early in this century, a physician, Archibald Garrod, noted that several hereditary human defects were produced by recessive mutations. Some of these defects could be traced directly to a metabolic defect, which led to the suggestion of "inborn errors" in metabolism. We know now, for example, that phenylketonuria, caused by an autosomal recessive, results from an inability to convert phenylalanine into tyrosine. Consequently, phenylalanine accumulates and is shunted to a toxic compound, phenylpyruvic acid. In another trait, it is the inability to convert tyrosine into melanin pigments that produces an albino. Anyway, the observations of Garrod focused attention on metabolic control by genes.

The experiments that clarified the actual function of genes were done with *Neurospora* by George Beadle and Edward Tatum, and for this work they received a Nobel Prize. After irradiation, ascospores were grown and tested for their ability to grow on minimal medium. Whereas wild types can grow on minimal medium, some ascospores were found to be unable to grow. However, the latter *could* grow when they were transferred to a medium containing certain specific chemical additives, showing that they were not dead. Some nongrowers responded to arginine, some to pyridoxine, and some to other specific chemicals. These auxotrophic requirements were all found to be inherited as single gene mutations, and gave 1 : 1 ratios when crossed with wild types. Beadle and Tatum concentrated on a group of independently isolated arginine-requiring auxotrophic mutants and set about mapping the chromosomal location of each. We'll simplify the story by telling you that all of their arginine (*arg*) mutants mapped into three different locations on separate chromosomes, which we'll call the *arg-1*, *arg-2*, and *arg-3* genes. Furthermore, they found that, as well as differing in their locations, auxotrophs for each of the three loci (genes) differed in their response to chemical compounds related to arginine, called ornithine and citrulline (Figure 9-5). Thus, *arg-1* strains could grow not only on arginine, but on either ornithine or citrulline; *arg-2* mutants could grow on either arginine *or* citrulline (but not ornithine); and *arg-3* mutants could grow only in the presence of arginine.

$$
\begin{array}{ccc}
NH_2 & NH_2 & \\
| & | & \\
C{=}NH & C{=}O & \\
| & | & \\
NH & NH & NH_2 \\
| & | & | \\
(CH_2)_3 & (CH_2)_3 & (CH_2)_3 \\
| & | & | \\
CHNH_2 & CHNH_2 & CHNH_2 \\
| & | & | \\
COOH & COOH & COOH \\
\text{Arginine} & \text{Citrulline} & \text{Ornithine}
\end{array}
$$

FIGURE 9-5

It was known that related compounds are interconverted in cells by biological catalysts called enzymes; so Beadle and Tatum and their coworkers formed the hypothesis that, in *Neurospora,* some precursor was enzymatically converted into ornithine, which in turn was converted into citrulline and finally into arginine.

$$\text{Precursor} \xrightarrow[\text{Enzyme X}]{} \text{Ornithine} \xrightarrow[\text{Enzyme Y}]{} \text{Citrulline} \xrightarrow[\text{Enzyme Z}]{} \text{Arginine}$$

This enables us to interpret the basis for the three types of *arg* auxotrophs. The *arg-1* mutants would have a defective enzyme X and consequently would lack ornithine for further conversion into arginine. Growth of *arg-1* strains by the addition of ornithine shows that they are quite capable of carrying out the ornithine-to-citrulline-to-arginine conversions (therefore, they have enzymes Y and Z). The same kind of reasoning applies to *arg-2* and *arg-3* mutants, which lack enzymes Y and Z, respectively. Thus, in the biosynthetic "pathway," a block can be circumvented by feeding the cells any compound that normally comes *after* the block and this supplement is taken to the final product, arginine. All this was inferred from the properties of the mutant classes and only later were the enzymes themselves actually shown to be defective by chemical tests.

These experiments led to a very exciting formulation: *the one-gene-one-enzyme hypothesis.* At last a great insight into the function of genes was possible: genes were somehow responsible for the function of enzymes, and one gene affected only one enzyme. The same kind of observations as those of Beadle and Tatum were soon obtained for other biosynthetic pathways, and the hypothesis became consolidated. It is perhaps one of the great unifying concepts in biology because genetics had now become melded with chemistry.

MESSAGE

Genes work by controlling chemical reactions by means of enzymes.

Soon, all "classical" genetic ratios could be explained simply by the one-gene-one-enzyme concept. For example, the $9:7$ F_2 dihybrid ratio obtained for flower pigment was derived as follows:

		Pigmented	×	White
P		*AABB*		*aabb*
F_1			*AaBb* Pigmented	
F_2	9		*A-B-* Pigmented	
	3		*A-bb* White	
	3		*aaB-* White	
	1		*aabb* White	

This ratio was easily explained by a biosynthetic pathway leading ultimately to a purple petal pigment that is preceded by colorless (white) precursors.

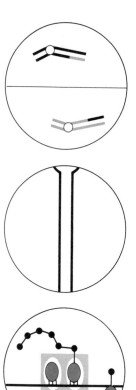

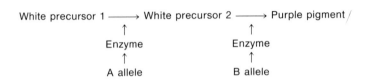

(Try 9:3:4, 13:3, etc., for yourselves.)

Also, the meaning of dominance and recessiveness became a little clearer: dominance probably represented enzyme function, and recessiveness a lack of enzyme function. A heterozygote would still have one allele producing a functional enzyme.

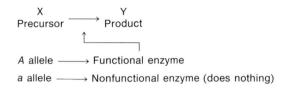

Along similar lines, the recessiveness of *arg* mutants in *Neurospora* can be demonstrated by making heterokaryons between two *arg* mutants. If an *arg-1* mutant is placed on minimal medium with an *arg-2* mutant, the two cell types fuse and form a heterokaryon composed of both nuclear types in a common cytoplasm (Figure 9-6). What happens is that the *arg-1* nuclei retain the ability to catalyze one step and the *arg-2* nuclei, the other step, so that arginine is made by their combined abilities in the shared cytoplasm. This is precisely analogous to complementation in a mixed phage infection of *rII* mutants in the A and B cistrons.

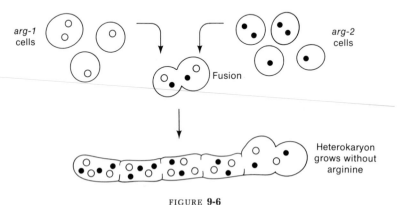

FIGURE 9-6

In humans, we now know the precise enzymes that are inactive in a large number of hereditary diseases, some of which are given in Table 9-1. Furthermore, in experimental organisms, literally hundreds of correlations between specific mutant genes and specific enzyme malfunctions are now known.

TABLE **9-1**

233

Enzymopathies: Inherited Disorders in Which Altered Activity (usually deficiency)[1] of a Specific Enzyme Has Been Demonstrated in Man

Condition	Enzyme with deficient activity
Acatalasia	Catalase
Acid phosphatase deficiency	Acid phosphatase
Adrenal hyperplasia I	20, 21-Desmolase*
Adrenal hyperplasia II	3-β-Hydroxysteroid dehydrogenase*
Adrenal hyperplasia III	21-Hydroxylase*
Adrenal hyperplasia IV	11-β-Hydroxylase*
Adrenal hyperplasia V	17-Hydroxylase*
Albinism	Tyrosinase
Aldosterone deficiency	18-OH-dehydrogenase
Alkaptonuria	Homogentisic acid oxidase
Angiokeratoma, diffuse (Fabry disease)	Ceramide trihexosidase
Apnea, drug-induced	Pseudocholinesterase
Argininemia	Arginase
Argininosuccinic aciduria	Argininosuccinase
Aspartylglycosaminuria	Specific hydrolase (AADG-ase)
Ataxia, intermittent	Pyruvate decarboxylase
Carnosinemia	Carnosinase
Norum disease	Lecithin cholesterol acetyltransferase (LCAT)
Citrullinemia	Arginosuccinic acid synthetase
Crigler-Najjar syndrome	Glucuronyl transferase
Cystathioninuria	Cystathionase
Disaccharide intolerance I	Invertase
Disaccharide intolerance II	Invertase, maltase
Disaccharide intolerance III	Lactase
Ehlers-Danlos syndrome, type V	Lysyl oxidase
Ehlers-Danlos syndrome, type VI	Collagen lysyl hydroxylase
Ehlers-Danlos syndrome, type VII	Procollagen peptidase
Fanconi panmyelopathy	Exonuclease*
Farber lipogranulomatosis	Ceramidase
Formininotransferase deficiency	Formininotransferase*
Fructose intolerance	Fructose-1-phosphate aldolase
Fructosuria	Hepatic fructokinase
Fucosidosis	α-L-Fucosidase
Galactokinase deficiency	Galactokinase
Galactose epimerase deficiency	Galactose epimerase
Galactosemia	Galactose-1-phosphate uridyl transferase

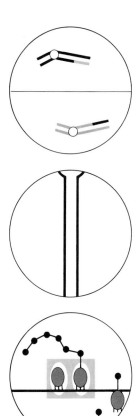

Source: *Mendelian Inheritance in Man*, 4th ed., By Victor A. McKusick (Johns Hopkins University Press, 1975).

Note: In some conditions marked * (as well as some which are not listed) deficiency of a particular enzyme is suspected but has not been proved by direct study of enzyme activity.

[1] The form of gout due to increased activity of PPRP is the only disorder listed here with *increased* enzyme activity.

TABLE 9-1 (Continued)

Condition	Enzyme with deficient activity
Gangliosidosis, generalized, type I GM	β-Galactosidase A, B, C
Gangliosidosis, GM$_1$, type II or juvenile form	β-Galactosidase B, C
Gangliosidosis, GM(3)	Acetylgalactosaminyl transferase
Gaucher disease	Glucocerebrosidase
Glycogen storage disease I	Glucose-6-phosphatase
Glycogen storage disease II	α-1-4-Glucosidase
Glycogen storage disease III	Amylo-1-6-glucosidase
Glycogen storage disease IV	Amylo (1-4 to 1-6)-transglucosidase
Glycogen storage disease V	Muscle phosphorylase
Glycogen storage disease VI	Liver phosphorylase*
Glycogen storage disease VII	Muscle phosphofructokinase
Glycogen storage disease VIII	Liver phosphorylase kinase
Gout	Hypoxanthine guanine phosphoribosyltransferase
Gout	PPRP synthetase (increased)
Granulomatous disease	NADPH oxidase
Hemolytic anemia	Adenosine triphosphatase
Hemolytic anemia	Adenylate kinase
Hemolytic anemia	Aldolase A
Hemolytic anemia	Diphosphoglycerate mutase
Hemolytic anemia	γ-Glutamylcysteine synthetase
Hemolytic anemia	Glucose-6-phosphate dehydrogenase
Hemolytic anemia	Glutathione peroxidase
Hemolytic anemia	Glutathione synthetase
Hemolytic anemia	Hexokinase
Hemolytic anemia	Hexosephosphate isomerase
Hemolytic anemia	Phosphoglycerate kinase
Hemolytic anemia	Pyrimidine 5' nucleotidase
Hemolytic anemia	Pyruvate kinase
Hemolytic anemia	Triosephosphate isomerase
Histidinemia	Histidase
Homocystinuria I	Cystathionine synthetase
Homocystinuria II	N (5, 10)-methylenetetrahydrofolate reductase
β-Hydroxyisovaleric-aciduria and methylcrotonylglysinuria	β-Methylcrotonyl CoA carboxylase*
Hydroxyprolinemia	Hydroxyproline oxidase
Hyperammonemia I	Ornithine transcarbamylase
Hyperammonemia II	Carbamyl phosphate synthetase
Hyperglycinemia, ketotic form	Propionyl CoA carboxylase*
Hyperglycinemia, nonketotic form	Glycine formininotransferase

TABLE **9-1** (*Continued*) 235

Condition	Enzyme with deficient activity
Hyperlipoproteinemia, type I	Lipoprotein lipase
Hyperlysinemia	Lysine-ketoglutarate reductase
Hyperprolinemia I	Proline oxidase
Hyperprolinemia II	δ-1-Pyrroline-5-carboxylate dehydrogenase*
Hypoglycemia and acidosis	Fructose-1, 6-diphosphatase
Hypophosphatasia	Alkaline phosphatase
Immunodeficiency disease	Adenosine deaminase
Immunodeficiency disease	Uridine monophosphate kinase
Intestinal lactase deficiency (adult)	Lactase
Isovalericacidemia	Isovaleric acid CoA dehydrogenase
Ketoacidosis, infantile	Succinyl CoA: 3-ketoacid CoA-transferase
Krabbe disease	A β-Galactosidase
Lactosyl ceramidosis	Lactosyl ceramidase
Leigh necrotizing encephalomyelopathy	Pyruvate carboxylase
Lipase deficiency, congenital	Lipase (pancreatic)
Lysine intolerance	L-lysine: NAD-oxido-reductase
Male pseudohermaphroditism	testicular 17, 20-desmolase
Male pseudohermaphroditism	testicular 17-ketosteroid dehydrogenase*
Male pseudohermaphroditism	α-Reductase*
Mannosidosis	α-Mannosidase
Maple sugar urine disease	Keto acid decarboxylase
Metachromatic leukodystrophy	Arylsulfatase A (sulfatide sulfatase)
Methemoglobinemia	NAD-methemoglobin reductase
Methylmalonicaciduria I (B12-unresponsive)	Methylmalonic CoA mutase
Methylmalonicaciduria II (B12-responsive)	Deoxyadenosyl transferase*
Methylmalonicaciduria III	Methylmalonyl-CoA racemase
Mucopolysaccharidosis I	α-L-Iduronidase
Mucopolysaccharidosis II	Sulfo-iduronide sulfatase
Mucopolysaccharidosis IIIA	Heparan sulfate sulfatase
Mucopolysaccharidosis IIIB	N-Acetyl-α-D-glucosaminidase
Mucopolysaccharidosis IV	6-Sulfatase*
Mucopolysaccharidosis VI	Arylsulfatase B
Mucopolysaccharidosis VII	β-Glucuronidase
Myeloperoxidase deficiency with disseminated candidiases	Myeloperoxidase (leukocyte)
Niemann-Pick disease	Sphingomyelinase
Ornithinemia	Ornithine ketoacid amino-transferase

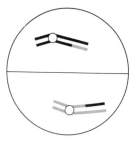

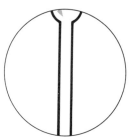

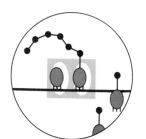

TABLE **9-1** (*Continued*)

Condition	Enzyme with deficient activity
Oroticaciduria I	Orotidylic pyrophosphorylase and orotidylic decarboxylase
Oroticaciduria II	Orotidylic decarboxylase
Oxalosis I (glycolic aciduria)	2-Oxo-glutarate-glyoxylase carboligase
Oxalosis II (glyceric aciduria)	D-glycerate dehydrogenase
Pentosuria	Xylitol dehydrogenase (L-xylulose reductase)
Phenylketonuria	Phenylalanine hydroxylase
Porphyria, acute intermittent	Uroporphyrinogen I synthetase
Porphyria, congenital	Uroporphyrinogen III cosynthetase
Pulmonary emphysema and/or cirrhosis	α-1-Antitrypsin
Pyridoxine-dependent infantile convulsions	Glutamic acid decarboxylase
Pyridoxine-responsive anemia	δ-Aminolevulinic acid synthetase*
Pyruvate carboxylase	Pyruvate carboxylase
Refsum disease	Phytanic acid oxidase
Renal tubular acidosis with deafness	Carbonic anhydrase B
Richner-Hanhart syndrome	Tyrosine aminotransferase
Rickets, vitamin-D-dependent	25-Hydroxycholecalciferol*
Sandhoff disease (GM$_2$-gangliosidosis, type II)	Hexosaminidase A, B
Sarcosinemia	Sarcosine dehydrogenase*
Sulfite oxidase deficiency	Sulfite oxidase
Tay-Sachs disease	Hexosaminidase A
Thyroid hormonogenesis, defect in, II	Peroxidase*
Thyroid hormonogenesis, defect in, IV	Iodotyrosine dehalogenase (deiodinase)
Trypsinogen deficiency	Trypsinogen
Tyrosinemia I	Para-hydroxyphenylpyruvate oxidase
Tyrosinemia II	Tyrosine transaminase
Valinemia	Valine transaminase
Wolman disease	Acid lipase
Xanthinuria	Xanthine oxidase
Xanthurenic aciduria	Kynureninase
Xeroderma pigmentosum	Ultraviolet specific endonuclease
Xylosidase deficiency	Xylosidase

All very fine, but *how* do genes control the functioning of enzymes? Enzymes belong to a general class of molecules called *proteins* and a knowledge of protein structure is needed to understand how that question was answered.

Put simply, proteins are macromolecules composed of *amino acids* attached end-to-end in a linear string. The general formula for an amino acid is $H_2N—CHR—COOH$, in which the R group can be anything from a hydrogen atom (as in the amino acid glycine) to a complex ring (as in the amino acid tryptophan). There are twenty common amino acids in living organisms, each having a different R group. They are linked together by covalent (chemical) bonds called *peptide bonds* through the reaction shown in Figure 9-7, in which condensation, or the removal of a water molecule, takes place.

$$H—N—C—C—OH + H—N—C—C—OH + \cdots$$

aa$_1$ Peptide bond   aa$_2$ Peptide bond   aa$_3$

$+ (H_2O)_n$

FIGURE **9-7**

A molecule that comprises several amino acids linked together by peptide bonds is called a *polypeptide*. The linear arrangement of amino acids in a polypeptide chain is called the *primary structure* of a protein. Many of the amino acid side groups attract or repel each other, resulting in a spiral or zigzag form that is called the *secondary structure,* and the spiral is often folded, resulting in a tertiary structure. Often several folded structures associate, resulting in a quarternary structure that is *multimeric,* being composed of several polypeptide monomers (Figure 9-8). Many proteins do look like blobs and are part of a class called *globular proteins.* Enzymes and antibodies are examples of globular proteins. On the other hand, unfolded (fibrous) proteins are components of such things as hair and muscle.

Proteins, acting as biological catalysts, as hormones, as structural elements in spindle fibers, hair, muscle, and so forth, are the single most important group of molecules in living systems. When we say that specific proteins differ from each other, what do we mean? If we purify a protein such as insulin, which is necessary for the metabolism of sugar, and then chemically break all of its peptide bonds, we find that each insulin molecule is characterized by having a specific number of each amino acid. But does that mean that specific proteins are

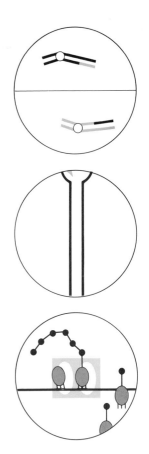

Primary ———————————————————— Amino acid chain

Secondary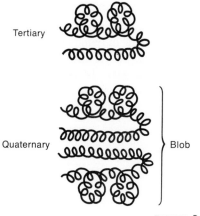

Tertiary

Quaternary } Blob

FIGURE **9-8**

simply polypeptides formed by a random hook up of set amounts of amino acids? No, the hook up is not random. It is possible to take small polypeptides and, by clipping off one terminal amino acid at a time, to determine the *amino acid sequence*. However, large polypeptides cannot be readily "sequenced" in this way. Frederick Sanger worked out a brilliant method for deducing the sequence of large polypeptides. There are several different *proteolytic enzymes* that break peptide bonds between specific amino acids in proteins. Thus, a large protein can be broken by such enzymes into a number of smaller fragments. These fragments can be separated according to their migration speeds on chromatographic paper. Because the speed of mobility of each fragment may differ depending on the solvent used, *two-way chromatograms* can be used to enhance separation of the fragments (Figure 9-9).

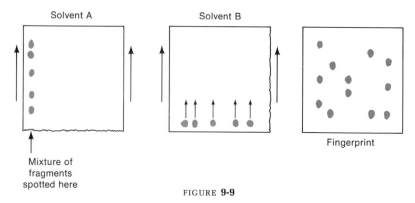

FIGURE **9-9**

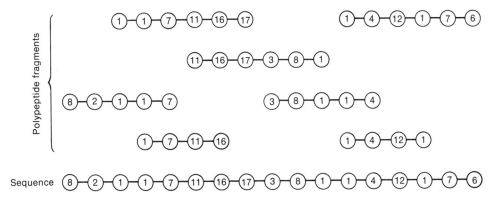

Polypeptide fragments

Sequence (8)-(2)-(1)-(1)-(7)-(11)-(16)-(17)-(3)-(8)-(1)-(1)-(4)-(12)-(1)-(7)-(6)

FIGURE **9-10**

On staining for polypeptides, they can be seen as spots in a characteristic chromatographic pattern called a *fingerprint*. Each of the spots can be eluted from the paper and, because each spot contains small polypeptides, their amino acid sequences can be derived. By using different proteolytic enzymes to cleave the protein at different points, polypeptide fragments from different treatments will overlap. The overlapping parts can be matched and, in crossword-puzzle fashion, the sequence of the large polypeptide inferred (Figure 9-10).

Using this elegant technique, Sanger was able to show that not only is the amino acid content specific for a protein, but the amino acid *sequence* is what makes insulin insulin, for example.

With such information in mind, we can examine the results of some studies by Vernon Ingram on the globular protein hemoglobin, which is the molecule that transports oxygen in red blood cells. Ingram compared hemoglobin from normal people (HbA) with hemoglobin from people carrying a mutant gene causing sickle-cell anemia (HbS. Proteolytic digestion of the hemoglobin showed that the fingerprints of polypeptides from the two genotypes differed in only one spot. In sequencing that spot, Ingram found that only one amino acid within that fragment differed in the two genotypes. So, out of the 150 amino acids known to make up a hemoglobin molecule, only one was different from normal in the anemic patients: at one point glutamic acid had been replaced by valine.

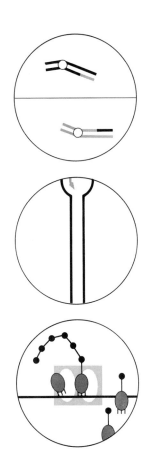

HbA Val-His-Leu-Thr-Pro-*Glu*-Glu-Lys . . .
HbS Val-His-Leu-Thr-Pro-*Val*-Glu-Lys . . .

This single substitution causes the death of anemic people unless they receive medical attention! But the main result obtained by Ingram is that a gene mutation results in an altered amino acid *sequence* in a polypeptide. Since then, numerous changes in hemoglobin have been detected, and each is the consequence of a single amino acid difference in the entire chain of 150 amino acids (Figure 9-11 shows a few examples).

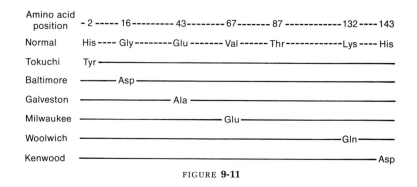

FIGURE **9-11**

MESSAGE

Genes therefore must determine the primary sequence of amino acids in specific proteins.

What is the relationship between gene structure inferred genetically and polypeptide sequence determined biochemically? Let's consider the work of Charles Yanofsky who was studying the enzyme tryptophan synthetase in *E. coli.* This enzyme catalyzes the conversion of indoleglycerol phosphate into tryptophan. The locus controlling this enzyme is called *trp* and happens to be composed of two cistrons, A and B; each produces a polypeptide, which then unites with the other to form the active enzyme tryptophan synthetase. We will concentrate on mutations in the A cistron. Yanofsky isolated many A mutants, which were mapped genetically relative to each other and, from each of these, he examined the A polypeptide. He found that each mutation could be shown to have a defective polypeptide associated with a specific amino acid substitution at a specific point (just as Ingram had shown for hemoglobin). This was an exciting study because it was found that there was an exact correlation between the order of the mutant sites in the gene map of the A cistron and the location of the respective changed amino acids in the polypeptide chain. Moreover, the farther two mutant sites are apart genetically, the more amino acids there are between the corresponding substitutions (Figure 9-12). Thus, there is *colinearity* between a gene and its corresponding polypeptide. In other words,

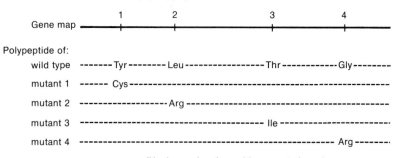

(Unchanged amino acids are not shown)

FIGURE **9-12**

MESSAGE

*The primary structure of a protein is a direct reflection of the linear
structure of the gene.*

It seems, therefore, that each amino acid is somehow "coded" by a specific
region of the gene.

In 1967, Yanofsky made additional analyses that enhanced our understanding
of the structure of the gene even more. He crossed two *trp A⁻* mutants— A23
and A46—*both* of which map very closely together and have an amino acid
substitution at position 47 of the polypeptide. When he plated the progeny on a
medium that did not contain tryptophan, colonies that were *trp A⁺* recombinants appeared. He then extracted enzyme from the recombinants and found
which amino acid was present at position 47. A summary of his results is given
in Figure 9-13. This demonstrates that the piece of gene that codes for an amino

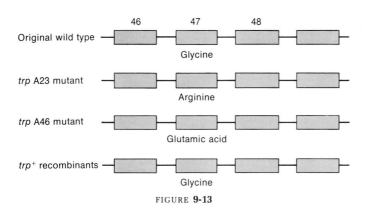

FIGURE **9-13**

acid (call it a *codon*) is composed of at least two recons. The chromosomal or
genic diagram of the cross shown in Figure 9-14 will make this a little clearer.
Thus, from an arginine and a glutamic acid codon, a glycine codon is obtainable
by crossing over.

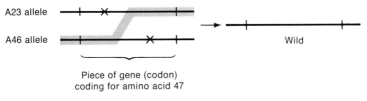

FIGURE **9-14**

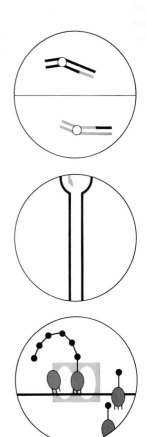

MESSAGE

Each codon is composed of at least two recons and at least two mutons.

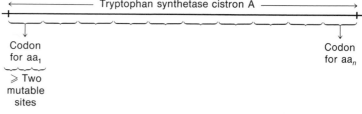

Well, now we know what a cistron is: it is a region of the genetic material that codes for one polypeptide. In fact, the one-gene–one-enzyme hypothesis can now be better expressed as the *one-cistron–one-polypeptide* hypothesis, which also emphasizes that genes (cistrons) code for proteins other than enzymes. We shall have to wait until the following chapters before discovering what codons, mutons, and recons are.

Having destroyed the bead theory, let's kick our new theory around a little bit before ending this chapter. It is known that enzymes do their job of catalysis by physically grappling with the substrates and twisting or bending them to break or make bonds. To do this they have notches in them called *active sites*, which are shaped very much like the substrates (Figure 9-15). Thus, much of the blob

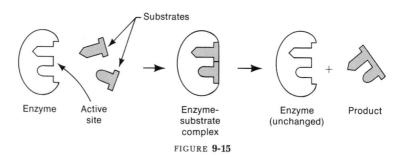

FIGURE **9-15**

structure of an enzyme is nonreactive material that simply supports the active site. One might predict that there would be considerable laxity in which amino acids are found in the support, and considerable specificity in the part that gives the precise shape to the active site. Hence the possibility arises that *unique* amino acid sequences are not necessary for enzyme function. Yanofsky did an experiment that bears on this point. Starting with mutant A23 (arginine at position 47), he detected "back mutations" by plating many A23 cells on medium with no tryptophan. These back mutants, or revertants, were examined and, whereas some did have the expected glycine at position 47, others had, for example, threonine or serine. Thus, position 47 can be filled by several alternative amino acids that are still compatible with enzyme function. At other sites, only a wild-type substitution will restore activity and these are most likely at very critical parts of the active sites.

The one-cistron–one-polypeptide hypothesis, as we have developed it in this chapter, was of paramount importance (and still is) in drawing attention to the fact that a genetic locus could comprise several different functional subunits (cistrons). However, complementation tests sometimes give results that cannot be interpreted neatly in terms of the cistron concept.

Let's examine an example and see how the confusion arises. In the adenine-3 (*ad-3*) locus of *Neurospora*, two adjacent regions can be recognized, the *ad-3A* and the *ad-3B* regions. All *ad-3A* mutants complement all *ad-3B* mutants; *ad-3A* mutants map genetically at one end of the locus and *ad-3B* mutants at the other end; and *ad-3A* mutants affect one enzyme function whereas *ad-3B* mutants affect another. No *ad-3A* mutant has ever exhibited complementation with another *ad-3A* mutant and so it looks like a good Benzer-type cistron. However, some (but not all) *ad-3B* mutants do complement *some* other *ad-3B* mutants. But before we start thinking of *ad-3B* as being composed of several cistrons, let's look at some typical complementation data. In the following grid, the results of testing four *ad-3B* alleles in all possible combinations of pairs are shown.

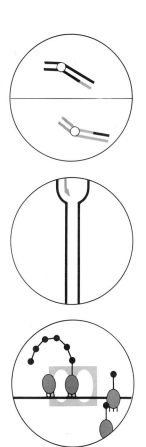

	1	2	3	4
1	−	+	−	−
2	+	−	+	−
3	−	+	−	+
4	−	−	+	−

A plus represents complementation (i.e., growth of the heterokaryon on minimal medium). What can be made of these data? To try to make sense of them in terms of several cistrons is impossible. The best that can be done is to draw a *complementation map* in which a failure to complement is indicated by overlapping "bars."

```
        3
   _____
            1
   _____
                   4
        _____
                       2
            _____
```

As you can see, if 1 and 3 were said to be in the same cistron they should *both* complement 4, yet only 3 does. The possibility that these are small deletions is ruled out by showing that the mutants are easily revertible. The complementation map looks like a deletion map, but, although derived by the same principle, they should not be confused.

What does the complementation map mean? This form of behavior has been observed in many genetic regions of many organisms; so it is not a rare exception. It is relevant to mention that the order of mutants in a complementation map (in our example, 3-1-4-2) is usually the same as the order found in the recombination-based allele map. The complementation maps are generally interpreted as follows: the bars represent regions of a polypeptide where distortion has occurred owing to the substitution of an amino acid in the mutant; two distorted polypeptides can come together and, if the distortions do not overlap, they can complement by "propping up" each other's distorted region (Figure 9-16). Thus, this kind of complementation might be expected to occur

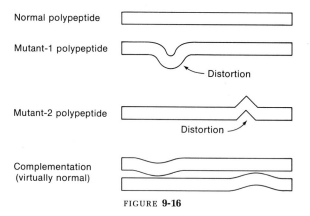

FIGURE **9-16**

only in proteins that normally have a quaternary structure; it should not occur at loci coding for monomeric proteins—these should behave like good cistrons.

In some systems, complementation products have been obtained in vitro by mixing extracts of both mutants. Not only is protein (enzyme) function detected, but a hybrid molecule can be detected by using the appropriate techniques.

So the word cistron cannot be applied to genetic regions showing complex complementation maps. What can they be called? There is no official word. Region will do.

MESSAGE

One cistron to one polypeptide is a useful general rule, but complementation between two mutations in the same cistron occurs at the level of polypeptide aggregates.

Temperature-sensitive Alleles

Sometimes a high or low temperature will turn a strain that is wild type at normal temperature into a mutant. Some amino acid substitutions, although functionally identical to wild type in phenotype at *permissive* temperatures, make the protein susceptible to heat- or cold-induced conformational distortion at *restrictive* temperatures. Such mutants are called temperature-sensitive; a heat-sensitive example is shown in Figure 9-17. As indicated in an earlier chapter,

 Normal protein   Functionally normal at 25°C   Mutant at 37°C

FIGURE **9-17**

conditional mutations such as temperature-sensitives can be very useful to geneticists. Much of this usefulness is in the ability to grow, in permissive conditions, a mutation that has drastic deleterious effects on an organism in restrictive conditions. Such mutations can be very useful in the genetic dissection of biological systems. For example, with a temperature-sensitive allele, the time at which a gene is acting can be found by shifting to restrictive temperatures at various times during development.

Genetic Dissection of Genes

In this chapter we have seen a good indication of the power of genetic analysis. Using the delicate forceps and scalpel of selective systems, by fine-structure mapping we have picked away at the bead theory until it crumbled. Out of it a much deeper understanding of the structure and function of a gene was obtained. Thus we have genetically dissected the gene. In later chapters we will genetically dissect other structures and functions of cells.

MESSAGE

Genetic analysis can be used to dissect the processes of heredity itself, or any other cellular or organismal process.

Problems

1. Mutagenic treatment in a haploid organism produced ten strains showing mutation in a hypothetical gene Q, which is concerned with the synthesis of the substance Q. All ten mutant strains require Q, and the position of the mutations can be located in the Q gene by mapping analysis. Luckily, a means for the detection of complementation is available and all mutants are tested against each other pairwise in a transarrangement to see if the pairs can complement to synthesize their own Q substance. The results are shown in the grid below, in which a plus indicates successful complementation.

	1	2	3	4	5	6	7	8	9	10
1	−	+	+	+	+	+	+	+	+	+
2		−	+	−	+	−	−	+	+	−
3			−	+	−	+	+	−	−	+
4				−	+	−	−	+	+	−
5					−	+	+	−	−	+
6						−	−	+	+	−
7							−	+	+	−
8								−	−	+
9									−	+
10										−

   a. How many cistrons are there in the Q gene? (Assume here that complementation means separate cistrons.)
   b. Which mutations belong in which cistron?
   c. How would you order the cistrons? Can you do it on the basis of the above data? If not, what procedure would be necessary?

2. Suppose that you have a map of the *rII* locus.

   You detect a new mutation, r_x, and find that it does not complement any mutants in the A or B cistrons. You find that wild-type recombinants are obtained in crosses with r_a, r_b, r_e, and r_f, but not with r_c or r_d. Suggest possible explanations for these results. Set up tests to distinguish between the explanations.

3. Various pairs of *rII* mutants of phage T4 were tested in *E. coli* in both the *cis* and *trans* positions and comparisons were made of the average number of phage particles produced per bacterium ("burst size"). One set of hypothetical results for six different *r* mutants—*rU*, *rV*, *rW*, *rX*, *rY*, and *rZ*—is:

Cis	Burst size	Trans	Burst size
rU rV/wild type	250	*rU*/*rV*	258
rW rX/wild type	255	*rW*/*rX*	252
rY rZ/wild type	245	*rY*/*rZ*	0
rU rW/wild type	260	*rU*/*rW*	250
rU rX/wild type	270	*rU*/*rX*	0
rU rY/wild type	253	*rU*/*rY*	0
rU rZ/wild type	250	*rU*/*rZ*	0
rV rW/wild type	270	*rV*/*rW*	0
rV rX/wild type	263	*rV*/*rX*	270
rV rY/wild type	240	*rV*/*rY*	250
rV rZ/wild type	274	*rV*/*rZ*	260
rW rY/wild type	260	*rW*/*rY*	240
rW rZ/wild type	250	*rW*/*rZ*	255

If we assign *rV* to the A cistron, what are the locations of the other five *rII* mutations in respect to the A and B cistrons?[1]

4. In *Salmonella,* a number of mutations affecting one enzyme involved in the histidine biosynthetic pathway are recovered. The mutants can be tested in all combinations of pairs by transduction. The recovery of wild-type transductants is indicated by a plus; zero means that no wild types are recovered. The results are:

		Mated to mutant stock			
		A	B	C	D
	A	0	+	0	+
Mutant	B		0	0	0
stock	C			0	+
	D				0

a. Considering that some or all of these mutations may be deletions, draw a possible topological map of this area.
b. If a point mutation produced wild-type recombinants with all the above mutations except C, at which position on this map would it most likely be located?

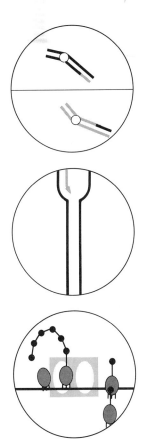

5. In a phage, a set of deletions was intercrossed in pairwise combinations and the following results were obtained (in which a plus indicates that wild-type recombinants were obtained from that cross).

	1	2	3	4	5
1	−	+	−	+	−
2		−	+	+	−
3			−	−	−
4				−	+
5					−

a. Construct a deletion map from this table.
b. If this were a complementation map instead of a deletion map, how many discrete complementation regions would be indicated?
c. The first geneticists to do a deletion-mapping analysis in the mythical schmoophage SH4 (which lyses schmoos) came up with this unique set of data:

	1	2	3	4
1	−	−	+	−
2		−	−	+
3			−	−
4				−

Show why this is a unique result by drawing the only deletion map that is compatible with this grid. (Don't let your mind be shackled by conventional expectations.)

6. The deletion map at the left below shows four deletions involving the *rIIA* cistron of phage T4. Five point mutations in *rIIA* were tested against these four deletion mutants for their ability to give r^+ recombinants. The results are given at the right below.

1 ——

2 ————————

3   ——————

4   ——

Point mutants					
	a	b	c	d	e
1	+	+	−	+	+
2	+	+	−	−	−
3	−	−	+	−	+
4	+	−	+	+	+

a. What is the order of the point mutants?
b. Another strain of T4 has a point mutation in the *rIIB* cistron. When this strain is mixed in turn with each of the above *rIIA* deletion mutants and the mixtures are used to infect *E. coli* K(λ) at a multiplicity of infection such that each host cell will be infected by at least one *rIIA* and one *rIIB*, it is found that a normal plaque is formed with deletions 1, 2, and 3, but none with 4. Given that the B

cistron is to the right of A, explain the behavior of deletion 4. Does your explanation affect your answer to part a?

7. In *Neurospora*, there is a gene controlling the production of adenine and mutants in this gene are called *ad-3* mutants. The *his-2* locus is 2.0 map units to the left and the *nic-2* locus is 3.0 units to the right of the *ad-3* locus (*his-2* controls histidine and *nic-2* controls nicotinamide). Thus, the genetic map is

his-2	*ad-3*	*nic-2*
2 m.u.	3 m.u.	

Three different *ad-3* auxotrophs are detected, *ad-3ᵃ*, *ad-3ᵇ*, and *ad-3ᶜ*. (Label them *a*, *b*, *c*). The following crosses are made:

Cross 1 *his-2⁺ a nic-2⁺* × *his-2 b nic-2*
Cross 2 *his-2⁺ a nic-2* × *his-2 c nic-2⁺*
Cross 3 *his-2 b nic-2* × *his-2⁺ c nic-2⁺*

The ascospores were then plated on minimal medium containing histidine and nicotinamide and *ad-3⁺* prototrophs were picked up. The following results were obtained:

Genotype of *ad-3⁺* recombinants	Number of *ad-3⁺* spores picked up		
	Cross 1	Cross 2	Cross 3
his-2 + nic-2	0	6	0
his-2⁺ + nic-2⁺	0	0	0
his-2 + nic-2⁺	15	0	5
his-2⁺ + nic-2	0	0	0
Total ascospores scored	41,236	38,421	43,600

What is the map order of the *ad-3* mutants and the genetic distance between them?

8. The common weed St.-John's-wort is toxic to albino animals, and causes blisters in animals with white areas of fur. Can you suggest a possible basis for this reaction?

9. In humans, the disease galactosemia causes mental retardation at an early age because lactose in milk cannot be broken down, which affects brain function. How would you provide a secondary cure for galactosemia? Would you expect it to be dominant or recessive?

10. The Notch locus of *Drosophila* is assumed to be a single cistron. William Welshons has recovered two classes of Notch mutants. Class I mutants have a dominant effect on the wings, bristles, and eyes and act as recessive lethals. All class II mutants are nonlethal and have recessive mutant phenotypes affecting eyes, bristles, or wings. Heterozygotes for a Class I and a Class II mutant are viable but exhibit the dominant phenotype of the Class I mutant and the recessive phenotype of the Class II mutant. Construct an explanation and ways to test your model.

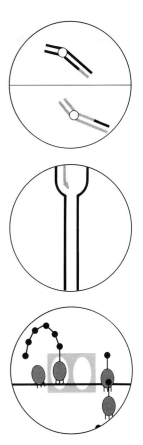

11. In sweet peas, the synthesis of purple anthocynanin pigment in the petals is under the control of two genes, B and D. Diagrammatically, the pathway is:

$$\underset{\text{intermediate}}{\text{White}} \xrightarrow{\text{Gene } B \text{ enzyme}} \underset{\text{intermediate}}{\text{Blue}} \xrightarrow{\text{Gene } D \text{ enzyme}} \underset{\text{anthocyanin}}{\text{Purple}}$$

   a. What color petals would you expect in a pure-breeding plant unable to catalyze the first reaction?
   b. What color petals would you expect in a pure-breeding plant unable to catalyze the second reaction?
   c. If the plants in parts a and b are crossed, what color petals would the F_1 plants have?
   d. What ratio of purple:blue:white plants would you expect in the F_2?
   e. In another kind of flowering plant, a cross between a pure-breeding red plant and a pure-breeding blue plant produces a purple F_1 and an F_2 of 9 purple:3 blue:3 red:1 white. How does this pathway differ from the one for peas?

12. Several mutants are isolated, all of which require compound G to grow. Compounds in the biosynthetic pathway are known and tested for their ability to support the growth of each mutant. The results are given in the following table, in which a plus indicates growth and a minus, none.

		Compounds					
		A	B	C	D	E	G
	1	−	−	−	+	−	+
	2	−	+	−	+	−	+
Mutant	3	−	−	−	−	−	+
	4	−	+	+	+	−	+
	5	+	+	+	+	−	+

   a. What is the order of these compounds in the pathway?
   b. At which points are the mutants blocked?
   c. Would a heterokaryon composed of double mutant 1,3 plus double mutant 2,4 grow on minimal medium? 1,3 plus 3,4? 1,2 plus 2,4 plus 1,4?

13. In *Drosophila*, the autosomal recessive *bw* causes a dark brown eye, the unlinked autosomal recessive *st* causes a bright scarlet eye. A homozygote for both genes has a white eye. Thus:

$$
\begin{array}{lll}
+/+ & +/+ & = \text{red eye (wild type)} \\
+/+ & bw/bw & = \text{brown eye} \\
st/st & +/+ & = \text{scarlet eye} \\
st/st & bw/bw & = \text{white eye}
\end{array}
$$

   Construct a hypothetical biochemical pathway showing how the gene products interact and why the different mutant combinations have different phenotypes.

14. George Beadle and Boris Ephrussi devised a means of transplanting eye imaginal discs of one larva into another larval host. When the host metamorphoses into an adult, the transplant can be found differentiated as a colored eye in its abdomen.

They took two strains of flies that were phenotypically identical in having bright scarlet eyes: one because of the sex-linked mutant vermillion (*v*); the other because of cinnabar (*cn*) on chromosome 2. If *v* discs are transplanted into *v* hosts or *cn* discs into *cn* hosts, the transplant develops mutant scarlet eyes. Transplanted *cn* or *v* discs in wild-type hosts develop wild-type eye colors. A *cn* disc in a *v* host develops a mutant eye color but a *v* disc in a *cn* host develops wild-type eye color. How would you explain these results and what experiments can you do to test your explanation?

15. Assume that, in *Neurospora* (a haploid), there are two genes that participate in the synthesis of valine. Their mutant alleles are called *val-1* and *val-2*, and their wild-type alleles called *val-1*$^+$ and *val-2*$^+$. These two genes are linked on the same chromosome such a distance apart that on the average one out of every two meioses produces a crossover between them.
    a. In what proportion of meioses are there no crossovers between them?
    b. Use the map function to determine the recombinant frequency between these two genes.
    c. If progeny from the cross *val-1 val-2*$^+$ × *val-1*$^+$ *val-2* are plated on medium containing no valine, what proportion will grow?
    d. The *val-1 val-2*$^+$ strains accumulate intermediate B, and the *val-1*$^+$ *val-2* strains accumulate intermediate A. The *val-1 val-2*$^+$ strains will grow on valine or A, but the *val-1*$^+$ *val-2* strains grow only on valine and not on B. Show the pathway order of A and B in relation to valine, and indicate which genes control which conversions.

16. All *zip* alleles result in defective enzyme Q and map at one genetic locus. Draw a complementation map from the following data. What kind of mutant might 3 be? What can you say about the structure of enzyme Q?

	1	2	3	4	5	6
1					+	
2					+	+
3						
4						+
5	+	+				+
6		+		+	+	

17. Two albinos marry and have a normal child. How is this possible? Suggest at least two ways.

18. Some genes in humans are known to have a multiple (or "pleiotropic") effect on phenotype. Does this constitute an invalidation of the one-gene–one-enzyme hypothesis?

19. The Martian bacterium *Martibacillus novellus* is green and can make phlizic acid. A mutation, *wht*, makes the bacteria white. A number of phlizic acid auxotrophs (*phl*$^-$) are selected: *wht* and *phl* are cotransducible by the phage pm-22, and we arbitrarily place *wht* on the right. A series of transductions are performed using donors that are *wht*$^+$ *phl-x* and recipients that are *wht phl-y*; the transductants are plated on minimal agar to select *phl* prototrophs, but no selection is made for color. The following table gives two kinds of data: first, the ratio of green/white colonies

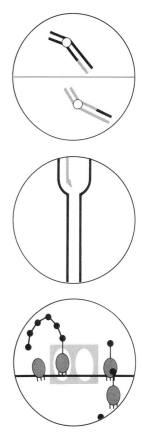

among the transductants; and, second, the presence (+) or absence (0) of micro-colonies from abortive transductions.* From these data

a. Place the six markers on a map in their proper order.
b. Divide them into complementation groups and mark the lines between cistrons on the map.[2]

| | Donors | | | | | |
Recipients	phl-1	phl-2	phl-3	phl-4	phl-5	phl-6
phl-1 wht	——	0.1 +	0.2 +	0.1 +	0.2 +	0.1 +
phl-2 wht	3.0 +	——	2.7 +	0.1 0	2.6 +	2.4 0
phl-3 wht	2.4 +	0.1 +	——	0.1 +	0.2 0	0.1 +
phl-4 wht	3.1 +	3.4 0	3.0 +	——	2.8 +	2.6 0
phl-5 wht	2.7 +	0.2 +	2.4 0	0.1 +	——	0.1 +
phl-6 wht	2.7 +	0.2 0	2.5 +	0.1 0	2.4 +	——

20. Consider the following table concerning two recessive metabolic diseases in humans. The figures represent units of enzyme activity and show the range found.

Disease	Patients	Parents of patients	Normal individuals
Acatalasemia (enzyme catalase)	0	1.2–2.7	4.3–6.2
Galactosemia (enzyme gal-1-P uridyl transferase)	0–6	9–30	25–40

This kind of information is available for many metabolic genetic diseases.

a. Of what use is it to a genetic counselor?
b. Indicate any possible sources of ambiguity.
c. Reevaluate the concept of dominance in the light of such data.

21. Amniocentesis is a technique in which a hypodermic needle is inserted through a pregnant woman's abdominal wall into the amnion, the sac that surrounds the developing embryo, to withdraw a small amount of amniotic fluid. This fluid contains cells from the embryo (not the woman), which can be cultured and which will divide and grow to form a population of cells on which enzyme analyses and karyotype analyses can be performed. What use would such a technique be to a genetic counselor? Name at least three specific conditions under which amnio-centesis might be useful. (*Note:* There is a small but finite risk for both woman and embryo; take this into account in your answer.)

---

*Abortive transduction is the failure of a transducing DNA segment to be incorporated into the recipient chromosome. The segment is functional but does not divide so only one daughter cell receives it and a microcolony is produced.

[2]Problem 19 is from Burton S. Guttman, *Biological Principles,* copyright © 1971, W. A. Benjamin, Inc., Menlo Park, California.

22. Assume that in a diploid organism squareness of cells is due to a threshold effect such that more than 50 units of "square factor" per cell will result in a square phenotype and less than 50 will result in a round phenotype. Allele s^f is a functional gene that actively synthesizes square factor. Each s^f allele contributes 40 units of square factor; thus $s^f s^f$ homozygotes will have 80 units and will be phenotypically square. A mutant allele, s^n, arose, which was nonfunctional, contributing no square factor at all. Which allele will show dominance, s^f or s^n? Are functional alleles necessarily always dominant? In a system such as this one, how might a specific allele become changed in evolution so that its phenotype shows recessive inheritance at generation 0 and dominant inheritance at generation t?

10 / DNA Structure

(Or: The unique trick that life has learned for achieving persistent order.)

We know now that genes or cistrons are linear arrays of recons, which are colinear with the primary sequence of amino acids in proteins. We assume that there is a molecular counterpart to the gene that embodies the linearity of the cistron. Can we determine what molecular species genes correspond to?

We have seen in Chapter 8 that Griffith's observations on transformation were explained by Avery's demonstration that the basis of these induced heritable changes is DNA. Although Avery's experiments were definitive, there was considerable reluctance on the part of many scientists to accept the conclusion that DNA and not proteins could be the genetic material. The final clincher was provided by Alfred Hershey and Martha Chase using the phage T2. It is obvious that phage infection must involve the introduction of the specific information that dictates viral reproduction. The phage is relatively simple in molecular constitution, most of its structure being protein with DNA contained within the protein sheath of its "head."

Phosphorus is not found in proteins but is an integral part of DNA; conversely, sulfur is present in proteins but never in DNA. Hershey and Chase, in 1952, incorporated the radioisotope of phosphorus, ^{32}P, into the DNA, and of sulfur, ^{35}S, into the proteins of phages. They then multiply infected bacteria with phages labelled with either isotope and, after a sufficient time for injection had elapsed, the empty phage carcasses (called "ghosts") on the bacteria were sheared off in a kitchen blender. The bacterial cells and phage ghosts were separated by centrifugation and radioactivity was measured in the two fractions. They found that most of the radioactive phage material injected into the bacterium was ^{32}P or DNA and that most of the proteins (^{35}S radioactivity) of

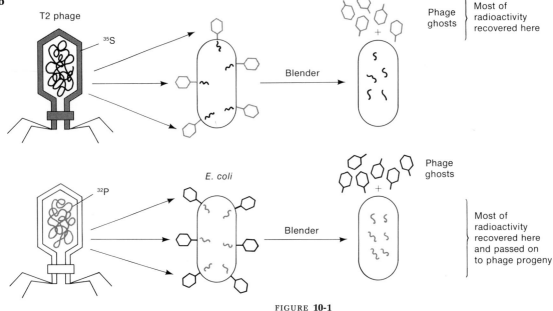

T2 phage

^{35}S

Blender

Phage ghosts

} Most of radioactivity recovered here

E. coli

^{32}P

Blender

Phage ghosts

Most of radioactivity recovered here and passed on to phage progeny

FIGURE **10-1**

the phage remained outside of the cell in the ghost. Again, this forces the inescapable conclusion that DNA is the hereditary material (Figure 10-1).

Does DNA have properties that are compatible with our genetic knowledge? What is DNA like?

The basic building blocks of DNA had been known for many years, but the way in which they were arranged to make DNA had not. By partly degrading purified DNA, the basic elements making up DNA could be isolated and determined. DNA was found to be rather simple in its constituent elements. In fact, it was this simplicity that encouraged its detractors to deny that DNA had the complexity required for the role of a gene.

DNA is made up of four basic molecules called *nucleotides,* which are identical except for the nitrogen "bases" they contain. All of them contain phosphates and sugars (of the deoxyribose type) and each has one of the four bases. The nitrogen bases are called *adenine* and *guanine* (related chemically and called *purines*) and *cytosine* and *thymine* (called *pyrimidines*). We'll use the abbreviations A, G, C, and T in referring to the four nucleotides. Their molecular structures are shown in Figure 10-2.

James Watson and Francis Crick were the first people to work out the unified structure of DNA. They had two kinds of clues to work from. First, a lot of X-ray crystallographic data on DNA structure had been amassed. The procedure for obtaining such data consists of firing X rays at DNA molecules and observing the scatter of the rays by catching them on photographic film on which they produce spots. The angle of scatter represented by each spot on the film gives information about the position of an atom or of groups of atoms in the target DNA molecule. This is not a simple procedure to do (or to explain)

PURINE NUCLEOTIDES

← Nitrogen base
(adenine)

← Deoxyribose sugar

Adenosine 5'-phosphate

← Guanine

Guanosine 5'-phosphate

PYRIMIDINE NUCLEOTIDES

← Cytidine

Cytosine 5'-phosphate

← Thymine

Thymidine 5'-phosphate

FIGURE **10-2**

and often a computer is needed to interpret the spots. These data suggested that DNA is long and skinny and is bipartite; that is, there are two parts to the molecule, which are parallel to each other and which run the length of the molecule.

Second, a lot of DNAs from different organisms had been examined by Erwin Chargaff, who arrived at several empirical rules about the amounts of each component of DNA. These were:

1. The amount of T + C (total pyrimidine nucleotides) always equals the amount of A + G (total purine nucleotides).

2. The amount of T always equals the amount of A, and C always equals G. But the amount of A + T is not necessarily equal to the amount of G + C.

The structure that Watson and Crick derived from these clues is now well known—a double helix looking rather like two interlocked bedsprings. Each bedspring is a chain of nucleotides held together by *phosphodiester bonds,* and the two bedsprings are held together by *hydrogen bonds* between the bases. A part of the structure of the two springs (uncoiled) is diagrammed in Figure 10-3. This

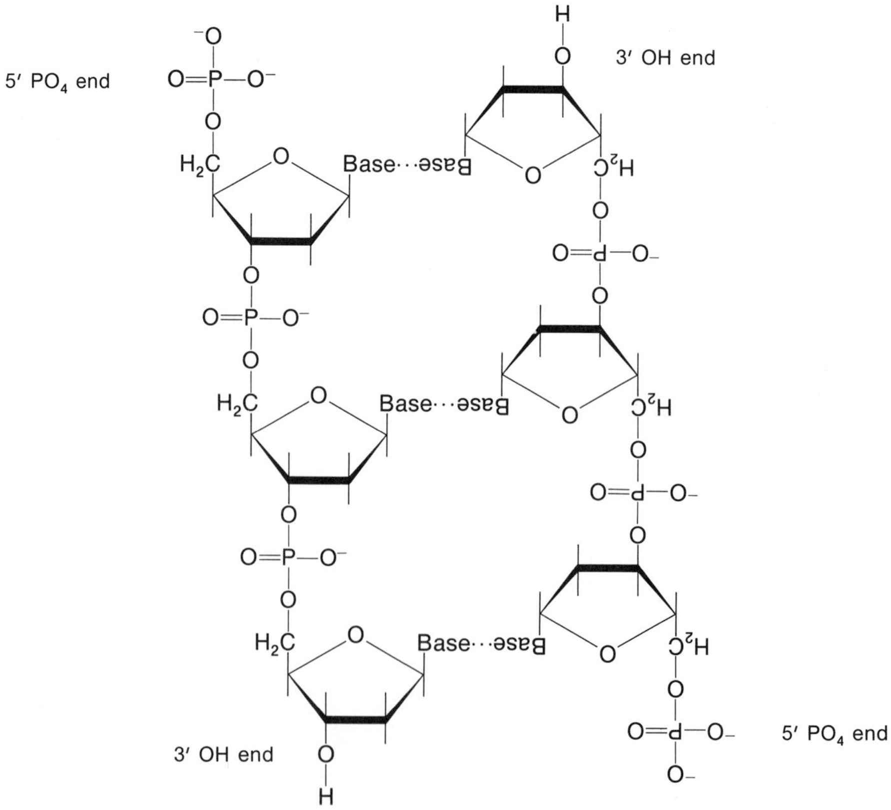

FIGURE 10-3

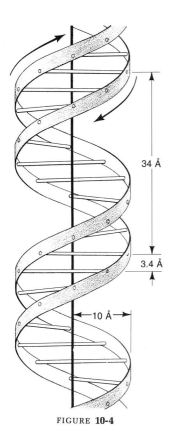

FIGURE **10-4**

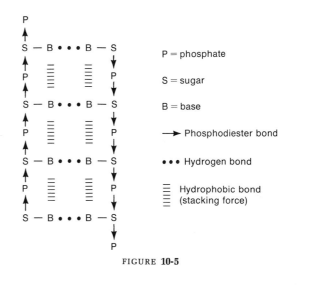

FIGURE **10-5**

P = phosphate

S = sugar

B = base

⟶ Phosphodiester bond

••• Hydrogen bond

☰ Hydrophobic bond
(stacking force)

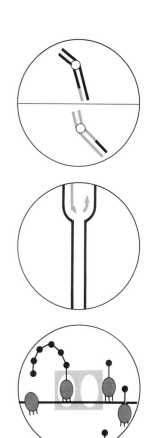

can be simplified by representing the successive phosphate-sugar elements by parallel lines (Figure 10-4). Note, in Figure 10-3, that the two springs, which are often called "sugar-phosphate backbones," are running in opposite directions: they are said to be *antiparallel* and one is called (for reasons that are apparent in the diagram) $5' \rightarrow 3'$ and the other $3' \rightarrow 5'$. If a three-dimensional model of this type of structure is made, the entire molecule becomes twisted so that the bases stack on top of each other. This stacking of bases adds tremendously to the stability of the molecule by excluding water molecules from the spaces between the flat, platelike base rings. (This phenomenon is very much like the stabilizing force you can feel when you are trying to separate two plates of glass that are under water.) The three kinds of forces hooking the nucleotides together are summarized in Figure 10-5.

Examinations of wire models of the structure of DNA led Watson and Crick to the realization that the thickness of the DNA double helix (revealed by X-ray crystallography) could be accounted for if a purine nucleotide was hydrogen bonded to a pyrimidine nucleotide as shown in Figure 10-6. This, of course, accounts for the Chargaff rule that states that A + G (purine nucleo-

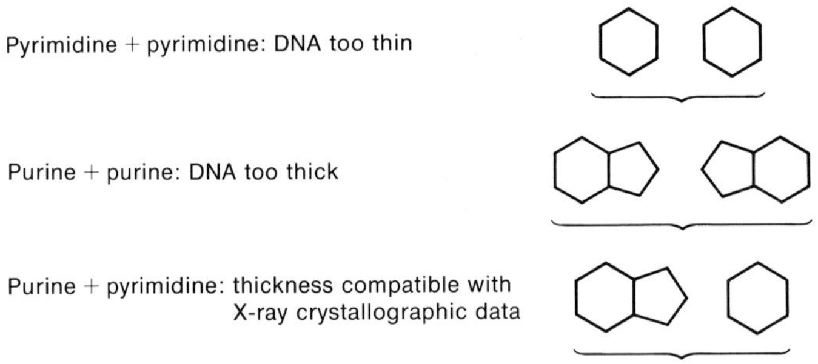

Pyrimidine + pyrimidine: DNA too thin

Purine + purine: DNA too thick

Purine + pyrimidine: thickness compatible with
X-ray crystallographic data

FIGURE **10-6**

tides) = T + C (pyrimidine nucleotides). But from this rule alone we might expect such pairings as:

$$T \cdots A$$
$$T \cdots G$$
$$C \cdots A$$
$$C \cdots G$$

However, only two of these possibilities are actually found in DNA, and the reason is that only two pairs have the necessary complementary "lock" and "key" shapes to allow efficient hydrogen bonding. Hydrogen bonds occur between hydrogen atoms with a small positive charge and acceptor atoms with a small negative charge. For example, this interaction forms a hydrogen bond.

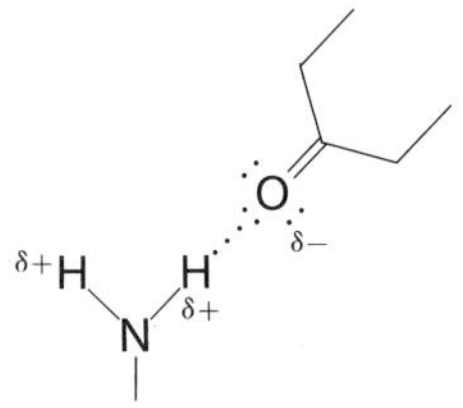

Each hydrogen atom in the NH_2 group is slightly positive because the nitrogen atom tends to "hog" the electrons involved in the N–H bond, thereby leaving the hydrogen atom slightly short of electrons, and hence "$\delta+$" (slightly plus). The oxygen atom has six unbonded electrons in its outer shell that form a $\delta -$ electron cloud around it. Hydrogen bonds are quite weak (about 3% as strong as covalent bonds), but this is, as we shall see, a necessity for the function of a hereditary molecule. One further important chemical point is that hydrogen bonds are much stronger if the participating atoms are "pointing" at each other.

If we look at the hydrogen-bonding potential between the various purine-pyrimidine pairs, we see that only two pairs have the necessary arrangement of (+) hydrogens and (−) acceptor atoms. These two are the T-A pair and the C-G pair, both of which show beautiful lock-and-key fit as shown in Figure 10-7 (the molecules have been stripped of everything except their hydrogen-

A ... T

G ... C

FIGURE **10-7**

bonding sites in order to emphasize them). Note that (1) the (+) and (−) charges are in a lock-and-key arrangement; and (2) the C-G pair has three strong hydrogen bonds, whereas the T-A pair has only two. (One would predict from this that DNA containing many G-C pairs would be more stable than that with many A-T pairs. In fact, this is found to be the case.) Once again, of course, the Chargaff rule that T = A and G = C is neatly explained (Figure 10-8).

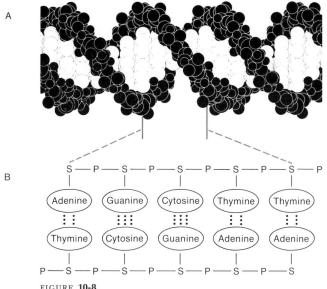

A

B

S — P — S — P — S — P — S — P — S — P

(Adenine)   (Guanine)   (Cytosine)   (Thymine)   (Thymine)

(Thymine)   (Cytosine)   (Guanine)   (Adenine)   (Adenine)

P — S — P — S — P — S — P — S — P — S

FIGURE **10-8**
Summary of DNA structure:
(A) view of an accurate three-dimensional model;
(B) diagram of a short part of a DNA
molecule showing complementary base pairing (the base
pair sequence shown is an invented, random one).

Elucidation of the structure of DNA caused a lot of excitement in genetics (and in all areas of biology) for two basic reasons:

1. The structure suggested an obvious way in which it could be *duplicated* (or *replicated*): an essential property of a genetic molecule and one that had been a mystery until then.

2. The structure suggested a possible reason for the colinearity of gene and polypeptide; that is, the *sequence* of nucleotide pairs in DNA could be dictating the sequence of amino acids in the protein organized by that gene. In other words, there is some sort of genetic code whereby information in DNA is written as a letter (nucleotide pair) sequence and is translated into protein amino acid sequences.

To those of us who have been studying biology for many years, these two basic points may seem trite and a trifle faded. But put yourself into the scene in 1953 and try to feel the excitement! Man could at last explain two of the biggest "secrets" of life. A very enjoyable personal account of the discovery is given in James Watson's book entitled *The Double Helix* (which is available in paperback).

Like every good model, the Watson-Crick model was ideally suited to being tested, and we will go through some of the predictions and tests.

Replication of DNA

The way in which DNA could replicate (as proposed by Watson and Crick) is illustrated diagrammatically in Figure 10-9, in which the sugar-phosphate backbones are represented by lines and the sequence of nucleotide pairs is random. Let's imagine that the double helix is like a zipper, which unzips starting at the top. You can see that, if this zipper analogy is valid, the unwinding of the two strands will expose single bases on either strand. Because the pairing requirements imposed by this structure are strict, each exposed base will pair only with its *complementary* base. Because of base complementarity, each of

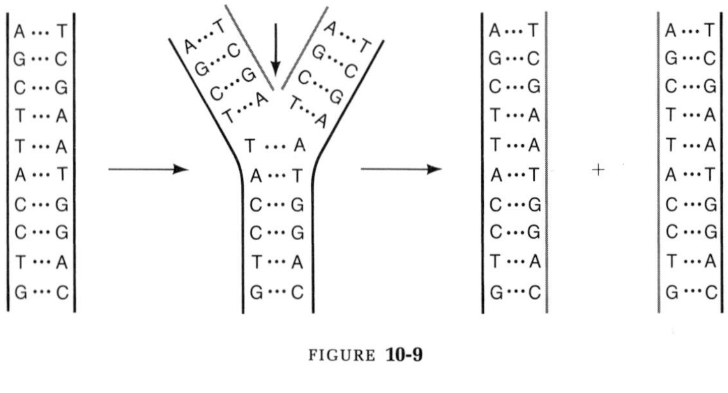

FIGURE **10-9**

the two strands will begin to reform a double helix identical with the one from which it was unzipped. The newly incorporated nucleotides come from a pool of free nucleotides that could be present in the cell. If this model is correct, four predictions should be testable:

1. The daughter molecules should have nucleotide compositions identical with the original parent molecule.
2. The daughter molecules should contain one parental (solid line in Figure 10-9) nucleotide chain and one newly synthesized (shaded line) nucleotide chain.
3. A replication "fork" should be visible during replication.
4. During replication, the information at the starting point of replication should be present in two copies more often than that at the other end.

As well as testing these four predictions, the following experiments give an indication of several of the technical devices that are used in molecular genetics.

TESTING PREDICTION 1

That daughter molecules have nucleotide compositions identical with the original parent molecule can be tested by actually replicating some DNA in a cell-free system in a test tube (an *in vitro* rather than an *in vivo* situation). Arthur Kornberg, in the late fifties, was successful in identifying and purifying an enzyme, DNA polymerase, which was catalytic in this reaction.

$$\text{Primer (parental) DNA + ATP} \xrightarrow{\text{DNA polymerase}} \text{Progeny DNA}$$
$$\text{GTP}$$
$$\text{CTP}$$
$$\text{TTP}$$

(The triphosphate forms of nucleotides are needed to make this reaction work.) As much as twenty times the original amount of primer DNA is recoverable in such a reaction. Most interestingly, the ratio of $(G + C)/(A + T)$ in the "progeny" DNA is identical with the ratio in the primer DNA, irrespective of the ratio of ATP:GTP:CTP:TTP that is put into the reaction pot. The $(G + C)/(A + T)$ ratio is a most interesting nucleotide ratio because it directly reflects the linear abundance of G-C and A-T nucleotide pairs along the DNA molecule. A primer DNA with a $(G + C)/(A + T)$ ratio of 2 must produce daughter molecules with a ratio of 2 if the model is correct. This is what is found for many primers having different ratios.

TESTING PREDICTION 2

That daughter molecules contain one parental nucleotide chain and one newly synthesized one has been tested in both prokaryotes and eukaryotes. Just by relaxing in your chair you can think of a couple of a priori ways in which a

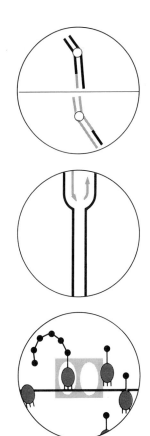

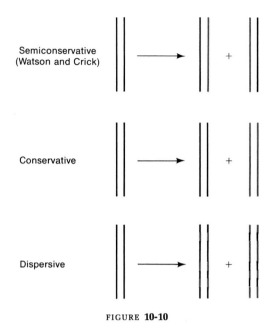

Semiconservative
(Watson and Crick)

Conservative

Dispersive

FIGURE **10-10**

parental DNA molecule may be structurally related to the daughter molecules. These hypothetical modes have been called semiconservative, conservative, and dispersive (Figure 10-10). An experiment that set out to distinguish between them in *E. coli* was done by Matthew Meselson and Franklin Stahl in 1958. They grew *E. coli* cells in a medium containing the heavy isotope of nitrogen, ^{15}N, instead of the normal light ^{14}N form. This isotope is inserted into the nitrogen bases, which can then be incorporated into newly synthesized DNA strands. After many cell divisions in ^{15}N, the DNA of the cells is well labelled with the isotope. These cells were then removed from the ^{15}N medium, put into ^{14}N medium, and sampled after one and two cell divisions. DNA was extracted from the cells in each of these samples and put onto a gradient of cesium chloride (CsCl) in an ultracentrifuge. If cesium chloride is spun in a centrifuge at astronomically high speeds (50,000 rpm) for many hours, the salt ions tend to be pushed by centrifugal force toward the bottom of the tube. Ultimately, a *gradient* of Cs^+ and Cl^- ions is established in the tube with the highest concentration at the bottom. Molecules of DNA will also be pushed toward the bottom by centrifugal force but, as they travel down the tube, they encounter an increasing salt concentration, which tends to push them back up because of DNA's buoyancy or tendency to float. Thus, the DNA will reach an equilibrium point in the tube at which the centrifugal forces are equal to the buoyancy of the molecules in the gradient. The buoyancy of DNA depends on its density (which in turn reflects the ratio of G-C to A-T base pairs). The presence of the heavier isotope of nitrogen changes the buoyant density of DNA.

Molecules of DNA extracted from cells grown for several cell cycles on either ^{15}N or ^{14}N medium can be easily distinguished on a CsCl gradient by their

Controls

First
generation

Second
generation

^{14}N

^{15}N

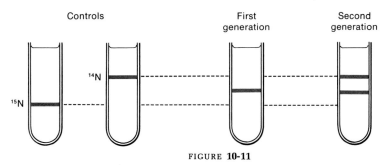

FIGURE **10-11**

equilibrium positions, and are referred to as "heavy" and "light," respectively. Meselson and Stahl found that, one generation after the incubation of "heavy" cells in ^{14}N, the DNA formed one band with density intermediate between ^{14}N and ^{15}N controls. After two generations in ^{14}N, two bands were formed, one at the intermediate position and one at the ^{14}N (or light) position (Figure 10-11). This result is compatible with *only* a semiconservative mode of replication, *if* one starts out with a single double-helix chromosome (Figure 10-12).

Semiconservative

Conservative

Dispersive

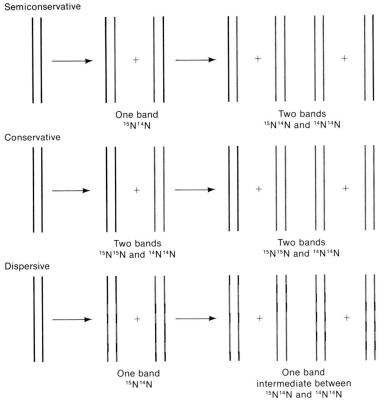

One band
^{15}N^{14}N

Two bands
^{15}N^{14}N and ^{14}N^{14}N

Two bands
^{15}N^{15}N and ^{14}N^{14}N

Two bands
^{15}N^{15}N and ^{14}N^{14}N

One band
^{15}N^{14}N

One band
intermediate between
^{15}N^{14}N and ^{14}N^{14}N

FIGURE **10-12**

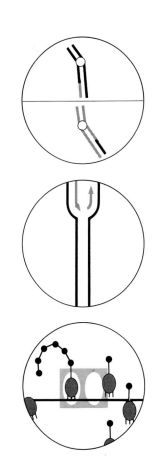

This experiment on *E. coli* was essentially duplicated by Herbert Taylor in 1958 on the chromosomes of bean root-tip cells, using a cytological technique. The root cells were put into a solution containing *tritiated* (^3H) *thymidine* (thymine nucleotide containing a radioactive hydrogen isotope called tritium). There they were allowed to undergo mitosis so that the ^3H-thymidine could be incorporated into DNA. The tips were then washed and transferred to a solution containing nonradioactive (cold) thymidine. If colchicine is added at this point, the spindle apparatus is inhibited and chromosomes in metaphase fail to separate so that sister chromatids remain "tied" together for a while by the centromere. The cellular location of ^3H can be determined by *autoradiography:* As ^3H decays, it emits a beta particle (electron). If a layer of photographic emulsion is spread over a cell that contains ^3H, a chemical reaction takes place wherever a beta particle is emitted and strikes the emulsion. The emulsion can then be developed, just like a photographic print, so that the emission track of the beta particle appears as a black spot or grain. The cell can also be stained so that one can tell exactly where the radioactivity lies. Thus autoradiography is a process whereby radioactive cell structures "take their own pictures." By adding colchicine right after the roots are removed from the ^3H and autoradiographing the cells after one or two cell divisions in cold medium, the results shown in Figure 10-13 can be observed. It is possible to interpret these results by

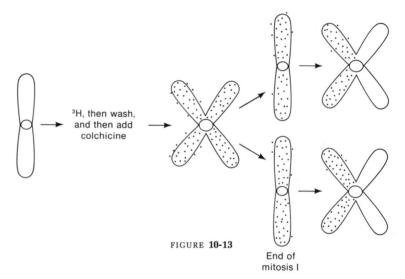

^3H, then wash, and then add colchicine

FIGURE **10-13**

End of mitosis I

representing each chromatid by a single DNA molecule that replicates semi-conservatively (Figure 10-14).

Taylor also showed, by similar techniques, that chromosome replication at meiosis was also semiconservative, thus putting another nail in the coffin of the copy-choice theory of crossing over (discussed in Chapter 4), which would require *conservative* chromosome replication at meiosis.

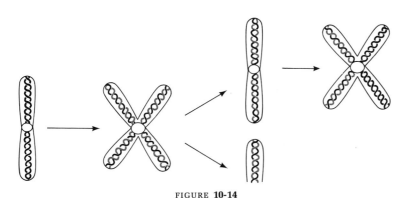

FIGURE **10-14**

Figures 10-13 and 10-14 bring up what is still one of the great unsolved questions of genetics: Is a chromosome basically one DNA molecule surrounded by a protein matrix? There are two things that strongly suggest that this is, in fact, the case. First, if there were many DNA molecules in the chromosome, whether side-by-side, or end-to-end, or randomly oriented, it would be almost impossible for the chromosome to replicate semiconservatively, with all of the label going into one chromatid as in Taylor's results. Look at Figure 10-15 and try to figure out how it could be done. Recent studies on isolated chromosomes and long DNA molecules are consistent with the suggestion that *each chromatid is a single molecule of DNA*. That makes a very long molecule, because there is enough DNA in a single human chromosome, for example, to stretch out to several feet. Second, genes are DNA and genes behave as though they are attached end-to-end in a single string or thread, which we call a linkage group. All genetic linkage data tell us that we need nothing more than a single linear array of genes to explain the facts.

One complication in this way of looking at chromosomes is that there is far too much DNA in a chromosome for it to be a single line down the middle so to speak. It has to be packed very efficiently, and the current thinking (for which there is good microscopic evidence) is that this is done by coiling and super-coiling. Twist a rubber band with your fingers and notice the way it coils. Chromosomes may be like this.

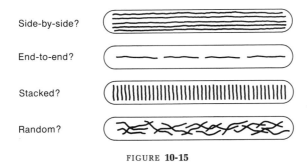

Side-by-side?

End-to-end?

Stacked?

Random?

FIGURE **10-15**

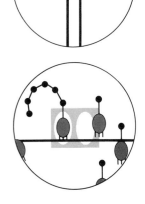

That a fork is found during replication was tested by John Cairns in 1963 by allowing replicating DNA in bacterial cells to incorporate tritiated thymidine. Theoretically, all newly synthesized daughter molecules should then contain one radioactive (hot) strand and another nonradioactive (cold) strand. After varying intervals and numbers of replication cycles in hot medium, the DNA was extracted from the cells, put on a slide, and autoradiographed for examination under the electron microscope. After one replication cycle in ^{3}H-thymidine, rings of dots were seen in the autoradiogram, which were interpreted as shown in Figure 10-16. During the second replication cycle, forks

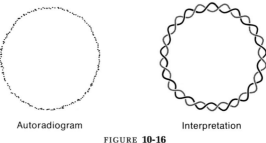

| Autoradiogram | Interpretation |

FIGURE **10-16**

were in fact seen as predicted by the model. Furthermore, the density of grains in the three segments was such that an interpretation such as that shown in Figure 10-17 could be made. All sizes of moon-shaped autoradiographic patterns were seen, corresponding to the progressive movement of the replication zipper, or fork. (These are referred to as theta (θ) structures.)

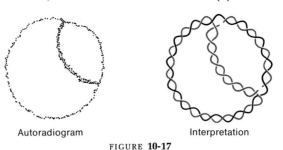

| Autoradiogram | Interpretation |

FIGURE **10-17**

It looks easy in Figure 10-17, but several problems remain:

1. The double helix must rotate in the process because the two strands are intertwined (or locked). Some have advocated molecular swivels to circumvent this problem in the bacterial ring chromosome.

2. *How* the fork is generated is still unclear. The simple notion of an unzipping molecule is inadequate because it has been shown that in all of the DNA polymerases now known growth is initiated only at the 3′ end, whereas in the zipper model, growth would have to start at both the 3′ and the 5′ ends

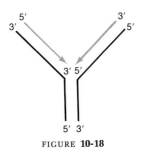

FIGURE **10-18**

(Figure 10-18). Several fascinating alternatives have been proposed, including one in which new chains are synthesized in short segments and then hooked together later by enzymes that are known to exist, called ligases (Figure 10-19).

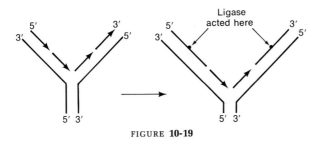

FIGURE **10-19**

3. Kornberg's original DNA polymerase does not seem to be the main DNA replication enzyme. It assists the main enzyme by completing the short fragments and also acts to repair DNA after radiation damage. This was shown by the recovery of a mutant that lacks the Kornberg enzyme activity but is still capable of replicating DNA.

For geneticists, the basic prediction of a fork structure is confirmed experimentally. The actual molecular mechanism that takes place in the crotch of the replicating fork remains to be solved by biochemists. It is best to leave the ultimate details clothed in a loin cloth (DNA polymerase) until later (Figure 10-20).

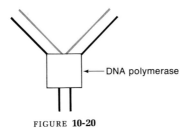

FIGURE **10-20**

TESTING PREDICTION 4

If DNA replication begins at one end and proceeds to the other end, then the initiation end should have more copies of genetic material than the terminal end. This was tested in 1963 by Toshio Nagata, who used *E. coli* lysogenic for two different phages, λ and 424. These two phages insert at different sites of the bacterial chromosome. Nagata synchronized his cells by filtering them through eighteen layers of filter paper. The cells that got through are all about the same size and are therefore at the same stage in the cell cycle. At different times, he took a sample of cells and irradiated them with ultraviolet light (which induces prophages into a lytic cycle). He then harvested the phages released and measured the relative proportions of the two. If we assume that the number of phages released is in direct proportion to the number of original prophages present on the bacterial chromosome, this ratio should vary with time of replication (Figure 10-21). The observations are consistent with a polarized replication from a set initiation point in the Hfr chromosome.

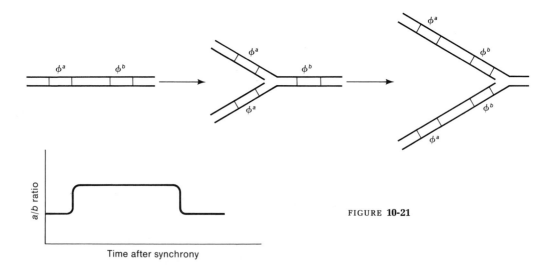

FIGURE **10-21**

In 1963, Hiroshi Yoshikawa and Noboru Sueoka used a different method to assay for gene dosage with *Bacillus subtilis,* which has no sexual process. They extracted DNA from a nonsynchronized population of cells and used the frequency of transformation for different markers as a measure of the relative proportions of each gene. They reasoned that, if replication does start at a unique point, then even in a nonsynchronized population, the relative proportion of genes will be highest at the initiation end and will decrease in direct proportion to distance from it (Figure 10-22).

Because the transforming ability of different markers may vary, they used number of transformants per amount of DNA extracted from nonreplicating

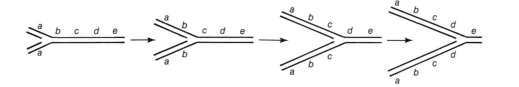

FIGURE **10-22**

spores as their basic value. Transformants per amount of DNA from dividing cells could then be standardized. They could then compare the ratio of trans-forming activity for each of the markers. You can see that the maximum ratio possible would be $a/e = 2.0$ and a minimum would be $d/e = 1.0$. We can now construct a new linkage map based on the *relative position of the loci on the replicating chromosome* as determined by transforming ability.

By growing cells from spores, synchronized cultures could be obtained and DNA extracted at different times and assayed for gene dosage by transforma-tion. Using this method later, Sueoka and his associates found that the gene order inferred from nonsynchronized populations was retained but a maximum ratio of 4 was obtained for a/e. This suggests that, even before one round of replication was completed, a second round began at the initiation end (Figure 10-23). In any case, these experiments confirm the polarized replication from one end.

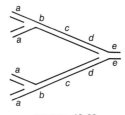

FIGURE **10-23**

If a eukaryotic cell is briefly exposed to ^{3}H-thymidine (*pulse*) and then provided an excess of cold thymidine (chase) and the DNA extracted and autoradiographs made, what appear to be successive replicating regions can be observed (Figure 10-24). Similarly, a pulse-and-chase study of DNA replication in polytene chromosomes of *Drosophila* by autoradiography reveals many

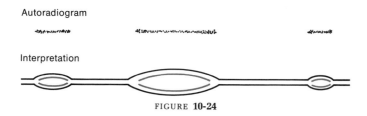

Autoradiogram

Interpretation

FIGURE **10-24**

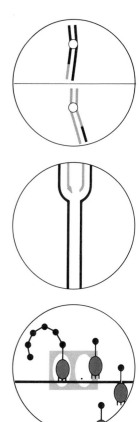

replication regions within single chromosome arms. Whether this indicates that there are many DNA molecules or many start points in one DNA molecule remains to be proved (Figure 10-25). The structure of the eukaryotic chromosome remains as one of the exciting problems yet to be resolved.

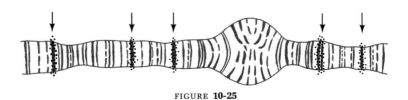

FIGURE **10-25**

So far, all the evidence we have reviewed is consistent with a unidirectional, polarized replication of DNA. However, autoradiography gives us a clue that something is occurring that is inconsistent with that idea. When autoradiographs of the labelled DNA loops are examined, the distribution of grains is not found to be uniform. In fact, the concentration of grains is lowest in the center of the loop and increases toward the ends. This is beautifully demonstrated in DNA from *B. subtilis* (Figure 10-26). How can we explain these observations? If the lightest area of grains represents the earliest replicating DNA (because the ^{3}H-thymidine has not had time to be concentrated within the cell), then it would appear the DNA is replicating *bidirectionally!*

Can we trust the grain patterns? Another way of testing bidirectionality was devised by Inman in 1966. He showed that, if λ DNA is treated with alkali (high pH) for short periods, the DNA strands in the regions rich in A-T tend to separate or denature (or "melt") before regions containing more G-C. You will recall that A-T pairs share two hydrogen bonds, whereas G-C pairs share three. By preventing the two complementary strands from hooking back up (using formaldehyde), he observed a highly reproducible pattern of "bubbles" along the length of the molecule. This is called a "denaturation" map (Figure 10-27).

Upon isolating θ structures of λ DNA, he made denaturation maps of them. Incidentally, he always found that the "melt profile" of the two segments of equal length were identical, thus proving that newly synthesized strands are identical with each other. If we take a specific bubble as a reference point in the molecule, then the unidirectional and bidirectional models of DNA replication lead to very different predictions (Figure 10-28). The unidirectional model predicts that the bubble should remain a constant distance from the origin of replication (one fork), whereas the bidirectional model predicts that the distance from the forks will change. Results supporting the latter model were obtained.

What about the *B. subtilis* transformation experiment discussed earlier? A transformation polarity is obtained that suggests unidirectional replication. These results could be obtained, however, if all the markers are on one side of the replication origin point.

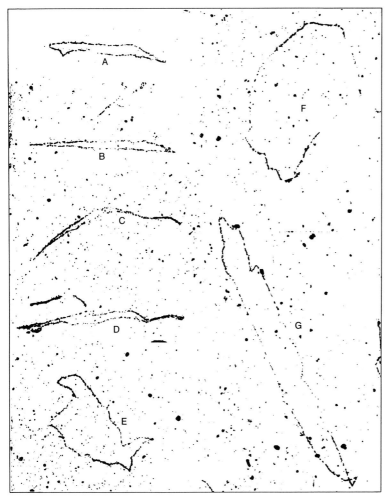

FIGURE **10-26**
Replication loops (labelled A through G) in *Bacillus subtilis* DNA molecules.

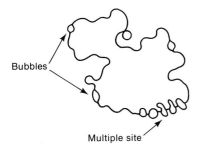

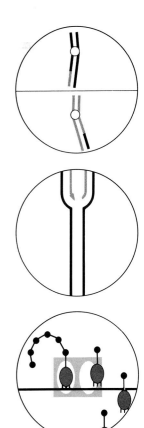

FIGURE **10-27**
Partly denatured DNA as seen under the electron microscope. Several regions rich in A-T base pairs are apparent, including one multiple site.

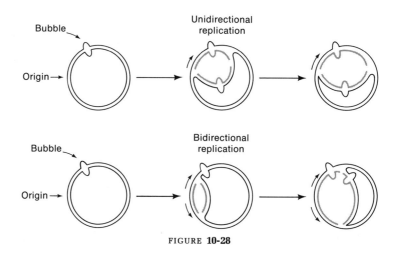

FIGURE **10-28**

Bidirectional replication probably occurs in linear or circular DNA molecules and has now been reported in phages, bacteria, yeast, and *Drosophila* DNA. Presumably the situation is something like that shown in Figure 10-29.

The relative number of copies of different chromosome regions during replication was measured by Lucien Caro's group in 1972. They took F⁻ cells lysogenic for λ. They then selected among these cells for lysogens for another phage called Mu-1. Mu-1 is very bizarre in that it can insert itself anywhere in the *E. coli* chromosome. If it inserts within a cistron, the function of that gene is lost. Thus, by selecting for auxotrophy for a variety of compounds, Mu-1 lysogens in specific loci can be recovered. Caro's group selected several strains with Mu-1 inserted at different positions in the bacterial chromosome. You can see that the closer Mu-1 is to the origin of replication, the more copies of Mu-1 there will be relative to λ, which remains at a single fixed site. The ratio of the two prophages should indicate the direction of replication.

The way in which this ratio was measured requires an analytical tool that has had enormous impact on molecular biology. In 1960, Paul Doty and Julius Marmur had observed that, when DNA is heated to 100°C, all of the hydrogen

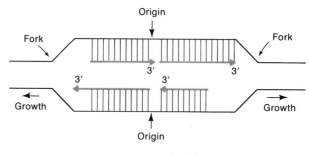

FIGURE **10-29**

bonds between the complementary strands are destroyed and the DNA becomes single-stranded (this is called *melting*, or denaturation). On cooling the solution slowly, some double-stranded DNA is formed that is biologically normal (for example, it may have transforming ability). This is presumed to result from the collision between single strands and the alignment of complementing base sequences to "reanneal" to the original double helix. The specificity and precision of this reannealing can be a very powerful tool because different DNAs with stretches of complementary nucleotide sequences will also anneal after melting and mixing. Thus the efficiency of annealing is a function of DNA similarity.

MESSAGE

The similarity between DNA molecules can be measured by melting them and examining the amount of intermolecular hybridization. The technique is called DNA/DNA hybridization.

Caro and his associates made use of hybridization by extracting DNA from mature λ and Mu-1 phages, denaturing it, and fixing it to separate filters. The average number of copies of λ and Mu-1 prophage DNA was then measured by adding denatured labelled DNA extracted from the double lysogen. The amount of label binding to the filter is a measure of the number of prophage gene copies in the replicating bacterial DNA. Their results show that the ratio of Mu-1 to λ prophages is lowest when Mu-1 is inserted at *trp* and increases in either direction from *trp* to a maximum at *ilv* (Figure 10-30). The interpretation of this curve is that there is a fixed point for initiation of replication around *ilv* and replication then proceeds bidirectionally. Each replication fork must proceed at the same rate because replication seems to terminate around *trp*, which is exactly 180° from *ilv* on the genetic map.

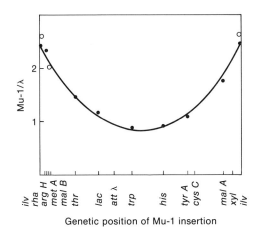

Genetic position of Mu-1 insertion

FIGURE **10-30**

MESSAGE

All the genetic, biochemical, and cytological evidence remains consistent with the semiconservative replication of a duplex DNA structure that begins at an initiation point and proceeds in both directions linearly.

You may encounter the term *replicon,* which is simply a unit of replication (from initiation to termination).

DNA and the Gene

We have shown that the gene is a linear sequence of recons that is colinear with the sequence of amino acids in a protein. Now we learn that DNA is the genetic material and consists of a linear sequence of base pairs. The obvious conclusion is that the allele maps are the genetic equivalent of the arrangements of the base pairs in DNA. This assumption can be validated if we can show that *the genetic maps are congruent with DNA maps.* This has been made possible, using some elegant genetic and biochemical tricks.

The DNA of the phage λ turns out to be a linear stretch of DNA that can be circularized because each 5′ end of the two strands (often called the Watson and the Crick strands) has an extra terminal extension of twelve bases that is complementary to the other 5′ end (Figure 10-31). Because they are complementary, they can pair to form a circle. These are called "cohesive," or "sticky," ends.

If linear objects such as DNA are subjected to shear stress (by pipetting or stirring), the mechanics of the stress produce breaks that occur primarily in the middle of the molecule. When DNA from the phage λ is sheared in half, the two halves differ in G-C ratio, which means that their buoyant density is different and that they can be separated in CsCl. When the two half molecules are separated, their genetic content can be assayed by introducing the DNA into bacteria in the following way. The DNA is introduced into the bacteria simultaneously upon infection with mutant phage λ. Recombination can occur between the phage DNA and the fragment as shown in Figure 10-32 in which *a* through *f* are phage genes. The introduced DNA can be incorporated into the normal λ DNA and "rescued" by inducing lysis. You can see that, by using different strains, the marker content of the DNA fractions can be analyzed. This is called a *marker-rescue* experiment. In this way, it was shown that one-half of the DNA molecule carries the information of one-half of the linkage map.

Dale Kaiser and his associates separated one of the DNA halves carrying a cohesive end and attached the other end to chromatographic material over which

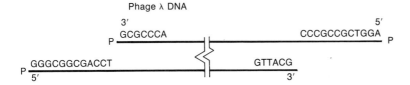

Phage λ DNA

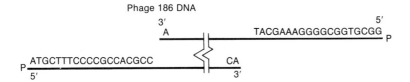

Phage 186 DNA

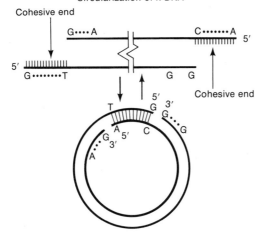

Circularization of λ DNA

FIGURE **10-31**

DNA could be passed (Figure 10-33). The single-stranded ends dangle free like fish hooks. If the other half of the DNA is sheared into smaller and smaller molecules and then passed over the sticky ends, the complementary sequences will stick. These stuck pieces can be easily detached by raising the temperature to break and hydrogen bonds in the sticky ends. In this way, a series of fractions of one end of the λ DNA that varied in size could be separated out and tested for content by marker rescue. Kaiser and his associates showed that an unambiguous arrangement of genes on the DNA could be determined that was completely congruent with the genetic map (Figure 10-34).

MESSAGE

We are now justified in concluding that the sequence of bases in the DNA is indeed congruent with the gene map.

The Genetic Code

If genes are segments of DNA and DNA is just a string of nucleotide pairs, how does the sequence of nucleotide pairs dictate the sequence of amino acids in protein? The analogy to a code suggests itself immediately: can there be such a thing as a *genetic code?* Seeing how the code was cracked will take us close to the end of Chapter 11. The experimentation was sophisticated and swift: we have known the precise nature of the code for some time now.

Armchair logic tells us that, if nucleotide pairs are the *letters* in a code, then a combination of letters could form words representing different amino acids. We need to answer the question, How many letters make up a word (codon) and which specific codons stand for (code for) specific amino acids?

There are only four nucleotide pairs: AT, TA, GC, and CG. So if the words are one letter long, only four words are possible. This is not enough because we need words for all twenty amino acids commonly found in cellular proteins. If the words are two letters long, sixteen words are possible (4^2); for example,

$$\begin{matrix} \text{-AT-} \\ \text{-TA-} \end{matrix} \quad or \quad \begin{matrix} \text{-CT-} \\ \text{-GA-} \end{matrix} \quad or \quad \begin{matrix} \text{-CC-} \\ \text{-GG-} \end{matrix}$$

This still is not enough. If the words are three letters long, 4^3 or sixty-four words are possible; for example,

$$\begin{matrix} \text{-ATT-} \\ \text{-TAA-} \end{matrix} \quad or \quad \begin{matrix} \text{-GCG-} \\ \text{-CGC-} \end{matrix} \quad or \quad \begin{matrix} \text{-TGC-} \\ \text{-ACG-} \end{matrix}$$

Common sense tells us that we can conclude that at least three nucleotide pairs are needed to constitute a code word. However, there are only twenty amino acids found in most proteins, so we have a considerable excess if all words are "triplets."

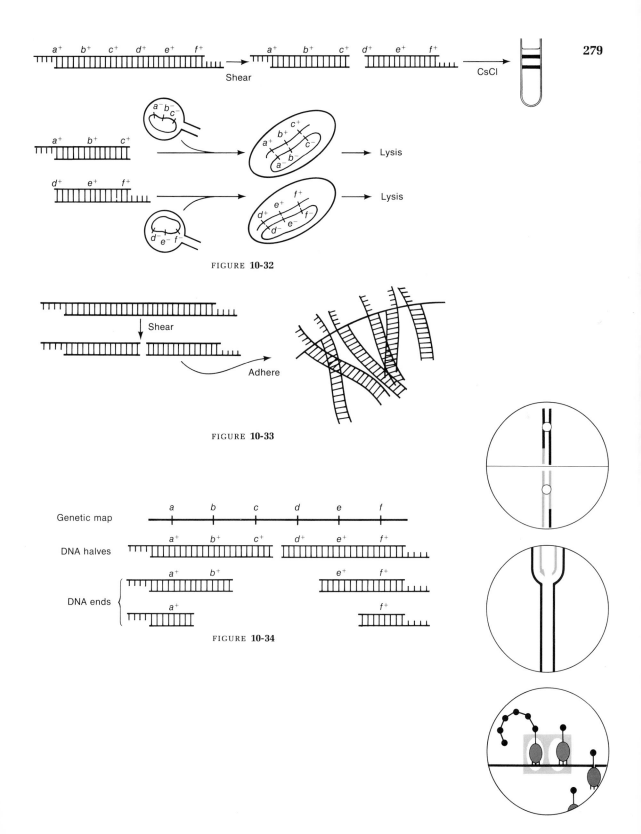

FIGURE **10-32**

FIGURE **10-33**

Genetic map

DNA halves

DNA ends

FIGURE **10-34**

Convincing proof that a codon is, in fact, three letters long (and no more than three) came from beautiful genetic experiments first reported in 1961 and extended later by Crick and his group, using mutants in the *rII* locus of T4 phage. Mutations causing the rII phenotype were induced using a chemical called proflavin, which was thought to act by the addition or deletion of single nucleotide pairs in DNA. This is an assumption based on experimental evidence that we will not go into. An example of the action of proflavin follows:

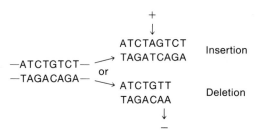

Then, starting with one particular proflavin-induced mutation (called FCO—Francis Crick O?), proflavin was again used to induce "reversions" that could be detected by their wild-type plaques on *E. coli* strain K(λ). The plaques of "revertants" were not identical to true wild types, thereby suggesting that the back mutation was not an exact reversal of the original forward mutation. In fact, the reversion was found to be caused by the presence of a *second mutation* at a different site from—but in the same cistron as—that of FCO, which suppressed mutant expression of the original FCO. More surprisingly, the *suppressor* mutation could be separated from the original forward mutation by recombination, and was shown to be an *rII* mutation by itself (Figure 10-35).

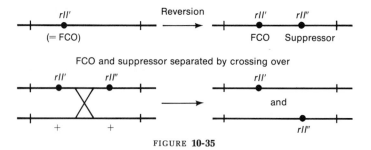

FIGURE **10-35**

How can we explain this? *If* the cistron is "read" from both ends, then the two proflavin-induced mutations should still give a mutant phenotype when combined.

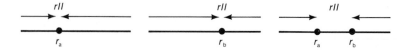

However, if reading is polarized—that is, the cistron is read from one end—then the original proflavin-induced addition or deletion could be mutant because it interrupts a normal reading mechanism that sets the number of bases to be read at a time. If this is indeed the case, then, by shifting the "reading frame" (proflavin-induced mutants are called *frame shifts*), the information from that point on would be read as *gibberish*. However, the proper reading frame could be restored by a compensatory insertion or deletion somewhere else, giving only a short stretch of gibberish between the two. Let us take an example of a sentence in which all the words are the same length:

```
                 THE FAT CAT ATE THE BIG RAT
     Delete C:   THE FAT ATA TET HEB IGR AT
     Insert A:   THE FAT ATA ATE THE BIG RAT
```

By itself, the insertion also disrupts the sentence:

```
                 THE FAT CAT AAT ETH EBI GRA T
```

If it is assumed that the FCO is caused by an addition, then the second suppressor mutant would have to be a deletion. This simply restores the "reading frame" of the gene as shown in the preceding diagram. In the following diagram (and from now on), we use a hypothetical nucleotide chain to represent DNA for simplicity, realizing that the complementary chain is automatically dictated by the specificity of the hydrogen-bonding affinities. We also assume that the code words are three letters long and that the words are read in one direction, from left to right:

```
 CAT      CAT      CAT      CAT      CAT             Wild-type message

 CAT      ACA      TCA      TCA      TCA      T      rII' message: distal words
 √        X        X        X        X              changed by frame-shift
          └─ Addition                               mutation

 CAT      ACA      TCT      CAT      CAT      CAT    rII' rII'' message:
 √        X        X│       √        √        √      few words wrong;
                    ↓                                reading frame restored
          Deletion
```

The few wrong words in the suppressed genotype could account for the fact that the "revertants" (suppressed types) that Crick and his associates recovered did not look exactly like the true wild types phenotypically.

We assumed here that the original frame-shift mutation was an addition, but the explanation would have worked just as well had the original FCO been a deletion and the suppressor an addition. If the FCO is defined as "+," then suppressor mutations are automatically "−" (hence, proflavin-induced mutations are also called "sign" mutants). It was found that a "+" cannot suppress a "+," nor can a "−" suppress a "−." In other words, two suppressors of the same sign never act as suppressors of each other. However, very interestingly, it was found that combinations of *three* pluses or *three* minuses could act together to restore a wild-type phenotype. This was the first experimental indication that a word in the genetic code consists of three successive nucleotide pairs, or a *triplet:* the reasoning being that three additions or three deletions within a gene would automatically restore the reading frame, as in the following diagram:

```
                       Delete these three
                        ↑ ↑              ↑
        CAT     CAT     CAT     CAT     CAT     CAT     CAT
         √       X               X       √       √       √
```

Crick's work also suggested that the code is *degenerate.* That is no moral indictment; it simply means that almost all sixty-four triplets must spell out an amino acid and so each amino acid must have *more than one* triplet. If only twenty triplets were actually used, then a frame shift would surely produce one of the forty-four other triplets, which might be expected to stop the reading mechanism. However, because reading continues, Crick reasoned that many or all of the amino acids must have nicknames or be signified by several triplets. This was later confirmed biochemically.

Proof that the genetic deductions about proflavin were correct came from an analysis of proflavin-induced mutants in a gene whose protein product could be analyzed. George Streisinger worked with the gene that controls an enzyme, lysozyme, whose amino acid sequence is known. He induced a mutation in the gene with proflavin and selected for proflavin-induced "revertants," which were shown genetically to be double mutants of opposite sign. When the protein of the double mutant was analyzed, a stretch of "gibberish" amino acids lay between two wild-type ends, just as predicted:

Wild type   -Thr-Lys-Ser-Pro-Ser-Leu-Asn-Ala-
Pseudowild   -Thr-Lys-Val-His-His-Leu-Met-Ala-

Thinking back to Yanofsky's work, it now looks as though we have tied down the nature of the recon. You will recall that Yanofsky had shown that the piece of gene coding for one amino acid (a codon) is divisible by recombination. We have now seen that this piece of gene is three nucleotide pairs long: a nucleotide pair must surely be the recon, the piece of a gene that is indivisible by recombination. Perhaps something like what is diagrammed in Figure 10-36 occurred in Yanofsky's experiment.

In talking about one nucleotide pair being "changed" (we have not specified how), it looks as though we have also identified the muton, that piece of a gene that *is* the unit of change. In other words, the nucleotide pair is both the recon and the muton. We will consider the mechanisms whereby a nucleotide pair can change in Chapter 12.

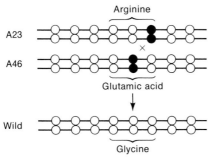

Arginine

A23

×

A46

Glutamic acid

Wild

Glycine

(plus a doubly wrong codon that is not
detected because it is mutant like the
parental DNA molecules)

FIGURE **10-36**

MESSAGE

*A codon is three nucleotide pairs. A muton and a recon are each one
nucleotide pair.*

We have now arrived at a stage of understanding that can be summarized
diagrammatically:

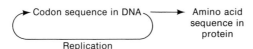

Codon sequence in DNA ⟶ Amino acid
sequence in
protein

Replication

At this stage, the game is essentially over as far as geneticists are concerned: the
rest of the details were for the most part worked out by people whose expertise
was in chemistry rather than genetic analysis. We will summarize the findings
because they are very interesting to geneticists, but geneticists can lay little
claim to their specific elucidation. However, as far as we have gone, it should be
pointed out that the main questions that many geneticists were sure would never
be answered in their generation are, in fact, already resolved.

1. What is the nature of the gene? It is a strip of DNA that can be viewed as a
   continuous sequence of triplets of nucleotide pairs, each triplet constituting a
   codon.
2. How do genes control phenotypes? Each codon stands for an amino acid in a
   polypeptide sequence.
3. What is the molecular nature of mutations? We shall see later that there are
   several kinds, but so far we have looked at two kinds: nucleotide-pair addition
   and deletion and nucleotide-pair change (or substitution).

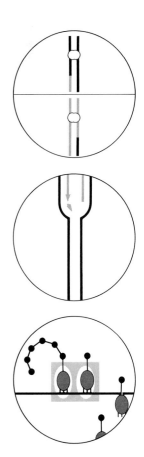

Problems

1. On Rama, the DNA is built of six nucleotide types, A, B, C, D, E, and F: A and B are called marzines; C and D, orsines; and E and F pirines. In all Raman DNAs, the following rules hold:

   Total marzines = total orsines = total pirines

   A = C = E

   B = D = F

   a. Prepare a model for the structure of Raman DNA.
   b. Bearing in mind the fact that on Rama mitosis produces three daughter cells, propose a replication scheme for your DNA model.
   c. Proflavin studies reveal that the code is (inevitably) a triplet code. How many different codons are possible in this system?
   d. Mutation at a single nucleotide site always leads to changes in *three* adjacent amino acids in a polypeptide, even though the one-gene-one-polypeptide relationship is found on Rama. What can be deduced about the way the code is read?
   e. Worry about meiosis on Rama.

2. Consider an *E. coli* "chromosome" in which every nitrogen atom is "labelled"; that is, every nitrogen atom is the heavy isotope ^{15}N instead of the normal isotope ^{14}N. The chromosomes are then allowed to replicate in an environment in which all the nitrogen is ^{14}N. Using a solid line to represent a heavy polynucleotide chain and a dotted line for a light chain, sketch the following:

   a. The heavy parental chromosome and the products of the first replication after transfer to ^{14}N medium, assuming that the chromosome is one DNA double helix and that replication is semiconservative.

b. Repeat part a, but assume conservative replication.

c. Repeat part a, but assume that the chromosome is in fact two side-by-side double helices, each of which replicates semiconservatively.

d. Repeat part c, but assume that each side-by-side double helix replicates conservatively and that the overall *chromosome* replication is semiconservative.

e. Repeat part d, but assume that the overall chromosome replication is conservative.

If the daughter chromosomes from the first division in ^{14}N are spun in a cesium chloride density gradient, and a single band is obtained, which of possibilities a through e can be ruled out? Reconsider the Meselson and Stahl experiment: What does it *prove*?

3. Assume that thymine makes up 15% of the bases in a typical DNA molecule. What percentage of bases are cytosine?

4. Draw a graph of DNA content against time, in a cell that undergoes mitosis and then meiosis.

5. You extract DNA from a small virus, denature it, and allow it to reanneal with DNA taken from other strains that carry a deletion, an inversion, or a duplication. What would you expect to see on inspection with an electron microscope?

6. DNA extracted from a mammal is heat-denatured and then slowly cooled to allow reannealing. The following curve is obtained:

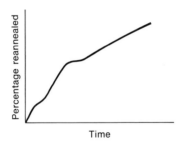

There are two "shoulders" in the curve. The first shoulder indicates the presence of a very rapidly annealing part of the DNA—so rapid in fact, that it occurs before strand interactions take place. What could it be? The second shoulder is a rapidly reannealing part as well. What does it suggest?

7. Design tests to determine the physical relationship between highly repetitive and unique DNA sequences in chromosomes. (*Hint:* It is possible to vary the size of DNA molecules by the amount of shearing they are subjected to.)

8. When plant and animal cells are given pulses of ^{3}H-thymidine at different times in the cell cycle, it is found that heterochromatic regions on chromosomes are invariably "late replicating." Can you suggest any biological significance this might have?

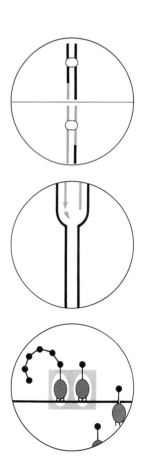

9. Ruth Kavenaugh and Bruno Zimm have devised an elegant technique to measure the maximum length of the longest DNA molecules in solution. They studied DNA samples from the following three *Drosophila* karyotypes:

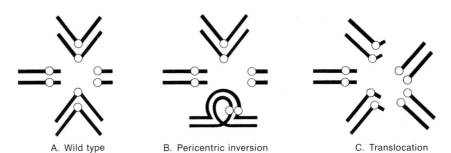

A. Wild type   B. Pericentric inversion   C. Translocation

They found the longest molecules in A and B to be of similar length and about twice the length of the longest in C. How would you interpret these results?

10. Plant root tips are exposed to ^{3}H-thymidine for several cell divisions. They are then washed and put into a nonradioactive thymidine-containing liquid for two cell divisions and prepared for autoradiography. James Peacock reported that, although one sister chromatid was labelled and one was unlabelled in most chromatid pairs, he sometimes observed:

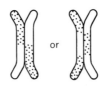

or

How could you explain these data? In addition, he occasionally found:

or

How would you explain such "iso-chromatid" or iso-labelling?

11. In mice there are viruses that are known to cause cancer. You have a pure preparation of virus DNA, a pure preparation of DNA from the chromosomes of mouse cancer cells, and pure DNA from chromosomes of normal mouse cells. Virus DNA will hybridize with cancer-cell DNA, but not with normal-cell DNA. Explore the possible significance of this observation genetically, at the molecular level, and also medically.

11 / DNA Function

(Or: How biological blueprints are turned into biological architecture.)

Now for some of the details of the fascinating molecular processes of "information transfer," whereby the "message" embodied in DNA is replicated into more DNA (the fundamentally unique characteristic of life) and the information in the message is translated into protein. Information is really an appropriate word here: it literally means "that which is necessary to give form" (to something), which is precisely what DNA does in being responsible for phenotypes. It gives form to organisms.

Very little of this part of our story was revealed by purely genetic analysis. Although mutants have been useful in much of the work, they have been tools to short-cut a lot of the biochemical work. Nevertheless, DNA function is important to genetics and the kind of reasoning used in investigating it illustrates the analytical approach characteristic of genetics.

Much of the work treated in this chapter took place in the past two decades and the many areas of uncertainty reflect the problems just now being posed. The spectacular rapidity with which new discoveries were made in molecular biology have brought us within striking range of genetic engineering with all its dazzling promise and horrendous possible consequences.

It soon became evident that information transfer was not directly from DNA into protein: another nucleic acid, ribonucleic acid (RNA) was necessary as an intermediary. There are good reasons for thinking this: For one thing, DNA is

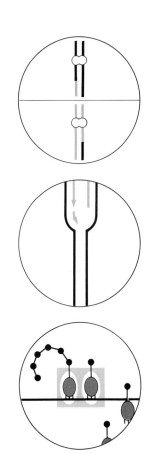

found in the nucleus (of eukaryotic cells), whereas protein is synthesized in the cytoplasm. If cells are fed radioactive RNA building-block precursors in an attempt to label RNA, the label shows up first of all in the nucleus, indicating that RNA is being synthesized there. However, if the cells are then put into a nonradioactive medium, any further RNA synthesized will be nonradioactive and then what happens to the labelled RNA can be followed (this is called a pulse-chase experiment, for obvious reasons). What is observed by means of autoradiography is that the labelled RNA first appears in the nucleus and, in later samples, the label appears in the cytoplasm. Thus RNA is a good candidate as an information-transfer intermediary (Figure 11-1).

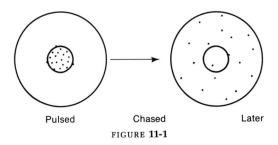

Pulsed   Chased   Later

FIGURE **11-1**

Another significant observation was made by Elliot Volkin and Lawrence Astrachan who found that, if *E. coli* is infected with the phage T2, one of the most striking molecular changes is a rapid burst of RNA synthesis. Furthermore, this phage-induced RNA "turns over" rapidly; that is, if the infected bacteria are pulsed with ^{3}H-uracil (as we shall soon see, a specific precursor of RNA that is abbreviated as U) and then chased with cold uracil, the RNA recovered shortly after the pulse is labelled, but it loses the label after the chase. Finally, when the nucleotide content of *E. coli* and T2 DNA is compared with the nucleotide content of the induced RNA, the latter is found to be very similar to that in the phage DNA.

The tentative conclusion: RNA is synthesized from DNA, and it passes into the cytoplasm where it is somehow used to synthesize protein. We now have three stages of information transfer: *replication, transcription,* and *translation* (Figure 11-2).

Although RNA is a long-chain macromolecule of nucleic acid like DNA, it has very different properties. First, it is single-stranded and not a double helix.

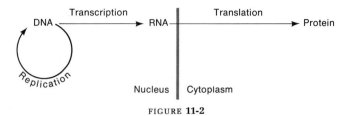

FIGURE **11-2**

Second, it has ribose sugar in its nucleotides instead of deoxyribose (hence the name).

Ribose   Deoxyribose

Third, it has the pyrimidine base *uracil* instead of thymine. However, uracil forms hydrogen bonds with adenine, just like thymine.

Uracil

Nobody is absolutely sure why RNA has uracil instead of thymine or ribose instead of deoxyribose, but the most important aspect of RNA is that it is single-stranded and very similarly constructed to DNA. This suggests a mode of transcription based on the complementarity of bases, which is the key to DNA replication. A transcription enzyme, RNA polymerase, is needed (Figure 11-3).

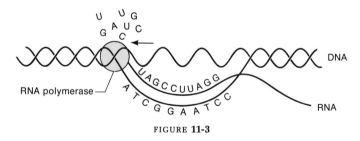

FIGURE **11-3**

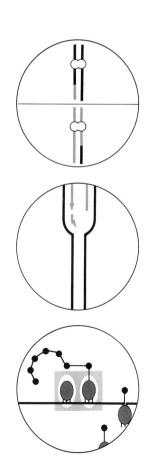

This picture of transcription in fact is observed cytologically (Figure 11-4). The fact that RNA is capable of being synthesized using DNA as a template can be shown by in vitro synthesis of RNA from nucleotides in the presence of DNA, using an extractable RNA polymerase. Whatever the source of DNA used, the RNA synthesized has an $(A + U)/(G + C)$ ratio similar to the $(A + T)/(G + C)$ ratio of DNA (Table 11-1). This does not tell us whether the RNA is synthesized from both DNA strands or just one, but it does tell us that the linear frequency of the A-T pairs in comparison with the G-C pairs in DNA is precisely mirrored in the relative abundance of $(A + U)$ in the RNA. (These points are difficult to grasp without drawing some imaginary nucleic acids: one

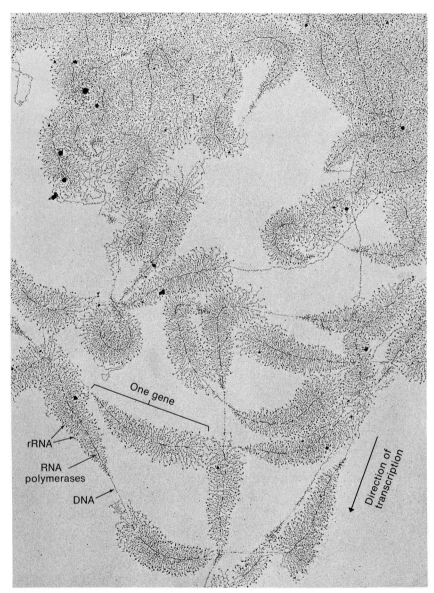

FIGURE **11-4**
Tandemly repeated ribosomal RNA genes being transcribed in the nucleolus of *Triturus viridescens* (an amphibian).

TABLE **11-1**

291

DNA Function

DNA source	$\dfrac{(A + T)}{(G + C)}$ of DNA	$\dfrac{(A + U)}{(G + C)}$ of RNA synthesized
T2 phage	1.84	1.86
Cow	1.35	1.40
Micrococcus (bacterium)	0.39	0.49

of the problems at the end of this chapter consists of an exercise in this kind of logic.)

What about the question of whether RNA is synthesized from only one or from both of the DNA polyncleotide strands? There is much sophisticated chemical evidence to prove that only one DNA strand acts as a template: one could sidestep all of it by reasoning that a gene that produced two "complementary" proteins would be under some very strange selection pressures in evolution. It seems more "sensible" to use only one strand and only one is used, but this is not necessarily the same one throughout the chromosome.

The specificity and precision of nucleic acid hybridization can be applied to determine the complementarity of DNA with RNA. For example, if DNA is labelled with a heavy isotope, as done by Meselson and Stahl, and denatured along with light RNA transcript, on slow cooling some of the RNA strands will anneal with complementary DNA to form a DNA:RNA hybrid, which differs in density from the DNA:DNA duplex. This discovery shows that RNA is indeed produced from a DNA template, and can be extended to show that RNA is synthesized from only one strand of DNA. The latter idea can be tested if RNA made by a DNA stretch can be purified and annealed to each of the two DNA strands separately. This is possible if the two DNA strands differ in their relative contents of purines and pyrimidines, which would affect their buoyant densities in cesium chloride. Marmur and his coworkers discovered that the DNA of *B. subtilis* phage SP8 does indeed have two strands of different buoyant densities. The single strands of DNA that has been denatured and then *fast-cooled*, do not reanneal but coalesce into individual lumps. In a gradient, then, each strand can be separated from the other. They showed that SP8 RNA hybridized to only one of the two strands, thereby proving that transcription is *asymmetrical;* that is, it occurs only on one DNA polynucleotide strand.

The information-bearing RNA is appropriately called *messenger RNA* (mRNA), and in a sense is comparable to a xerox copy of a DNA sequence. The details of how mRNA information is translated into protein are still not fully understood, but we do have a very good general picture. If you mix mRNA and all twenty amino acids in a test tube and hope to make protein, you will be disappointed. Other components are needed, and the discovery of what these are was the key to the elucidation of the mechanism of information transfer in the translation of mRNA into protein.

At this point, we should mention another important tool for the separation of molecules and molecular aggregates: sucrose density-gradient centrifugation.

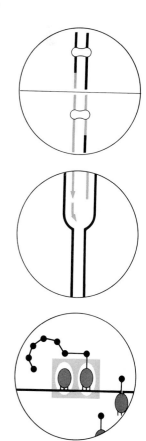

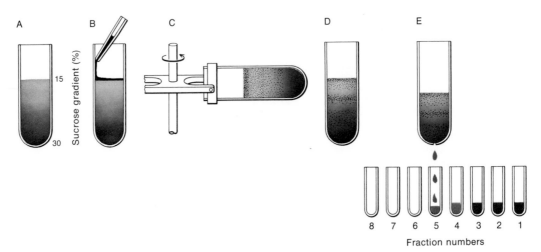

FIGURE **11-5**
A. Sucrose density gradient in ultracentrifuge tube.
B. Introduction of mixture of cell components on top of gradient.
C. Centrifugation causing differential sedimentation of various cell components.
D. The gradient is fractionated.
E. Different bands will be collected in different fractions.

Unlike the cesium chloride gradient, a gradient of sucrose is created in a test tube by layering successively lower concentrations of sucrose solution one on top of the other. The material to be studied is carefully placed on top. On spinning the solution in a machine that allows the test tube to swivel freely, the sedimenting material travels through the gradient at different rates that are related to the size and shape of the molecules. Larger molecules migrate further in a given length of time than do smaller molecules. The separated molecules can be collected individually by puncturing the bottom of the test tube and collecting the drops sequentially (Figure 11-5). The distance that a fraction moves indicates its sedimentation, or S, value, which is a measure of its size.

The two most important and interesting additional components needed are a class of small RNA molecules called *transfer RNA* (tRNA), with a sedimentation value of 4S, and *ribosomes,* with a value of 70S. The complete nucleotide sequences of some tRNA molecules have been determined; they appear to have some hydrogen-bonded and some single-stranded regions and form variations of a clover leaf structure.

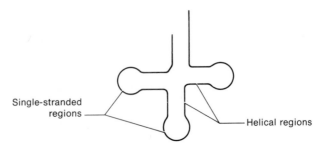

Ribosomes, on the other hand, are very complex aggregations of protein and *ribosomal RNA* (rRNA) components. There are at least three separate RNA molecules distinguishable by size in ribosomes (28S, 18S, and 5S). The precise function of ribosomes is still the subject of much research, and the story of how we have come to know what we do now is a fascinating one that lies outside the scope of this book. We do not know precisely how rRNA is bound up with the protein components, or how each component functions. Under the electron microscope, a ribosome is seen as a "blob." On chemical treatment, the blob splits into two main sub-blobs (50S and 30S) and this is how we will represent it:

Both rRNA and tRNA molecules can be shown to be DNA transcripts because they form RNA:DNA hybrids in vitro. Other important items needed to make protein synthesis work in vitro are: several enzymes (aminoacyl-tRNA synthetases and peptidyl transferase), several mysterious protein "factors" whose role is probably enzymic, and a chemical source of energy. The energy donor is necessary because an ordered structure is being created out of a mess of components, a process that requires energy because the system must lose entropy (disorder or randomness).

We can regard protein synthesis as a chemical reaction (which we will do at first), but then we will take a three-dimensional look at how the components interact physically.

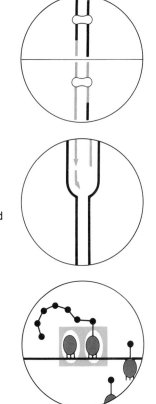

1. Each amino acid is hooked to a tRNA molecule that is specific to that amino acid by a high-energy bond derived from ATP. The process is catalyzed by a specific enzyme (the tRNA is said to be "charged" when the amino acid is hooked up).

$$aa_1 + tRNA_1 + ATP \xrightarrow{\text{Synthetase}_1} aa_1\text{---}tRNA_1 + ADP$$

2. The energy of the charged tRNA is converted into a peptide bond linking the amino acid to another on the ribosome.

$$aa_1\text{---}tRNA_1 + aa_2\text{---}tRNA_2 \xrightarrow[\substack{\text{Peptidyl transferase} \\ \text{on a ribosome}}]{} \underbrace{aa_1\text{-}aa_2\text{---}tRNA_2}_{\substack{\text{Small} \\ \text{polypeptide}}} + tRNA_1 \text{ released}$$

3. New amino acids are linked by means of a peptide bond to the growing chain.

$$aa_3\text{---}tRNA_3 + aa_1\text{-}aa_2\text{---}tRNA_2 \longrightarrow \underbrace{aa_1\text{-}aa_2\text{-}aa_3\text{---}tRNA_3}_{\substack{\text{Larger} \\ \text{polypeptide}}} + tRNA_2 \text{ released}$$

4. This continues until aa_n, the final amino acid, is added. Of course, the whole thing works only in the presence of mRNA.

We can visualize this incredible process if we recognize the main interacting sites of the components. First, tRNA. A very convincing experiment showed that it was tRNA rather than the amino acid itself that recognizes the codon on mRNA. In the experiment, cysteinyl tRNA (the tRNA specific for cysteine) charged with cysteine was reacted chemically with nickel hydride. This had the effect of converting cysteine, while still bound to $tRNA_{Cys}$, into another amino acid, alanine, without affecting the tRNA.

$$\text{Cysteinyl tRNA}_{Cys} \xrightarrow{\text{Nickel hydride}} \text{Alanyl tRNA}_{Cys}$$

Protein synthesized with this "Rube Goldberg" species had alanine wherever you would expect cysteine, thereby showing that amino acids are illiterate and are inserted at the proper position through recognition of the mRNA codons by tRNA "adaptors." Thus, we can delineate several sites of tRNA in terms of function as shown in Figure 11-6. The site that recognizes an mRNA codon is called an *anticodon* and works by base complementarity of codon and anticodon segments. This drawing in Figure 11-6 is supported by very sophisticated

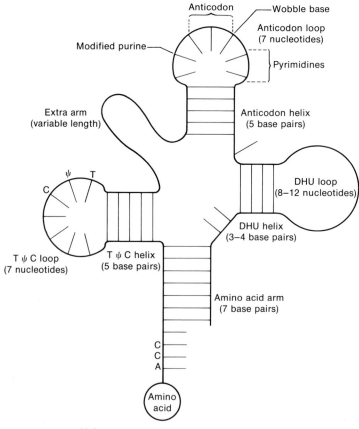

FIGURE **11-6**

chemical analysis of tRNA nucleotide sequences and X-ray crystallographic data on the overall shape of the molecule.

Where does tRNA come from? If radioactive tRNA is put into a cell nucleus in which the DNA has been partly denatured by heating, the radioactivity determined by autoradiography appears in localized regions of the chromosomes. These regions probably reflect the location of tRNA genes; they are regions consisting of DNA whose job is to make tRNA rather than code for a protein product. Why does the labelled tRNA "home in" on these sites? Because the complementary base pairs of the tRNA and of the DNA strand from which it is synthesized allow a stable bonding or hybridization of tRNA to its parent gene. So it appears that even the one-gene–one-polypeptide idea is not 100% watertight. Some genes do not code for protein; rather they specify RNA components of the translational apparatus.

MESSAGE

Some genes code for proteins; others have RNA (tRNA or rRNA) as their final product.

How does tRNA get its fancy shape? It probably folds up spontaneously into a conformation that produces maximum stability. Transfer RNA contains many "odd" or modified bases in its nucleotides (e.g., pseudouracil), which might be important in folding. They have also been implicated in other tRNA functions.

Ribosomes have several sites that can be delineated a priori. By adding radioactive tRNA to a solution of ribosomes and measuring how much of it becomes bound to the ribosomes, one can calculate that there are two tRNA binding sites per ribosome. There is also an mRNA binding site, and the enzyme peptidyl transferase is built into the ribosome (Figure 11-7). All the components

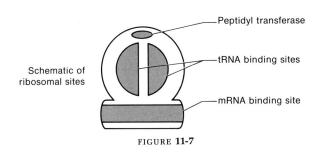

Schematic of ribosomal sites

Peptidyl transferase

tRNA binding sites

mRNA binding site

FIGURE 11-7

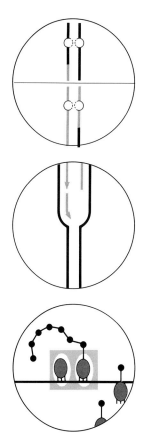

can now be assembled in a diagram that shows an amino acid being added to a growing polypeptide (Figure 11-8). The ribosome "moves along" the mRNA and, as each new codon enters the ribosome, conditions are right for the insertion of the appropriate aa-tRNA complex in the amino acid binding site (or A site) and for the formation of a bond between the amino acid in the A site and polypeptide being held to a tRNA in the polypeptide binding site (or P site).

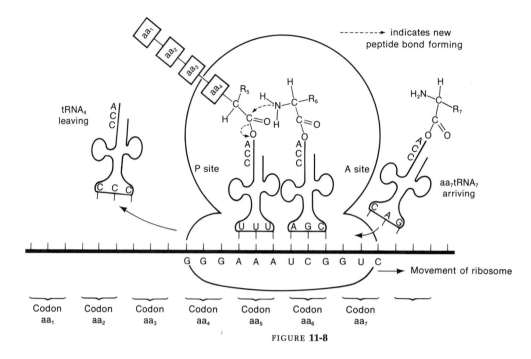

FIGURE **11-8**

When the peptide bond is formed, the tRNA in the P site is liberated, and the new tRNA-polypeptide complex moves into the unoccupied P site as the ribosome moves further along the mRNA.

Much of what is shown in Figure 11-8 was constructed by inference based on genetic and chemical analyses. Its validity was gratifyingly demonstrated recently by electron micrographs of the process of protein synthesis in an *E. coli* gene (Figure 11-9). As can be seen, the protein synthesis by this prokaryotic gene does not wait for the completion of mRNA synthesis: as soon as part of the mRNA is synthesized, the ribosomes latch on to it and begin translation! Protein synthesis in eukaryotes is different because mRNA is made in the nucleus and is shipped to the cytoplasm for translation; so transcription and translation are separated in time and space.

In Figure 11-8, a few imaginary codons have been inserted for demonstration. Of course this was the general notion in the minds of the investigators at the time, too, but knowledge of *which* specific codons represented *which* specific amino acids had to await another round of sophisticated investigations in chemistry, which came to be known as "cracking the code."

Code Cracking

The actual deciphering of the codons—that is, determining the triplets that specify each amino acid—is one of the most exciting breakthroughs of the past

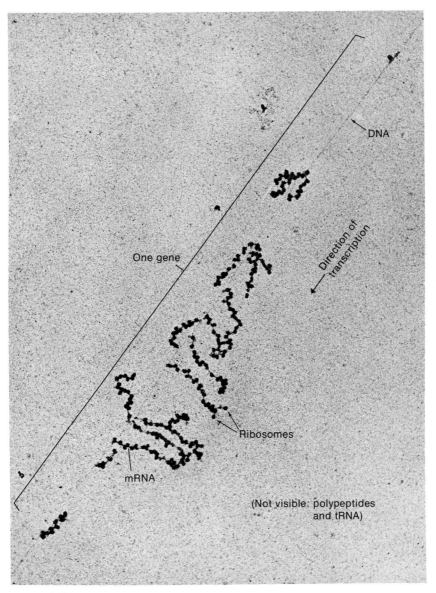
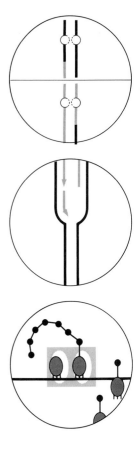

FIGURE **11-9**
A gene of *E. coli* being simultaneously transcribed and translated.

two decades. Once the techniques for this determination became available, the biological code was broken in a rush.

The first breakthrough came as a result of the discovery of how to make artificial or synthetic mRNA. If the nucleotides of RNA are mixed with a special enzyme (polynucleotide phosphorylase), a single-stranded RNA is formed in the reaction. No DNA is needed for this synthesis, and so the nucleotides are incorporated at random. The exciting aspect of this RNA was that it could be used in a test-tube protein-synthesizing system to see what proteins are made by specific mRNAs. The first synthetic messenger obtained was U-U-U-U (or poly-U) and this was made by reacting uracil nucleotides only with the RNA-synthesizing enzyme. In 1961, Marshall Nirenberg and Heinrich Mathaei mixed poly-U with the protein-synthesizing machinery of *E. coli* (ribosomes, amino-acyl tRNAs, a source of chemical energy, several enzymes, and a few other things) in vitro and *observed the formation of a protein!* Of course the excitement centered on the question of which amino acid(s) made up this protein. The answer to that question provided the first code word because the protein was polyphenylalanine—in other words, a chain of amino acids, all of them phenyl-alanine, strung together (Figure 11-10); thus the triplet U-U-U must stand for phenylalanine.

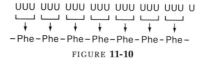

FIGURE **11-10**

This type of analysis was extended by mixing nucleotides in a known fixed proportion when making synthetic mRNA. In one experiment, the nucleotides uracil and guanine were mixed in a ratio of $3:1$. If they are incorporated at random into synthetic messenger, we can calculate the relative frequency that each triplet will appear in the sequence, as shown in Table 11-2, and the amino acids produced by this RNA in the in vitro protein-synthesizing system should

TABLE **11-2**

	Probability		Ratio
$p(UUU) = 3/4 \times 3/4 \times 3/4 =$	$27/64$		100
$p(UUG) = 3/4 \times 3/4 \times 1/4 =$	$9/64$		33
$p(UGU) = 3/4 \times 3/4 \times 1/4 =$	$9/64$		33
$p(GUU) = 3/4 \times 3/4 \times 1/4 =$	$9/64$		33
$p(UGG) = 3/4 \times 1/4 \times 1/4 =$	$3/64$		11
$p(GGU) = 3/4 \times 1/4 \times 1/4 =$	$3/64$		11
$p(GUG) = 3/4 \times 1/4 \times 1/4 =$	$3/64$		11
$p(GGG) = 1/4 \times 1/4 \times 1/4 =$	$1/64$		3

be expected to reflect the distribution of probabilities given. In fact, the protein produced had the following amino acid ratios:

Phenylalanine	100
Leucine	37
Valine	36
Cysteine	35
Tryptophan	14
Glycine	12

From this we can deduce that codons consisting of one guanine and two uracils (G + 2U) code for valine, leucine, and cysteine, although the specific sequences cannot be distinguished; similarly one uracil and two guanines (U + 2G) code for tryptophan, glycine, and perhaps one other. In all, it looked as though the precise sequence was of paramount importance, just as the Watson-Crick model had predicted. Many provisional assignments (such as those for G and U above) were soon obtained primarily by groups working with Nirenberg or with Severo Ochoa.

Specific code words were finally deciphered by means of two kinds of experiments. The first entailed making what might be thought of as "mini" mRNAs, each only three nucleotides in length. These, of course, are too short to promote translation into protein, but what makes them very interesting and useful is that they stimulate the binding of aminoacyl-tRNAs (aa-tRNAs) to ribosomes in a kind of abortive attempt at translation. One can make specific mini mRNAs and ask, *What* aminoacyl-tRNA will they bind to ribosomes?

For example, the G + 2U dilemma we discussed above can be resolved by making the following mini mRNAs:

GUU stimulates binding of valyl $tRNA_{Val}$

UUG stimulates binding of leucyl $tRNA_{Leu}$

UGU stimulates binding of cysteinyl $tRNA_{Cys}$

Analogous minis provided a virtually complete cracking of all the $4^3 = 64$ possible codons.

The second kind of experiment that was useful in deciphering the code will be presented as a problem at the end of the chapter; in solving it, you can pretend that you are H. Gobind Khorana, the worker who received a Nobel Prize for directing the experiments.

The code dictionary of sixty-four words is given in Figure 11-11. Inspection of this figure (and inspect it you should, for long periods of time in which you ponder the miracle of molecular genetics) reveals several points that require further explanation.

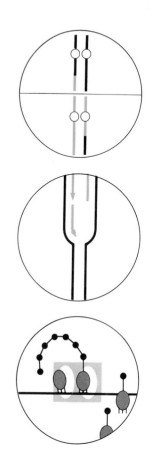

Second letter

		U	C	A	G	
First letter	U	UUU ⎱ Phe UUC ⎰ UUA ⎱ Leu UUG ⎰	UCU ⎱ UCC ⎱ Ser UCA ⎰ UCG ⎰	UAU ⎱ Tyr UAC ⎰ UAA Stop UAG Stop	UGU ⎱ Cys UGC ⎰ UGA Stop UGG Trp	U C A G	Third letter
	C	CUU ⎱ CUC ⎱ Leu CUA ⎰ CUG ⎰	CCU ⎱ CCC ⎱ Pro CCA ⎰ CCG ⎰	CAU ⎱ His CAC ⎰ CAA ⎱ Gln CAG ⎰	CGU ⎱ CGC ⎱ Arg CGA ⎰ CGG ⎰	U C A G	
	A	AUU ⎱ AUC ⎰ Ile AUA ⎰ AUG Met	ACU ⎱ ACC ⎱ Thr ACA ⎰ ACG ⎰	AAU ⎱ Asn AAC ⎰ AAA ⎱ Lys AAG ⎰	AGU ⎱ Ser AGC ⎰ AGA ⎱ Arg AGG ⎰	U C A G	
	G	GUU ⎱ GUC ⎱ Val GUA ⎰ GUG ⎰	GCU ⎱ GCC ⎱ Ala GCA ⎰ GCG ⎰	GAU ⎱ Asp GAC ⎰ GAA ⎱ Glu GAG ⎰	GGU ⎱ GGC ⎱ Gly GGA ⎰ GGG ⎰	U C A G	

FIGURE 11-11

First, the number of codons for different amino acids varies from one (tryptophan—UGG) to as many as six (serine—UCU, UCA, UCG, UCC, AGU, AGC). Why? The answer is complex but not difficult, and can be divided into two parts:

1. Certain amino acids can be brought to the ribosome by several *alternative* tRNA types (species), having different anticodons, whereas certain other amino acids are brought to the ribosome by only one tRNA.
2. Certain tRNA species can bring their specific amino acids in response to several codons, not just one, by a loose kind of base pairing at one end of the codon and anticodon. This sloppy pairing is called *wobble*.

We had better discuss wobble first, and it will lead us into a discussion of the various species of tRNA. Wobble is caused by a third nucleotide of the anticodon (5′ end) that is not quite aligned. This out-of-line nucleotide can sometimes form hydrogen bonds not only with its normal complementary nucleotide in the third position of the codon, but also with a different nucleotide in that position. There are certain "wobble rules," established by Crick, that dictate which nucleotides can and which cannot form new hydrogen-bonded associations through wobble.

5′ anticodon	3′ codon
G	U or C
C	G only
A	U only
U	A or G

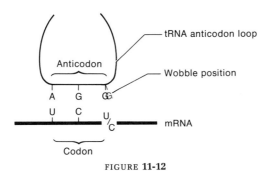

FIGURE **11-12**

Figure 11-12 shows the possible codons that one tRNA serine species can recognize. As seen from the wobble rules, G can pair with U or C. Let's list all the codons for serine and show how different tRNAs can service these codons, for this will be a good example of the various kinds of tRNA species possible.

Codon	tRNA	Anticodon
UCU UCC	$tRNA_{Ser_1}$	AGG + wobble
UCA UCG	$tRNA_{Ser_2}$	AGU + wobble
AGU AGC	$tRNA_{Ser_3}$	UCG + wobble

Sometimes there can be an additional tRNA species, which we will represent as $tRNA_{Ser_4}$; it has an anticodon identical with either 1, 2, or 3, but differs from these three in its nucleotide sequence elsewhere in the tRNA molecule. All these tRNAs are called *isoaccepting tRNAs* because they accept the same amino acid, but they are probably all transcribed from different tRNA genes.

The second point is that some codons do not specify an amino acid at all. They are labelled *stop* and can be thought of as analogous to punctuation marks—a comma or a period—that terminate a phrase of sentence. One of the first indications of the existence of stop codons came from an analysis by Sidney Brenner of certain mutants (m_1-m_6) in a single cistron that controls the head protein in the T4 phage. These mutants had two things in common: First, each mutant produced a head protein polypeptide that was shorter than the wild type. Second, the mutant phenotype of these mutants was suppressed (i.e., made wild type) in the presence of another mutation called a suppressor (*su*) at a separate chromosomal location. In the suppressed mutants, the normal length head protein was restored (Figure 11-13). The ends of the shortened proteins were examined and compared with wild-type protein, and the next amino acid that *would* have been inserted in a wild type was recorded for all six mutants. They were glutamine, lysine, glutamic acid, tyrosine, tryptophan, and serine. Initially, there is no apparent similarity between them, but Brenner brilliantly deduced

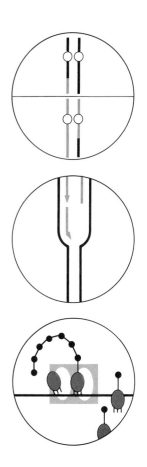

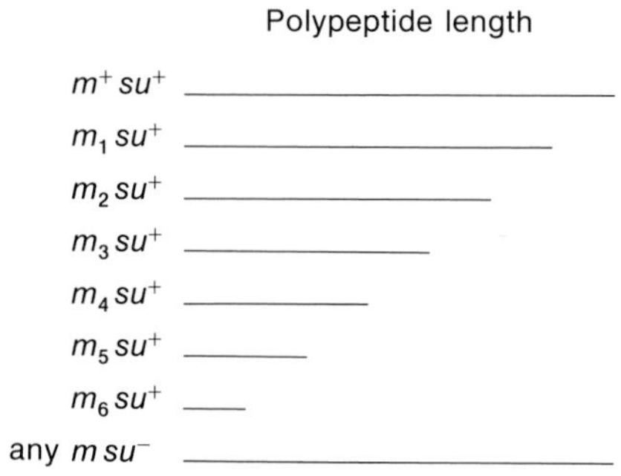

Polypeptide length

$m^+ su^+$ _____

$m_1 su^+$ _____

$m_2 su^+$ _____

$m_3 su^+$ _____

$m_4 su^+$ _____

$m_5 su^+$ _____

$m_6 su^+$ ____

any $m\ su^-$ _____

FIGURE **11-13**

that the codons for each of these amino acids were related in that all of them could mutate to the codon UAG by a single change in a DNA nucleotide pair. It was postulated therefore that UAG is a stop, or termination, codon that signals to the translation mechanism that the protein is now complete. UAG was the first stop codon deciphered and it is called the *amber* codon. Mutants that are defective because of the presence of an amber codon are called amber mutants, and their suppressors are amber suppressors. UGA (*opal*) and UAA (*ochre*) are also stop codons, and also have suppressors. Stop codons are often called nonsense codons because they designate no amino acid, but their sense is real enough and this is a misnomer that unfortunately persists. (*Missense* mutations are those in which a single amino acid is changed in the translated polypeptide). Not surprisingly, stop codons do not act as mini mRNAs in binding aa-tRNA to ribosomes.

MESSAGE

A missense mutation changes a codon so that it stands for a different amino acid. A nonsense mutation changes a codon so that it means stop to the translation system.

It is interesting to consider the nonsense suppressor mutations, which are now known to be mutations in the anticodon loop of specific tRNAs that allow recognition of a nonsense codon in mRNA. This results in the insertion of an amino acid at that point, thereby allowing continuation of translation past that triplet. An example, for m_1 in the T4 head protein analysis, is given in Figure 11-14.

The $tRNA_{Tyr}$ thus suppresses by inserting tyrosine instead of glutamine at that point of the polypeptide. Obviously, the insertion of tyrosine is not the critical factor that restores protein activity (probably a lot of amino acids would suffice); the important thing is that an amino acid *is* inserted and the translation apparatus trundles by and completes the rest of the protein.

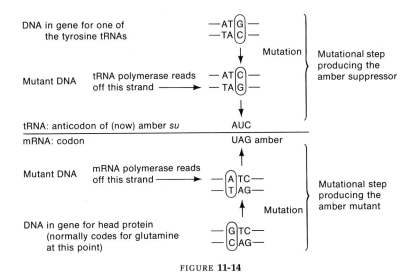

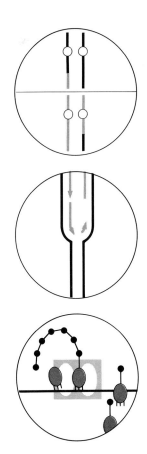

DNA in gene for one of the tyrosine tRNAs — AT⟨G⟩ —
— TA⟨C⟩ —

Mutation

Mutational step producing the amber suppressor

Mutant DNA   tRNA polymerase reads off this strand ⟶ — AT⟨C⟩ —
— TA⟨G⟩ —

tRNA: anticodon of (now) amber *su*   AUC

mRNA: codon   UAG amber

Mutant DNA   mRNA polymerase reads off this strand ⟶ —⟨A⟩TC—
—⟨T⟩AG—

Mutation

Mutational step producing the amber mutant

DNA in gene for head protein (normally codes for glutamine at this point) —⟨G⟩TC—
—⟨C⟩AG—

FIGURE **11-14**

Two questions are often asked at this point:

1. What is doing the job of putting tyrosine where tyrosine should be when the gene for tyrosine tRNA becomes a nonsense suppressor? Remember that there are often several tRNA forms (and hence genes) and that two of these may have the same anticodon loop. The "other" (one or more) wild-type tyrosine tRNA gene takes over when a suppressor locus is made out of the first.

2. What happens to normal stop signals, which indicate the ends of normal proteins, when nonsense suppressors are present? There is evidence that the normal ends of proteins are signalled by two different consecutive stop codons so that one suppressor could not gum up the works. It is not known how widespread this situation is.

The tRNA$_{Tyr}$ nonsense suppressor was only an example. Try to invent another nonsense suppressor that will suppress UAG. Also, try to figure out a suppressor that will suppress amber *and* ochre mutants.

The third point regarding our dictionary of code words is that there is one, AUG, that is thought to be the "capital letter" for a gene, or the "start" word. It is also the codon for the amino acid methionine. In *E. coli* and in some other organisms, the first amino acid in any newly synthesized polypeptide is always methionine. It does not get put in by tRNA$_{Met}$ however, but by initiator tRNA called tRNA$_{N-f-Met}$. This initiator tRNA has the normal methionine anticodon, but puts in *N*-formylmethionine rather than methionine (Figure 11-15). The formyl group apparently mimics an amino acid chain; at least the ribosome recognizes it as such and shifts one codon to amino acid position 2. Methionine won't cause this shift: if Met-tRNA$_{Met}$ inserts, nothing happens. The system has to wait until Met-tRNA$_{Met}$ diffuses away and an *N*-f-Met-tRNA$_{N-f-Met}$ chances

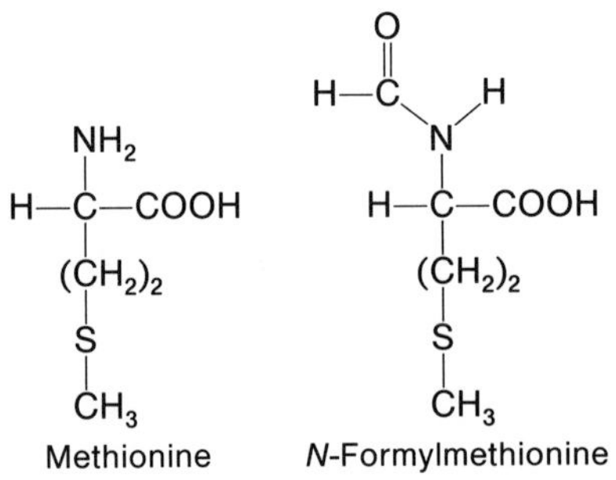

Methionine:
$$NH_2$$
$$H-C-COOH$$
$$(CH_2)_2$$
$$S$$
$$CH_3$$

N-Formylmethionine:
$$O$$
$$H-C \diagdown \quad H$$
$$N$$
$$H-C-COOH$$
$$(CH_2)_2$$
$$S$$
$$CH_3$$

Methionine   N-Formylmethionine

FIGURE 11-15

by to get things going. Similarly, when AUG occurs in the middle of a protein, N-f-Met-tRNA$_{N\text{-f-Met}}$ cannot form a peptide bond with the growing chain; so Met-tRNA$_{Met}$ has to be awaited.

Numerous problems remain. What initiates and regulates replication, transcription, and translation? How does the RNA get from the nucleus to the cytoplasm? How is the ribosome structured and how does it actually mediate translation? Where are the tRNAs activated and how do they actually swing into place in the ribosome? Certainly there is still lots of work to do in a field that has moved at lightning speed in the past fifteen years.

DNA Engineering

It might be instructive to cite some interesting feats of DNA tinkering at this point. Jonathan Beckwith and his group have intensively studied the *lac* locus in *E. coli* using a special episome called F*lac*, which carries the fertility factor and *lac*$^+$. A temperature-sensitive derivative, F$_{ts}$*lac*, was recovered, which is lost at high temperatures because of its inability to replicate autonomously. If F$_{ts}$*lac* is transferred to F$^-$*lac*$^-$ cells that are then grown at high temperatures, some *lac*$^+$ colonies grow. They are found to have F$_{ts}$*lac* integrated into the genome at the lac region. If F$_{ts}$*lac* is transferred to an F$^-$ carrying a deletion for the *lac* locus, *lac*$^+$ colonies are again found on plating at high temperatures, but at a lower frequency. In this case, F$_{ts}$*lac* is found to be inserted in a number of places throughout the genome.

Using this technique, two different *lac* insertion strains were recovered, one adjacent to the insertion site for λ, the other adjacent to the $\Phi 80$ insertion. The phages λ and $\Phi 80$ are related and share a considerable amount of homology. Beckwith's group recovered specialized transducing phages of λ and $\Phi 80$ that carry *lac*$^+$. Diagrammatically, they were:

Each line in the diagram represents a strand of DNA with symbolic phage genes represented by the letters. The 5′ end is represented by the arrow. The two strands differ in density and the "heavy" strand is indicated by the primed symbols. You can see that the *lac* locus in the two strains is inverted relative to the phage "markers."

By heat denaturation and cesium chloride density centrifugation, the light and heavy strands in each strain were separated. Thus, the heavy strands are:

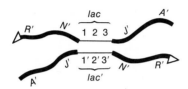

You can see that the two heavy strands have only one region, *lac*, in which there is complementarity that permits duplex formation. Thus, on mixing under annealing conditions, they obtained:

Using an enzyme that cleaves only single-stranded DNA, they were able to clip away the "whiskers" of unpaired strands on either side of *lac*. Under the electron microscope, the pre- and post-enzyme-treated complexes are shown in Figure 11-16. The picture on the right is a remarkable one: it is a picture of an isolated gene.

Using this technique it is possible, therefore, to prepare a test tube full of one kind of gene that is free of the rest of the genome—a very handy facility for the genetic engineer. Now let's look at another way of obtaining an isolated gene.

If we know the sequence of nucleotides in tRNA, then, by applying the rules of base complementarity, we also know the base sequence in the DNA coding for that tRNA. Indeed, knowing the genetic code, the gene for any protein whose amino acid sequence is known can be written down on paper. The big question is, Can such a gene *be made*, starting with just the nucleotides? Khorana succeeded in constructing a completely man-made gene from scratch in the following way. Nucleotides can be chemically linked in vitro. So if you want to make

A T G G C

you begin by taking A + T and making A T (this is simplified of course). But the reaction has to be stopped before all of the substrate is utilized or you might get ATA or ATT. At each step, then, some substrate is lost. When you are dealing with fifteen or sixteen nucleotide chains, there is very little substrate left

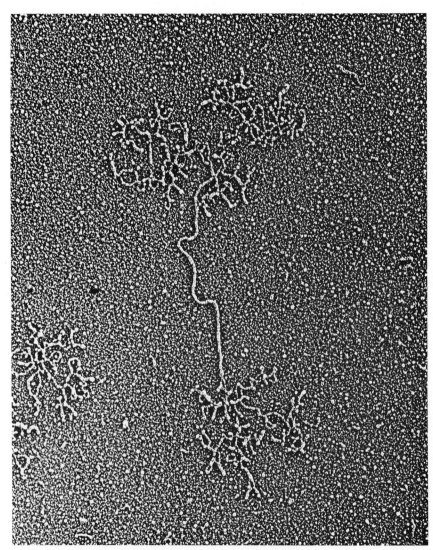

FIGURE **11-16**
Electron micrographs of *lac* gene: (above) with unpaired, adjacent, single-stranded
DNA "whiskers;" (facing page) isolated (whiskers enzymatically removed).

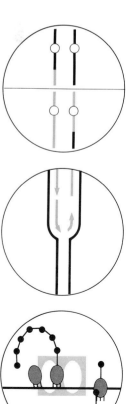

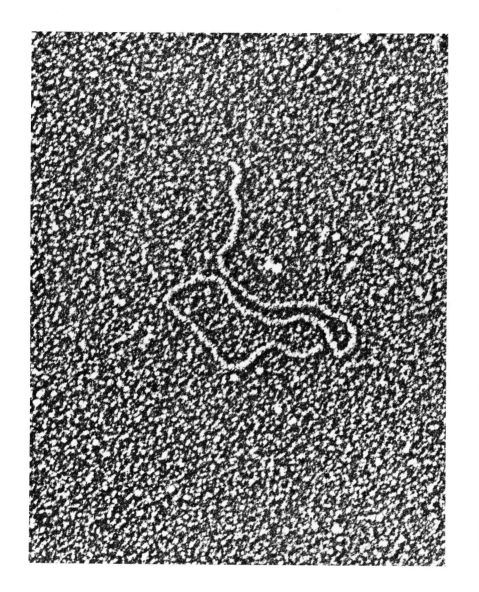

when the last nucleotide is added. Khorana made a brilliant move. Suppose we want to make a DNA sequence:

ATGGCTAATTCCGATA

TAGCGATTAAGGCTAT

He made short stretches of each strand with overlapping complementarity:

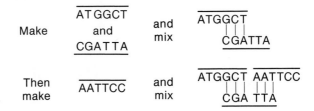

You can see that, in this way, the desired sequence can be patched together using short stretches. The completed "gene" can then be treated with ligase, which stitches together the ends of the patches:

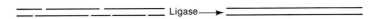

In this way, it should be possible to make very long stretches of DNA.

The question then is, Will such a molecule function (i.e., be transcribed) in vivo? How could this be tested? First of all, DNA could be made for a tRNA that differs from any known in *E. coli* (such as a yeast tRNA) or for a suppressor tRNA, which would allow recognition of its product (as a different tRNA or by suppression of a nonsense mutant). But how can one get it into a cell? Remember the sticky ends of the λ chromosome? By attaching a base sequence complementary to the λ cohesive end on the in vitro synthesized DNA, it could be attached to the λ chromosome and possibly "rescued" in a bacterial cell.

In a technique which is potentially very useful in medicine, Carl Merrill and his associates have succeeded in conferring wild-type function on a tissue culture of mutant *human* cells from patients with galactosemia, using a *bacterial* gene donor. *E. coli* has a gene that codes for an enzyme that is functionally similar to the one that is deficient in humans with galactosemia. Phage λ was used to pick up this gene (the *gal*$^+$ gene) from the bacterial donor cells and then "infect" the recipient diseased human cell culture. A persistent "cure" was achieved! If these methods can be applied to other diseases, the possibilities are very exciting. Of course, the "transduced" human cells would have to be put back into their owner where they could supply the missing bodily function. The new trait would not be passed on to the patient's offspring, but, as a medical cure, it would be very effective and permanent.

Very recently, another exciting technique for tampering with DNA has been developed. It utilizes the existence of bacterial enzymes that play a role in protecting bacterial cells from infection with foreign DNA. We cannot get into the elegant genetics and biochemistry that led to the understanding of the system; however, these enzymes are found to be endonucleases (that is, they cleave the DNA within the molecule in contrast to exonucleases, which start at the ends). They're called *restriction* enzymes and degrade foreign DNA in a very specific way, but do not attack the cell's DNA. The enzymes recognize specific nucleotide sequences that have a symmetry and cleave either in the center of the symmetry or on either side (Table 11-3). The asymmetrically displaced breaks

TABLE **11-3**

Restriction enzyme	Cleavage sites
Eco RI	↓ -C-T-T-A-A-G- -G-A-A-T-T-C- ↑
Eco RII	↓ -C-G-G-A-C-C-G- -G-C-C-T-G-G-C- ↑
Hin dII	↓ -C-A-Pu-Py-T-G- -G-T-Py-Pu-A-C- ↑
Hin dIII	↓ -T-T-C-G-A-A- -A-A-G-C-T-T- ↑
Hae	↓ -C-C-G-G - -G-G-C-C - ↑

leave molecules with single-stranded ends. For example, Eco-RI–treated DNA will be

AATTC	G
G	CTTAA

These free ends are complementary and will allow circularizing (Figure 11-17).

There is an episome called R that confers tetracycline resistance in bacteria. R DNA can be purified as a ring, cleaved with a restriction enzyme, and then annealed with any fragment of DNA treated with the same restriction enzyme

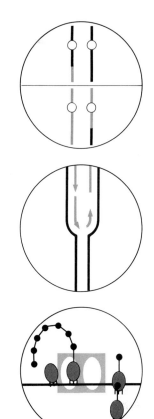

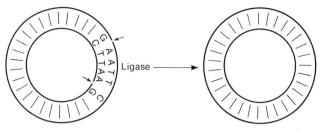

FIGURE **11-17**

(Figure 11-18). In the presence of tetracycline, bacteria that receive R DNA can be selected and thus also carry the extra DNA piece inserted into R. In this way a segment of DNA can be "cloned" in the bacteria. This affords the possibility of recovering specific parts of a eukaryotic genome and studying them in a much more genetically and biochemically manipulable system, the bacterial cell. At the same time, in a historically unprecedented step, a number of molecular geneticists have warned of the potential dangers of this kind of engineering (constructing highly infectious, indestructible bacteria, for example) and have called for stringent controls and limits on all experiments involving their construction. The dangers are clear: new parasitic genomes would be created to which the human (and other) species have had no evolutionary exposure and hence had no opportunity to develop resistance.

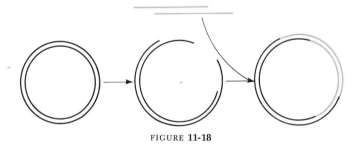

FIGURE **11-18**

In this chapter we have come a long way. We hope that you now have some understanding of the molecular nature of the genetic apparatus. The basic information-transfer, coding, and translation processes are virtually identical in all organisms that have been studied. All the different amino acid substitutions known in human hemoglobins are obtainable by single nucleotide pair substitutions using the genetic code derived from *E. coli!* This strongly suggests that the genetic code is universal. Also, the interchangeability of translational components between organisms points to a common mode of action through all life forms. Some minor differences exist, but not enough to distract from the wonderful unity of fundamental biological processes. Are these the *only* ones compatible with life on Earth? Or are they simply a reflection of a common ancestry? We may never know the answers to these questions.

MESSAGE

The processes of information storage, replication, transcription, and translation are fundamentally similar in all living systems. In showing this, molecular genetics has provided a powerful unifying force in biology. We now know the tricks that life uses to achieve persistent order in a randomizing universe.

A summary of the picture we have developed in the preceding chapter and this one of the relationship between DNA and protein is shown in Figure 11-19 (on the next two pages).

Now it is time to backtrack and look at some familiar concepts in light of what we know of the molecular level.

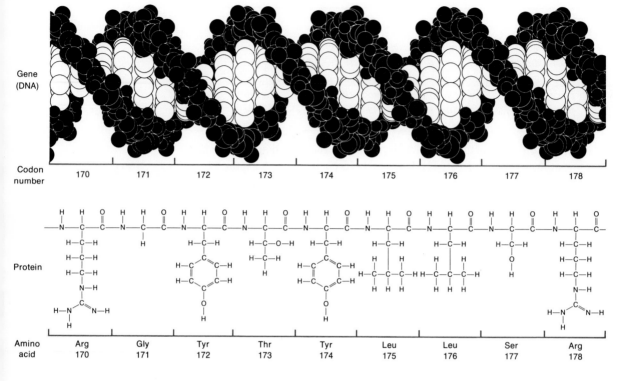

Gene (DNA)

Codon number: 170 171 172 173 174 175 176 177 178

Protein

Amino acid: Arg 170, Gly 171, Tyr 172, Thr 173, Tyr 174, Leu 175, Leu 176, Ser 177, Arg 178

FIGURE 11-19

Problems

1. Complete the following table, using the code dictionary on page 300:

C					T	G	A				DNA double helix
	C	A			U						mRNA transcribed
								G	C	A	Appropriate tRNA anticodon
		Trp									Amino acids incorporated into protein

Assume that reading is from left to right and that the columns represent transcriptional and translational alignments.

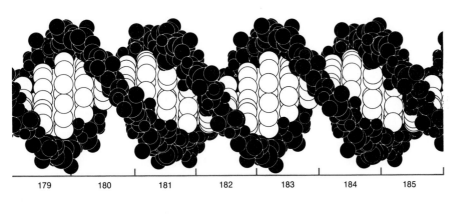

| 179 | 180 | 181 | 182 | 183 | 184 | 185 |

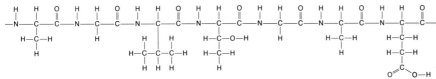

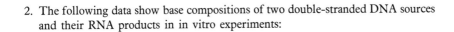

| Ala | Gly | Val | Thr | Gly | Ala | Glu |
| 179 | 180 | 181 | 182 | 183 | 184 | 185 |

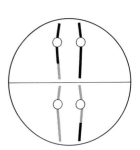

2. The following data show base compositions of two double-stranded DNA sources and their RNA products in in vitro experiments:

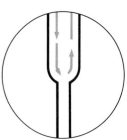

Species	DNA base ratio	RNA base ratios	
	A + T/G + C	A + U/G + C	A· + G/U + C
B. subtilis	1.36	1.30	1.02
E. coli	1.00	0.98	0.80

a. Can you distinguish from these data whether the RNA of these species is copied from one or both strands of the DNA?

b. Explain how you can tell whether the RNA itself is single- or double-stranded.[1]

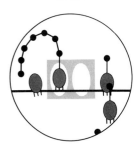

[1] Problem 2 is reprinted with permission of Macmillan Publishing Co., Inc. from *Genetics* by M. Strickberger. Copyright ©, Monroe W. Strickberger, 1968.

3. Suppose that you have synthesized 3 different messenger RNAs with bases incorporated at random in the following ratios:

   a. 1U:5C
   b. 1A:1C:4U
   c. 1A:1C:1G:1U

   In an in vitro protein-synthesizing system, indicate the proportions and types of amino acids that will be incorporated into proteins when each messenger is tested. (The codons are given on page 300.)

4. If a polyribonucleotide contained equal amounts of randomly positioned adenine and uracil bases, what proportion of triplets would code for:

   a. Phenylalanine?
   b. Isoleucine?
   c. Leucine?
   d. Tyrosine?

5. The comparison of physical distances with map distances is facilitated by our knowledge of the genetic code; the code permits us to "translate" numbers of amino acids in a protein into numbers of nucleotides in a corresponding stretch of DNA.

   a. Consider two mutant forms of a particular enzyme in *Neurospora*. The two proteins are known to differ at only two positions in their polypeptide chains. The two positions are separated by forty other amino acids. Crosses of the two mutants regularly give 1.0×10^{-5} recombinants between them. Approximately how many nucleotides in *Neurospora* lie between a pair of markers giving 10^{-5} recombinants?

   b. Consider two other mutant forms of the enzyme. Each has the normal (wild-type) number of amino acids, but they differ from each other and from the wild-type enzyme at amino acid 68 in the polypeptide chain. The wild-type protein has arginine at position 68, whereas the mutants have glutamine and serine, respectively. When mutants 1 and 2 were crossed, six kinds of offspring differing in the amino acid at position 68 were observed. Two of these were parental types in which glutamine and serine were present at position 68 and one had arginine like the wild type. Three novel types—having lysine, asparagine and histidine, respectively—made up the remainder. The types and frequencies of each class produced in the cross are tabulated here:

Amino acid at position 68	Frequency among offspring of the cross
Gln	~ 0.50
Ser	~ 0.50
Arg	4.0×10^{-7}
His	2.0×10^{-7}
Asn	1.0×10^{-11}
Lys	2.0×10^{-7}

Deduce the nucleotide sequences at codon number 68 for each of the six types.

6. One of the techniques that was used to decipher the genetic code was to synthesize polypeptides in vitro, using synthetic mRNA with various repeating base sequences—for example, $(AGA)_n$, which could be spelled out as AGAAGAAGAAGAAGA . . . , and so on. It was found that sometimes the synthesized polypeptide contained just one amino acid (homopolymer) and sometimes more than one (heteropolymer), depending on the repeating sequence. Furthermore, sometimes different polypeptides were made from the same synthetic mRNA, suggesting that the initiation of protein synthesis in the in vitro system does not always start on the end nucleotide of the messenger. For example, from $(AGA)_n$, three polypeptides may have been made: $(aa_1\text{-}aa_1)$, $(aa_2\text{-}aa_2)$, and $(aa_3\text{-}aa_3)$, which is shorthand for aa_1 homopolymer, aa_2 homopolymer, and aa_3 homopolymer. These probably correspond to a reading AGA, AGA, AGA, AGA, . . . , etcetera; GAA, GAA, GAA, GAA, . . . , etcetera; and AAG, AAG, AAG, AAG, . . . , etcetera. The actual results in the experiment done by Khorana were:

Synthetic mRNA	Polypeptide or polypeptides synthesized
$(UC)_n$	(Ser-Leu)
$(UG)_n$	(Val-Cys)
$(AC)_n$	(Thr-His)
$(AG)_n$	(Arg-Glu)
$(UUC)_n$	(Ser-Ser) and (Leu-Leu) and (Phe-Phe)
$(UUG)_n$	(Leu-Leu) and (Val-Val) and (Cys-Cys)
$(AAG)_n$	(Arg-Arg) and (Lys-Lys) and (Glu-Glu)
$(CAA)_n$	(Thr-Thr) and (Asn-Asn) and (Gln-Gln)
$(UAC)_n$	(Thr-Thr) and (Leu-Leu) and (Tyr-Tyr)
$(AUC)_n$	(Ile-Ile) and (Ser-Ser) and (His-His)
$(GUA)_n$	(Ser-Ser) and (Val-Val)
$(GAU)_n$	(Asp-Asp) and (Met-Met)
$(UAUC)_n$	(Tyr-Leu-Ser-Ile)
$(UUAC)_n$	(Leu-Leu-Thr-Tyr)
$(GAUA)_n$	None
$(GUAA)_n$	None

Note: The order in which the polypeptides or amino acids are given is not significant except for $(UAUC)_n$ and $(UUAC)_n$.

a. Why do $(GUA)_n$ and $(GAU)_n$ code for only two homopolypeptides?
b. Why do $(GAUA)_n$ and $(GUAA)_n$ fail to stimulate synthesis?
c. Assign an amino acid to each triplet in the following list. Bear in mind that there are often several codons for each amino acid, that the first two letters in a codon are usually the important ones, but that the same first two letters

---

[2]Problem 5 is from *The Mechanics of Inheritance*, 2d ed., by Franklin W. Stahl, © 1969, p. 192. Reprinted by permission of Prentice-Hall, Inc. Englewood Cliffs, New Jersey.

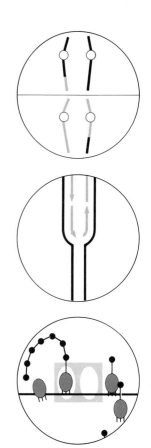

can sometimes mean a different amino acid, depending on the third letter. Also remember that some very different looking codons can sometimes code for the same amino acid. Try it without consulting the dictionary of code words.[3]

AUG, GUG, GUU, GUA, UGU, CAC, ACA, GAU, UUC, CUC, CUU, CUA, UCU, AGU, UUG, UUA, AUC, UAU, UAC, ACU, AAG, AAC, CAA, AGA, GAG, GAA, UAG, UGA

(To solve this problem requires both logic and trial and error. Don't be disheartened: Khorana received a Nobel Prize for doing it. Good luck!)

7. Suppose that you have a gene in *E. coli* that specifies a protein, part of whose sequence is:

Ala-Pro-Trp-Ser-Glu-Lys-Cys-His

You recover a series of mutants in this gene that lose enzyme activity. On isolation of the mutant products, you find the following sequences:

Mutant 1	Ala-Pro-Trp-Arg-Glu-Lys-Cys-His
Mutant 2	Ala-Pro
Mutant 3	Ala-Pro-Gly-Val-Lys-Asn-Cys-His
Mutant 4	Ala-Pro-Trp-Phe-Phe-Thr-Cys-His

What is the molecular basis for each mutation? What is the DNA sequence specifying this part of the protein?

8. The two strands of phage λ differ from each other in their G-C content. This property allows them to be separated in an alkaline (which denatures the double helix) cesium chloride gradient. When λ-induced RNA is isolated from infected cells, it is found to form DNA-RNA hybrids with both strands of λ DNA. What does this tell you? Formulate some testable predictions.

9. Suppressors of frameshift mutations are now known. Propose a mechanism for their action.

10. If you extract the DNA of the coliphage φX174, you will find that its composition is 25% A, 33% T, 24% G, and 18% C. Does this make sense in terms of Chargaff's rules? How would you interpret this? How might such a phage replicate its DNA?

11. The genetic material of some cancer-causing viruses is RNA instead of DNA. An infected cell produces an enzyme called *reverse transcriptase*, which makes DNA using the RNA as template. If you had some reverse transcriptase in a jar, you could theoretically *make* any gene in any organism. How would you go about this God-like task?

12. A single nucleotide addition and a single nucleotide deletion approximately fifteen sites apart in the DNA caused a protein change in sequence form

Lys-Ser-Pro-Ser-Leu-Asn-Ala-Ala-Lys

to

Lys-Val-His-His-Leu-Met-Ala-Ala-Lys

---

[3] Problem 6 is from *Genetics: Questions and Problems* by J. Kuspira and G. W. Walker. McGraw-Hill, 1973.

a. What are the old and new mRNA nucleotide sequences? (Use the code dictionary on page 300.)

b. Which nucleotide has been added and which deleted?[4]

13. Eukaryotic cells can be lysed without breaking down the nuclear membrane. This permits separation of cytoplasmic from nuclear material. In competition studies, cold nuclear RNA is found to compete with the binding of labeled cytoplasmic RNA to DNA. However, cold cytoplasmic RNA fails to interfere with the binding of a considerable amount of labelled nuclear RNA to DNA. What do these studies show? Can you suggest a possible biological interpretation of this phenomenon?

14. In in vitro protein synthesizing systems, the addition of a specific human mRNA to *E. coli* translational apparatus (ribosomes, tRNA, etc.) stimulates the synthesis of a protein very much like that specified by the mRNA. What does this show?

15. It is possible to separate highly redundant DNA sequences from unique ones because of their different rates of renaturation. Knowing the techniques of restriction enzyme digestion, and denaturation mapping, suggest ways to determine the order in which such sequences are arranged.

---

[4] Problem 12 is from *Theory and Problems of Genetics* by W. D. Stansfield. McGraw-Hill, 1969.

12 / Mechanisms of Genetic Change

*(Or: Mutation and crossing over
revisited at the molecular level.)*

We have been using the concepts of mutation and intrachromosomal recombination (crossing over) extensively in this book without really coming to grips with the precise events behind these processes at the molecular level. It is, in fact, an interesting historical paradox that mutation and crossing over have proven to be two of the most cryptic of genetic mechanisms. Geneticists used the two phenomena as tools for many years without knowing anything about their underlying mechanisms. It is only recently that some glimmer of understanding of the processes at the molecular level has been achieved, although this understanding is very limited, even today. In this chapter we will discuss some molecular mechanisms of mutation and crossing over. The models described should not be regarded as the final truth but as frameworks on which to hang future findings.

Mutation and recombination, as the main generators of change, can be thought of as the "mainsprings" of evolution. In Chapter 14, we will discuss the role of the selection of mutations and see its importance in the evolution that has occurred on Earth, but, for the present, the point must be made that evolution will occur *without* the aid of selection. Any system that has a built-in mechanism for change will evolve. There are many definitions of evolution, but it is usually associated with an increase in the diversity of forms. In other words, it is basically a *divergent* process. This divergent nature springs from the primary property of change: once a change has occurred, there are more potential

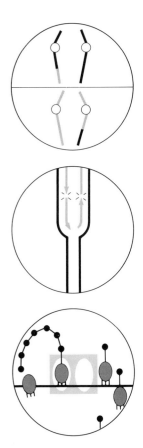

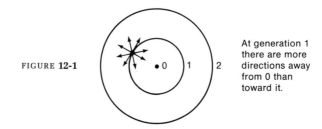

FIGURE 12-1

At generation 1
there are more
directions away
from 0 than
toward it.

directions of further change away from the origin than toward it. Figure 12-1 illustrates the principle, which incidentally is the same one that causes gases to diffuse outward. It is for this reason, also, that evolution is considered irreversible. To be otherwise would be like asking all the gas molecules in a room to assemble in a group in the corner. It just won't happen. So we will be discussing the mainsprings of evolution.

As we mentioned in Chapter 7, in general, reverse mutation rates are lower than forward mutation rates. This is because the normal wild-type gene can be mutated in a variety of ways to produce an inactive protein. However, usually mutant genes can be mutated in only one (or a very few) specific ways to produce a protein with wild-type function again. The last statement needs to be partly qualified because it is now known that many revertants are not genuine reversals of the original event. Two types of observations are important:

1. Several different amino acids can fulfill the role of the amino acid found in the wild-type protein at many polypeptide locations. Reversion can be to one of these alternatives rather than to the precise original (we discussed this in Chapter 9).

2. In some systems, second-site reversions are very common. Thus, an altered amino acid further down the polypeptide can often restore function at the first site, which remains mutant. This is a form of suppression called *intragenic supression by second-site change*.

> Wild-type sequence $\ldots aa_1 \ldots\ldots\ldots\ldots\ldots aa_2 \ldots$
>
> Mutant sequence $\ldots aa_3 \ldots\ldots\ldots\ldots\ldots aa_2 \ldots$
>
> Suppression sequence $\ldots aa_3 \ldots\ldots\ldots\ldots\ldots aa_4 \ldots$

We have already encountered intragenic suppression in another form in the discussion of restoration of the reading frame by two sign mutants.

MESSAGE

Reversion can be due to
1. Replacement of the original amino acid.
2. Replacement of a functionally similar amino acid.
3. Replacement of an amino acid at a second site.

Despite these qualifications, however, it is undoubtedly true that there are more polypeptide conformations and sequences that are *incompatible* with function than are compatible, and this inevitably gives rise to an excess of forward over reverse mutations.

The Causes of Mutation

Mutations can arise either spontaneously or by induction. Spontaneous mutations occur at quite low but constant rates in individuals. Induced mutations are produced as a result of the treatment of organisms with mutagenic agents or *mutagens* and are usually recovered at much higher frequencies.

SPONTANEOUS MUTATION

Once believed to be caused by cosmic rays or other natural mutagenic agents in the environment, spontaneous mutation is now believed to be attributable mainly to mistakes made by normal cellular enzymes. This change in thinking has come about because the mutations occur at too high a rate to be attributable to background radiation (Figure 12-2). How a mistake might be made can be seen by taking an example—say, a polymerase that on occasion "accepts" an illegitimate nucleotide pair (e.g., A-C), thereby generating an altered codon. Recombination enzymes or recombination itself by unequal exchange may also

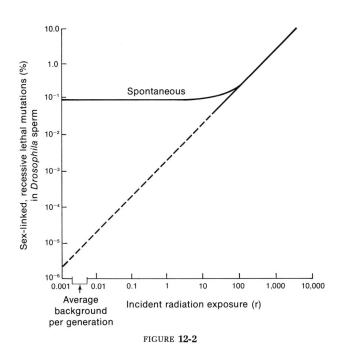

FIGURE **12-2**

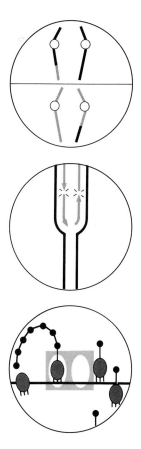

create heritable changes. There are several instances in which mutant genes increase the frequency of spontaneous mutation; they are called *mutator* alleles. The existence of mutator alleles further emphasizes the role of proteins or enzymes (possibly concerned with nucleic acid metabolism) in producing mutations.

Interestingly, if the spontaneous mutation frequency is dictated by the genome itself, it means that mutation frequency is capable of being molded (and probably has been) by natural selection. The spontaneous rates in present-day organisms probably represent a balance between the input of advantageous new alleles needed to retain evolutionary flexibility and the input of disadvantageous new alleles, which will have to be disposed of ultimately as dead or infertile individuals (genetic death).

INDUCED MUTATION

The principal mutagenic agents used by geneticists today are radiation and chemicals. The mode of action of some mutagens (e.g., ultraviolet light) is now relatively well understood.

Ultraviolet Light. Our present understanding of the way in which ultraviolet (UV) light causes mutation has arisen out of the elucidation of the cell's defenses against it. The whole story is a classic example of how genetic analysis has been used to dissect important biological functions at the subcellular level. We will follow only a skeletal form of the story, and try to convey an idea of the kinds of reasoning employed.

Ultraviolet light, as you know, is an efficient germicidal agent; in other words, it kills bacteria. Bacteria irradiated by ultraviolet light are killed in direct proportion to the radiation dose *so long as the cells are kept in the dark.* If, after irradiation, the cells are exposed to visible light, a large proportion of them survive! This is evidence of a light-induced system for repairing radiation-induced damage. This also says that the UV-induced lesion is not itself lethal.

That the primary target of ultraviolet light is DNA was inferred from a demonstration that the most potent killing was obtained with light whose wavelength is 260 millimicrons, precisely the wavelength most strongly absorbed by DNA. What happens to DNA after UV treatment? Chemical analysis of UV-irradiated DNA revealed a variety of modifications, the most prominent one being covalent linkage between adjacent pyrimidine nucleotides located on the same strand. The most common products are called *thymine dimers* and are usually represented by T͡T (Figure 12-3). As you may imagine, this T͡T structure produces a great bulge or distortion in the DNA helix. Apparently this bulge is recognized by a patrolling enzyme system, which then cleaves the dimer to the original single-base condition in the presence of light. Few, if any, mutations are recovered after this cleavage process, called "light repair"; it seems to do a very efficient clean-up job.

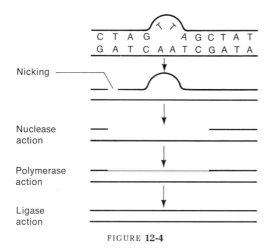

FIGURE **12-3**

However, some DNA repair can still go on in the dark, and it is only after dark repair that mutations are found. The existence of a dark-repair system was shown by the isolation of UV-sensitive mutants in *E. coli* that lack this dark-repair ability; we will call them *uvrA* mutants (standing for UV repair). In the dark, *uvrA*⁺ strains actually excise the entire dimer and some adjacent nucleotides, resynthesize a correct stretch of polymer, and finally close up the strand with ligase (Figure 12-4). *uvrA* mutants are defective in the nicking function

FIGURE **12-4**

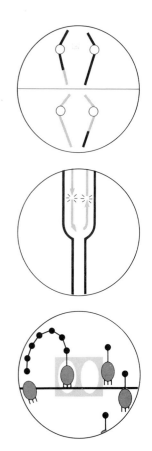

only. Their polymerase function is normal and is provided by pol1 (the enzyme utilized by Kornberg in his in vitro DNA syntheses). Thus the mutants are deficient in *excision repair,* a second UV-repair system.

Some mutations are produced during normal excision repair because the repair system is prone to make errors. The errors result in mutation probably through illegitimate nucleotide pairing (Figure 12-5). (*Xeroderma pigmentosum* is a recessive hereditary disease in humans and results in the formation of skin cancers on exposure to UV rays in sunlight. It is caused by a deficiency in one of the human excision-repair enzymes.)

In addition, in *E. coli,* mutants were found that resulted in the inability of an exogenote to recombine with an endogenote. They were called *recA* mutants.

FIGURE 12-5

Interestingly, *recA* mutants are also UV sensitive. Thus there seems to be a recombinationlike process associated with repair. Is this a third repair system? The answer is yes. It is confirmed by the following sorts of evidence: *uvrA* mutants still show some repair in the dark, so a third system is indicated. Also pol1 mutants have been isolated (mutants of the *polA* gene) and these are also capable of repair, again pointing to a third repair system. However, *polA recA* double mutants are a lethal combination in the dark suggesting that in these strains both repair systems have been knocked out. The third system is called recombination repair. (Incidentally, *polA* mutants are perfectly happy in a normal non-UV environment: they can replicate and undergo cell division. This suggests that the pol1 DNA polymerase is used only during excision-repair polymerase activities. It is now known in fact that there are other important DNA polymerases that carry out the major replication polymerase functions.) Most interestingly, in relation to mutation, in *recA* strains of *E. coli* it is difficult to recover UV-induced mutations at any locus in the genome. This implicates the *recombination-repair* system as the major contributor to mutation induction by ultraviolet light. Apparently, the recombination-repair system is very prone to errors, and the UV-induced damage is not repaired accurately. Precise ways in which the system errs have not been discovered.

The details of the process of recombination repair have been worked out chemically. The repair occurs after replication (it is sometimes called *post-replication repair*): the DNA replication system "runs over" dimers, leaving single-strand gaps. Reshuffling of the genomes by recombinationlike processes leads to whole strands that produce whole DNA double helices upon further replication. One possible scheme is shown in Figure 12-6.

MESSAGE

UV-induced mutations are produced not by the thymine dimers themselves, but by errors incurred during their repair in the dark.

X Rays. Although X rays were one of the first agents shown to be mutagenic, their mode of action is still largely obscure. Undoubtedly, free radicals and ions formed by the rays are involved, but the biological effects of X rays are so complex that this is probably not the whole story. X rays are renowned for their ability to produce chromosome breaks and hence gross

1. Light repair (not error prone)

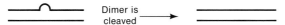

Dimer is
cleaved →

2. Excision repair (dark, error prone)

3. Postreplication (recombination) repair (dark, error prone)

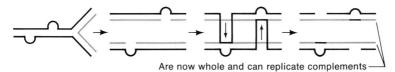

Are now whole and can replicate complements

FIGURE **12-6**

structural rearrangements. But they also produce point mutations (such as nucleotide pair substitutions) at high frequencies.

It is expected that the frequency of point mutations, m_1, should be proportional to the dose, d, whereas the frequency of two-break rearrangements, m_2 (such as deletions), should be proportional to the square of the dose. In these two situations we can write

$$m_1 = k_1 d$$
$$m_2 = k_2 d^2$$

where k_1 and k_2 are constants. Logarithmically,

$$\log m_1 = \log k_1 + \log d$$
$$\log m_2 = \log k_2 + 2 \log d$$

Thus, if m is plotted against d on a logarithmic scale, a slope of 1 or 2 is expected for these kinds of mutational events. A convincing demonstration of this has been shown in the production of *ad-3* mutants of *Neurospora* by X rays. The mutants can be shown by genetic analysis to be either point mutations within the *ad-3* region or deletions of part or all of the *ad-3* region (Figure 12-7). It can be seen that the slopes are 1 and 2 as expected. Often, however, two-break rearrangements in other systems show an exponent between 1 and 2, for example 1.5. This is due to the rejoining of broken ends, in some cases, to reconstitute the original chromosomal configurations. An exponent, and slope, of say 1.5 in an untested situation could also mean a mixture of one and two hit events.

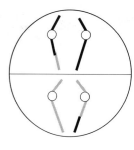

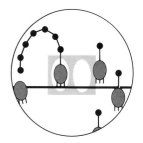

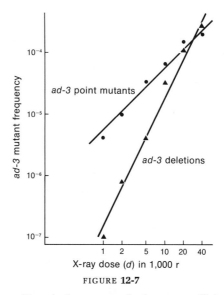

FIGURE **12-7**

Base Analogs. Chemical compounds that are sufficiently similar to the normal nitrogen bases of DNA that they occasionally get incorporated into DNA as if they were the right ones are called base analogs. Once in, however, their pairing properties are unlike the normal bases and they therefore can produce mutations by causing incorrect nucleotides to be inserted opposite them. On replication, an incorrect insertion ends up as a *nucleotide pair substitution,* as we shall soon see in specific examples. Such nucleotide pair substitutions are categorized as either *transitions,* in which a purine on a strand is replaced by the other purine (or a pyrimidine by the other pyrimidine), or *transversions,* in which a purine is replaced by a pyrimidine (or a pyrimidine by a purine).

$$
\begin{array}{ll}
\text{Transitions} & \begin{array}{l} \text{AT} \longrightarrow \text{GC} \\ \text{GC} \longrightarrow \text{AT} \end{array} \\
\text{Transversions} & \begin{array}{l} \text{AT} \longrightarrow \text{CG or TA} \\ \text{CG} \longrightarrow \text{AT or GC} \end{array}
\end{array}
$$

As an example, 5-bromouracil is an analog of thymine but instead of having a CH_3 group at the C_5 position, it has bromine. Although the hydrogen-bonding atoms are exactly the same as in thymine, the bromine atom has a significant effect on the distribution of electrons in the base so that the normal structure (the *keto* form) undergoes a relatively frequent *tautomeric shift* to the *enol* form (Figure 12-8). The enol form has the hydrogen-bonding properties of cytosine! Depending on when this enol form occurs, two kinds of transitions can be predicted (Figure 12-9). The compound 2-aminopurine acts in a similar manner. It is an analog of adenine that sometimes bonds with cytosine, and hence is expected to cause transitions.

Adenine … N—H⁺•••⁻O (Br) — 5 BU$_T$ keto form

Guanine … O⁻•••⁺H—O (Br) — 5 BU$_C$ enol form

FIGURE **12-8**

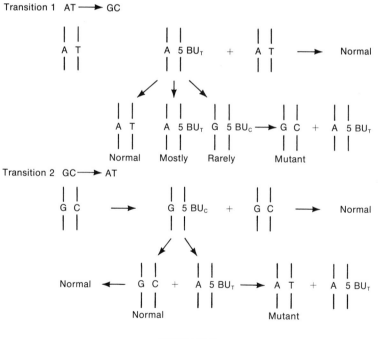

Transition 1   AT ⟶ GC

| | | |
A T   A 5 BU$_T$ +   A T   ⟶   Normal
| | | |

| | | | | |
A T   A 5 BU$_T$ G 5 BU$_C$ ⟶ G C   +   A 5 BU$_T$
| | | | | |
Normal   Mostly   Rarely   Mutant

Transition 2   GC ⟶ AT

| | | | | |
G C   ⟶   G 5 BU$_C$ +   G C   ⟶   Normal
| | | | | |

Normal ⟵ G C   +   A 5 BU$_T$ ⟶ A T   +   A 5 BU$_T$
Normal   Mutant

FIGURE **12-9**

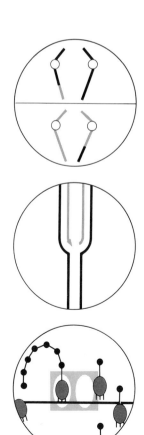

DNA Modifiers. DNA modifiers are agents that chemically react with DNA and change the hydrogen-bonding properties of the bases. Nitrous acid, for example, reacts with the C_6 amino groups of cytosine and adenine, replacing the amino groups (positive in hydrogen bonding) with oxygen (negative in hydrogen bonding). Consequently, the hydrogen-bonding properties of the modified base are changed and what was cytosine now bonds like thymine, and what was adenine now bonds like guanine. The transitions GC $\longrightarrow$ AT and AT $\longrightarrow$ GC are therefore predicted chemically.

Alkylating agents are potent mutagens. They are reactive compounds that can add an alkyl group (e.g., ethyl or methyl) at a variety of positions, thereby changing the hydrogen-bonding potential in a number of ways and producing transitions and transversions. Some commonly used ones are ethyl methane-sulfonate, methyl methanesulfonate, and nitrosoguanidine.

Hydroxylamine (HA) is a very handy mutagen because it appears to be very specific and produces only one kind of change; namely, the GC $\longrightarrow$ AT transition. The reason for the specificity is that HA reacts with only the cytosine molecule, probably right in the DNA. The C_6-bound amino nitrogen is hydroxylated and this causes it to bond like a thymine molecule.

The last DNA modifiers we shall discuss are the frameshift mutagens—for example, acridines such as proflavin and ICR-170. These compounds intercalate between the stacked bases at the heart of the DNA helix. In this position, they can cause either the insertion or the deletion of a single nucleotide, perhaps as shown in Figure 12-10.

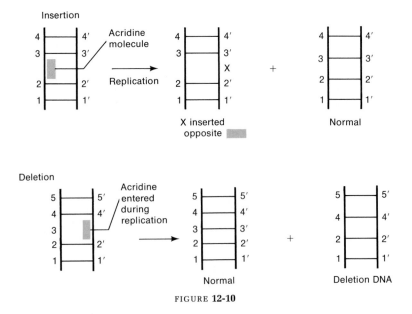

FIGURE **12-10**

Many of the actions of chemical mutagens we have described must still be considered speculative. From a knowledge of chemistry, reasonable predictions can be made about the ways in which specific compounds should affect DNA and cause mutation. The predictions can then be tested in two ways: first, by checking to see if the compound is mutagenic; and second, by checking the specificity of the mutagen by the properties of its induced mutations. For example, HA-induced mutations should *not* be revertible with HA. Similarly, only *some* transitions induced by nitrous acid should be revertible with HA, whereas all HA-induced mutations should be revertible with nitrous acid. In the same way, mutations induced by a supposed frameshift mutagen should not be revertible with transition- or transversion-inducing mutagens, and vice versa.

As a rule, these predictions do hold true and give geneticists enough confidence in the specificity of the postulated actions that the chemicals are used routinely to produce specific kinds of mutations. Probably no mutagen is completely specific, but the question will ultimately be answered only after an extensive investigation of mutant proteins produced at a locus by specific compounds. The relative potencies of some commonly used mutagens are given in Table 12-1, concerning *ad-3* mutations in *Neurospora*.

TABLE **12-1**

Mutagen	Exposure time (min)	Survival (%)	Number of *ad-3* mutants per 10^6 survivors
Control (spontaneous rate)		100	~0.4
X rays (2,000r/min)	18	16	259
UV (600 erg/mm²/min)	6	18	375
Nitrous acid (0.05 M)	160	23	128
Ethyl methanesulfonate (1%)	90	56	25
Amino purine (1–5 mg/ml)	During growth	100	3
Methyl methanesulfonate (20 mM)	300	26	350
Nitrosoguanidine* (25 μM)	240	65	1,500
ICR-170 acridine mustard (5 μg/ml)	480	28	2,287

*Also a potent carcinogen (cancer-causing agent).

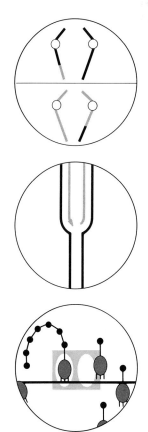

Mutant Detection in Drosophila. To give you an idea of how useful chemical mutagens are, we can follow an example in *Drosophila*. Until the mid-sixties, most drosophilists used radiation to induce mutations. At 4,000 r, perhaps from 5% to 6% of all X chromosomes of irradiated males carry a lethal. The problem is that, at higher doses, there is greater infertility of the treated flies. Furthermore, many of the induced mutations are associated with chromosome rearrangements, which can considerably complicate matters.

The alkylating agent EMS, in contrast, induces mutations that are almost exclusively point mutants. It is very easily administered by simply placing adult flies on a filter pad saturated with a mixture of sugar and EMS. Simple ingestion of EMS produces whopping numbers of mutations. For example, males fed 0.025 M EMS produce sperm carrying lethals on more than 70% of all X chromosomes and on virtually every second and third chromosome! At these levels of mutation induction, it becomes quite feasible to screen for many mutations at specific loci or for defects with unusual phenotypes.

It is possible to screen mutagenized chromosomes immediately in the F_1 generation if the region of interest is hemizygous. The X chromosome can be tested this way by crossing EMS-fed males with females carrying attached-X chromosomes and a Y ($\widehat{X}X/Y$). All F_1 males carry a mutagenized paternal X chromosome and each fly represents a different treated sperm. Thus, if individual F_1 males are crossed with $\widehat{X}X/Y$ females, each culture will represent a single cloned X. All F_1 zygotes carrying a sex-linked lethal will die but any newly induced visible mutant will be expressed. In this way, it has been possible to "saturate" regions for mutations and to select a wide range of behavioral and visible mutants (Figure 12-11). If the F_1 flies are reared at 22°C and then clones

$$\widehat{X}X/Y \,♀ \quad \times \quad X/Y \,♂ \text{ (EMS-treated)}$$

Sperm

		X*	Y*
Eggs	$\widehat{X}X$	$\widehat{X}X$ X* (dies)	$\widehat{X}X/Y*$ ♀
	Y	X*/Y ♂	YY* (dies)

* Denotes the sex chromosome exposed to the mutagen.

FIGURE **12-11**

of each established at 22°, 17°, and 29°C, heat- or cold-sensitive lethals can be readily detected and are found to constitute from 10% to 12% of all EMS-induced lethals. The utility of conditional mutant expression has already been discussed and, in *Drosophila*, many laboratories now routinely screen for the temperature sensitivity of mutants in screening tests.

A spin-off of the recovery of temperature-sensitive (ts) lethals has been their use for simplifying collecting procedures. One of the nuisances of any large-scale *Drosophila* experiment is separating males from females or collecting virgin females (because newly emerged males and females do not mate for 12 to 14 hours, the collection of all females within a 12-hour interval ensures their purity). This tiresome procedure can be eliminated by using ts lethals (represented as l^{ts}) to produce unisexual cultures at will. For example, $\widehat{X}X/Y$ females crossed with l^{ts}/Y males yields both sexes at permissive temperature. On shifting a culture to restrictive temperature, the l^{ts}/Y males die and only females hatch. Similarly, homozygosis of an l^{ts} in an $\widehat{X}X$ chromosome will produce only wild-type males at restrictive temperatures. For fun, you might try to incorporate ts recessive and dominant lethals into inversions or translocations to construct autosomal unisexual stocks. That's chromosome mechanics.

MESSAGE

The mutagen EMS has revolutionized the genetic versatility of Drosophila *by providing a potent method for the recovery of a wide range of point mutants.*

Crossing Over

A normal crossover or reciprocal intrachromosomal recombination event is really a miraculous process. Somehow the genetic material from one parental chromosome and the genetic material from the other parental chromosome are cut up and pasted together during each meiosis, and this is done with complete reciprocality. In other words, no genes are lost by one chromosome or gained by another: in fact, it is probably correct to say that no *nucleotides* are lost or gained either. How is this remarkable precision attained? The answer is that we do not know for sure, but many interesting phenomena are pertinent to this question and provide us with some important clues.

BREAKAGE AND REUNION OF DNA MOLECULES

Throughout our analysis of linkage, it was implicitly assumed that crossing over occurred by means of a process of breakage and reunion of chromatids. The evidence already presented against the copy-choice hypothesis is good *indirect* evidence in favor of breakage and reunion. Furthermore, there is good genetic and cytological evidence that crossing over occurs during prophase of meiosis rather than interphase when chromosomal DNA is replicating. There is a small amount of DNA synthesis during prophase but certainly chromosome replication is not associated with crossing over. But one of the first direct proofs that chromosomes (albeit viral chromosomes!) could break and rejoin came from experiments on λ done by Matthew Meselson and Jean Weigle.

They multiply infected *E. coli* with two strains of λ. One of the strains had the genetic markers *c* and *mi* at one end of the chromosome and this chromosome was "heavy" because the phages had been produced from cells growing in heavy isotopes of carbon (^{13}C) and nitrogen (^{15}N). The other strain was + + and had "light" DNA because it had been harvested from cells grown on the normal light isotopes ^{12}C and ^{14}N. The two DNAs (i.e., chromosomes) can be represented as:

The progeny phage released were spun in a cesium chloride density gradient. A wide band was obtained, indicating that the viruses ranged in density from heavy parental to light parental with a whole lot of intermediate densities.

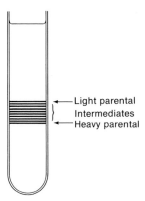

Very interestingly, some recombinant phages were recovered from very close to the heavy-parental position. They were of genotype $c +$ and must have arisen from an exchange event between the two loci.

They were, appropriately, mostly heavy because the c and mi loci are close to the right end. When the cross heavy $+ +$ × light $c\,mi$ was made, the heavy recombinants were found to be $+\,mi$. The mostly light recombinants were the reciprocal genetic types. There is only one way to explain these results: the recombination event must have occurred by the physical breakage and reunion of DNA.

Of course, it is unwise to extrapolate from viral to eukaryotic chromosomes, but at least we know that the breakage and reunion of DNAs is a chemical possibility.

Now let's examine the genetic information on the mechanism of crossing over that has been collected from eukaryotes. It is of a different kind. Much of the information on the mechanism of intrachromosomal recombination has come from bacteria and phage, especially at the chemical level. However, we will concentrate on eukaryotes not because these data are any more useful than from the prokaryotes, but because they are based more on genetic analysis and less on chemistry. We shall study what can be called the genetics of genetics! The clues are enumerated below.

Clue 1: Chromatid Conversion and Half-chromatid Conversion.   Much of the data on recombination in eukaryotes has come from fungi. There is one good reason for this: the ascus. All products of a single meiosis can be recovered and examined so that records of each meiosis can be kept in close tally in terms of the total genetic information being lost or gained. By being able to recover all four products of a specific meiosis, the confidence with which inferences can be made about the events occurring during that meiosis is immeasurably increased. The ascus represents a tight system of internally self-consistent controls.

In some asci, non-Mendelian allele ratios are detectable. Mendel would have predicted 4:4 segregations for all monohybrid crosses, but very rarely (depending on the fungus, 1/100 to 1/1,000 of all asci), a non-4:4 ratio is obtained. Figure 12-12 gives the most common aberrant ratios obtained from, for example, the cross $+ \times m$. It seems as though some genes in the cross have

+	+	m	+	+	+
+	+	m	+	+	+
+	+	m	+	+	+
+	+	m	+	m	m
m	+	m	+	m	+
m	+	m	m	m	m
m	m	+	m	m	m
m	m	+	m	m	m
4:4	6:2	2:6	5:3	3:5	3:1:1:3

(*Note:* 3:1:1:3 = aberrant 4:4.)

Normal	Chromatid conversion	Half-chromatid conversion

FIGURE **12-12**

been "converted" to the opposite allele, and the process became known as *gene conversion* and has, as an absolute requirement, *heterozygosity* for two different alleles. In some asci, it appears as though the whole chromatid in meiosis has converted (chromatid conversion), in others only half of the chromatid seems to have converted (half-chromatid conversion). In half-chromatid conversion, different members of a spore pair (which, remember, are produced by mitosis) have different genotypes! It should be emphasized, first, that alleles heterozygous at other loci in the same cross usually segregate 4:4 so these asci are not accidents of isolation, and, second, that the process of conversion is not mutation because it is directional: the allele converted always changes to the other specific allele of the locus, and this has been established by molecular studies on gene products of converted alleles. Figure 12-13 shows an example of chromatid conversion and of half-chromatid conversion.

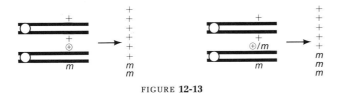

FIGURE **12-13**

Clue 2: Polarity. In genes for which accurate allele maps are available, it is possible to compare the conversion frequencies of alleles at various positions within the gene. It is nearly always found that sites closer to one end show higher frequencies than those further away from that end: in other words, there is a gradient or *polarity* of conversion.

Clue 3: The Association of Conversion with Crossing Over Between Flanking Markers. In heteroallelic crosses in which the locus under study is closely flanked by other genetically marked loci, it has been observed that the conversion event is very often (to keep the argument simple we will say 50% of the time) accompanied by an exchange in one of the flanking regions. This exchange is nearly always on the side of the allele that has converted. Furthermore, it nearly always involves the chromatid in which conversion has occurred. For example, in a locus m in which the left-hand alleles (as drawn) convert more often than those to the right, the following hypothetical case summarizes the association of conversion and crossing over. The cross is $a^+ m_2 b^+ \times a m_1 b$, where m_1 and m_2 are different alleles of the m locus, and a and b represent closely linked flanking markers, and may be diagrammed as follows:

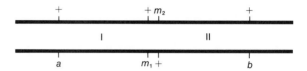

If we look at asci in which there has been conversion at the m_1 site (the most frequent kind of conversion in this locus), one-half of these asci will have a crossover in region I and one-half will have no crossover. In the minority of asci in which there has been conversion at the m_2 site, one-half will have a crossover in region II and one-half will have no crossover. Such events are detected in ascus genotypes like that shown in Figure 12-14, which can be interpreted as a

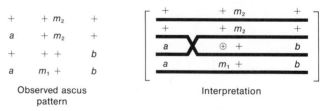

Observed ascus
pattern

Interpretation

FIGURE **12-14**

conversion of $m_1 \rightarrow +$, accompanied by a crossover in region I. (If this were a nutritional locus—for example, for arginine requirement—can you figure out how the genotypes $m_1 +$, $m_2 +$, and $m_1 m_2$ would be distinguished from each other?)

Clue 4: Double Conversion. In some asci, a single conversion event seems to include several sites at once: in a heteroallelic cross (like that shown in Figure 12-15), this is called double conversion.

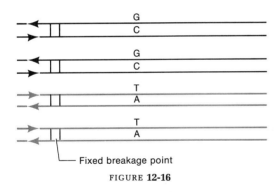

$$m_1 \times m_2$$

$m_1 +$	$m_1 +$	$m_1 +$
$+ +$	$m_1 +$	$+ m_2$
$+ m_2$	$m_1 +$	$+ m_2$
$+ m_2$	$+ m_2$	$+ m_2$
Single conversion of m_1	Double conversion of bottom genotype	Double conversion of top genotype

FIGURE **12-15**

A recombination model can be formulated that accounts for all these observations. It is complex because the phenomena it attempts to accommodate are also complex. It is called a *hybrid DNA model* and is typical of the kinds of recombination models that have been put forward both for eukaryotes and for prokaryotes. It may well be wrong, but it is simply a way of expressing the present state of knowledge.

We can set up a hypothetical cross, $+ \times m$, in which the $+$ site corresponds to a G-C nucleotide pair, and the m site corresponds to a transition mutant A-T as shown in Figure 12-16. The four chromatids are represented as four DNA

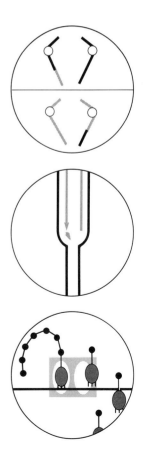

Fixed breakage point

FIGURE **12-16**

double helices: an impertinent assumption perhaps. The arrows represent the direction of antiparallelity ($5' \rightarrow 3'$ or $3' \rightarrow 5'$). Also represented is a *fixed breakage point*: perhaps a recognition site for an endonuclease enzyme (its significance will become clear). The fixed breakage points of two nonsister molecules now become the site of phosphodiester-bond breakage. The broken strands unravel away from the fixed breakage point for some distance (Figure 12-17). The unravelled strands are identical in type (indicated by the antiparallelity arrows) and can now rejoin at one another's breakage point by means

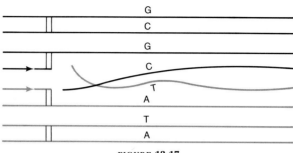

FIGURE **12-17**

of an enzyme, *ligase* (Figure 12-18). What we have now are two sets of hybrid DNA (the solid-and-shaded parts) containing illegitimate purine-pyrimidine pairs, G-T and C-A. We will return to these later.

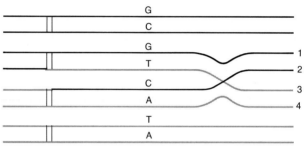

FIGURE **12-18**

The exchange point now resolves itself and we can assume that, 50% of the time, strands 2 and 3 break and rejoin (Figure 12-19) and that, 50% of the time,

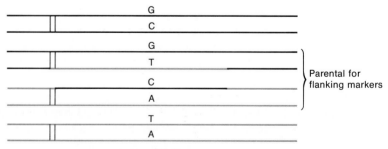

FIGURE **12-19**

1 and 4 break and rejoin (Figure 12-20). Thus we see that either a crossover or a noncrossover situation (with respect to outside markers) is produced, but both contain two regions of hybrid DNA.

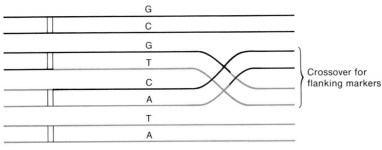

FIGURE **12-20**

Hybrid DNA is unstable: the mispaired nucleotides could produce a distortion in the DNA at these points that is recognized by a repair system much like the excision-repair system that removes thymine dimers produced by ultraviolet light. For example, the G-T pair could be repaired in one of two ways (Figure 12-21). Thus, the correction of hybrid DNA by such a system could result in

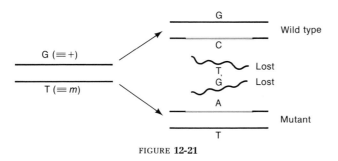

FIGURE **12-21**

either a wild-type or a mutant allele in a chromatid. Failure to correct the mismatched nucleotides would lead to a "hybrid" chromatid that, on replication, would generate two different daughter chromatids and hence the two different spore pair members found in half-chromatid conversion.

The various aberrant ratios can be produced by events similar to those in Table 12-2, which deals only with correction (and hence conversion) from $m/+ \rightarrow +$. The symbols m and $+$ are used for simplicity instead of the nucleotides used in Figures 12-15–12-21. Similar ratios can be developed for the correction of hybrid DNA to m.

Thus, the hybrid DNA model we have described accounts for most of the phenomena that we called "clues":

1. Aberrant ratios are accounted for by correction of the hybrid DNA.

2. Polarity is accounted for because the closer a monohybrid site is to the fixed breakage point, the more often it will be included in hybrid DNA.

3. The association of crossing over with gene conversion is explained by the resolution of the exchange point in two equally frequent ways. The ascus that

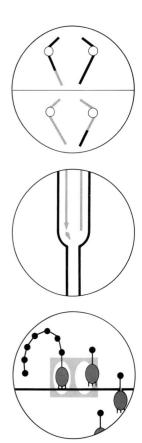

TABLE 12-2

DNA strands at start of meiosis	Strands at hybrid DNA stage	No correction	One hybrid DNA corrected to +	Both hybrid DNAs corrected to +
_m_	_m_	} 3 m	} 3 m	} 2 m
_m_	_m_			
_m_	_m_			
_m_	_+_	} 1 +		} 6 +
_+_	_m_	} 1 m	} 5 +	
_+_	_+_	} 3 +		
_+_	_+_			
_+_	_+_			
		Aberrant 4:4	5:3	6:2

we used to illustrate this clue would be accounted for as shown in Figure 12-22. (Conversion of the right-hand allele must be because hybrid DNA sometimes filters in from some distant breakage point to the right.)

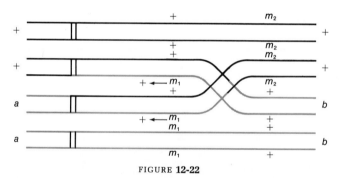

FIGURE 12-22

4. Double conversion is accounted for because both sites are in the region of hybrid DNA and both are excised in the same excision-repair act. This obviously produces a unidirectional conversion toward one parental type.

MESSAGE

It appears that, in concentrating on the phenomenon of gene conversion, we have gratuitously found ourselves in the middle of a crossover. Mendelian (1 : 1) allele ratios are normally observed in crosses because it is only rarely that a heterozygous locus is the precise point of chromosome exchange. Remember that asci showing gene conversion at a heterozygous locus are relatively rare.

Although the model formally accounts for many of the phenomena of recombination, it is still only a model and research on the molecular nature of recombination is currently active. Models at the molecular level are possibly premature, given the present knowledge of the structure of the eukaryote chromosome. Can we infer that hybrid DNA is the force behind all chiasma formation in all eukaryotes? It is a tempting idea, but not enough information is available at present. Nevertheless, it is one of the best models we have, and it is based on an interpretation of genetic data.

Recombination itself is a biological process and as such is amenable to analysis by genetic dissection. This approach is currently being used by many geneticists. By isolating mutants defective in some stage of recombination (the hybrid DNA model points to how many stages there *could* be!), it is hoped that the whole process will be pieced together someday. However, it is complex (especially in eukaryotes) and the controls must surely be polygenic and so progress will be slow. Chromosome exchange remains one of the most mysterious of processes and much exciting work remains to be done in this area.

One final note: don't forget that interchromosomal recombination (independent assortment of chromosomes) is also an important producer of genetic change. Much information on the cytology and chemistry of spindle fibers, centrioles, centromeres, and synaptinemal complexes is available, but little genetic analysis of these structures had been done.

Interestingly, although nonhomologous chromosomes assort independently, they are not without their effect on each other during meiosis. Thus, heterozygosity for a variety of chromosomal rearrangements such as inversions and translocations can enhance crossover values in chromosomes not involved in the aberrations.

In his classic study of nondisjunction, Bridges noted that the X chromosomes of secondary exceptional females were always noncrossovers. One might expect, then, that in the presence of an enhancer of crossing over, such as a heterozygous autosomal inversion, nondisjunction in X/X/Y females would decrease. In fact, this is *not* the case; both crossing over and nondisjunction increase and all of the X chromosomes in the nondisjunctional females are noncrossovers. How can this apparent contradiction be resolved?

Rhoda Grell has interpreted elegant genetic experiments as showing that, in fact, there are two types of chromosome pairing that occur sequentially. Initially, "exchange" pairing occurs only between homologs and is a prerequisite for crossing over. If chromosomes do cross over, they remain paired (perhaps held by chiasmata) and disjoin normally. If, on the other hand, crossing over does not occur, the chromosomes enter a "distributive pairing pool" in which *any* exchange chromosomes may pair, regardless of homology but with a distinct preference for pairing partners of similar size. Because more autosomes will enter the distributive pairing pool (crossovers in the inversions are selectively eliminated) there are more potential pairing partners for the noncrossover X's.

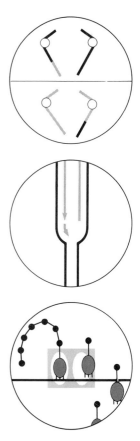

Try thinking of ways to test Grell's hypothesis genetically. What is the genetic result of nonhomologs distributively pairing?

MESSAGE

Nonhomologous chromosomes are not independently behaving elements during meiosis. They can affect crossing over interchromosomally and undergo two types of pairing interactions: exchange *and* distributive.

Problems

1. Mutations in locus 46 of the Ascomycete fungus *Ascobolus* produce light asco-spores. We'll call them *a* mutants. In the following crosses between different *a* mutants, asci were observed for the appearance of dark wild-type spores. In each cross, all such asci were of the genotypes indicated.

$$
\begin{array}{ccc}
a_1 \times a_2, & a_1 \times a_3 & a_2 \times a_3 \\
\downarrow & \downarrow & \downarrow \\
a_1 + & a_1 + & a_2 + \\
+ + & a_1 + & a_2 + \\
+ a_2 & + + & + + \\
+ a_2 & + a_3 & + a_3
\end{array}
$$

Interpret these results in light of recent models.

2. In the cross $A\, m_1 m_2\, B \times a\, m_3\, b$, the order of the mutant sites m_1, m_2, and m_3 is unknown in relation to each other and to A/a and B/b. One nonlinear conversion ascus is obtained; it is

$$
\begin{array}{lll}
A & m_1 m_2 & B \\
A & m_1 & b \\
a & m_1 m_2 & B \\
a & m_3 & b
\end{array}
$$

Interpret this with respect to the hybrid DNA theory and derive as much information about order as possible.

3. In the cross $a_1 \times a_2$ (alleles of one locus), the following ascus was obtained. Deduce what events may have led to it at the molecular level.

$$
\begin{array}{ll}
a_1 & + \\
a_1 & a_2 \\
a_1 & a_2 \\
+ & a_2
\end{array}
$$

4. a. Why is it not possible to induce nonsense mutations (represented at the mRNA level by the triplets UAG, UAA, UGA) by treating wild-type strains with mutagens that cause only AT-to-GC or TA-to-CG transitions in DNA?
   b. Hydroxylamine causes only GC-to-AT or CG-to-TA transitions in DNA. Will hydroxylamine produce nonsense mutations in wild-type strains?
   c. Will hydroxylamine treatment revert nonsense mutations?

5. Devise clever screens for selecting very rare
   a. Nerve mutants in *Drosophila*.
   b. Flagella-less mutants in a haploid unicellular alga.
   c. Supercolossal-sized mutants in bacteria.
   d. Mutants leading to overproduction of the black compound melanin in haploid fungus cultures (which are normally white).
   e. People with natural polarized vision in large human populations. (Do they exist? How could you locate them?)
   f. Negatively phototropic *Drosophila* or unicellular algae.
   g. UV-sensitive mutations in haploid yeast.

6. Strain A of *Neurospora* contains an *ad-3* mutation that reverts spontaneously at a rate of 10^{-6}. Strain A is crossed with a newly acquired wild-type isolate and *ad-3* strains are recovered from the progeny. When twenty-eight different *ad-3* progeny strains were examined, it was found that thirteen lines reverted at the rate of 10^{-6}, but the remaining fifteen reverted at the rate of 10^{-3}. Formulate a hypothesis to account for these findings and indicate how to test it.

7. In *E. coli*, how would you expect the following double mutants to behave with respect to UV sensitivity and to UV mutability in the light and in the dark?
   a. *uvrA recA*.
   b. *uvrA polA*.
   c. *polA recA*.

8. Noreen Murray crossed α and β, two alleles of the *me-2* locus in *Neurospora*. Also in the cross were two markers, *trp* and *pan*, which flank *me-2* each at a distance of 5 m.u. The ascospores were plated onto a medium containing tryptophan, pantothenate, and no methionine. The methionine prototrophs growing were isolated and scored for the flanking markers. The following results were obtained:

Genotype of me-2⁺
prototrophs

		trp +	+ pan	trp pan	+ +
Cross	$\dfrac{trp\alpha\ +}{+\ \beta pan}$	26	59	16	56
	$\dfrac{trp\beta\ +}{+\ \alpha pan}$	84	23	87	15

Interpret these results in the light of the models presented in this chapter. Be sure to account for the asymmetries in the classes.

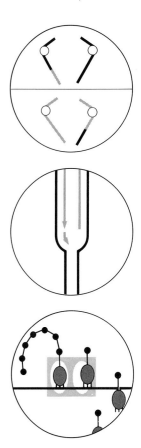

9. In *Neurospora*, the cross A *his-1*$_x$ $\times$ a *his-1*$_y$ is made in which x and y are alleles of the *his-1* locus, and A/a are sex alleles. The recombination frequency between the *his* alleles is measured by prototroph frequency on histidineless medium and comes out to 10^{-5}. Progeny of parental genotype were backcrossed to the parents and the results were as follows:

a. All a *his-1*$_y$ progeny backcrossed to A *his-1*$_x$ parent showed frequencies of 10^{-5}.

b. A *his-1*$_x$ progeny backcrossed to a *his-1*$_y$ parent showed two prototroph frequencies: half of the crosses showed 10^{-5} but the other half showed a very much higher frequency of 10^{-2}.

Propose an explanation for these results. How would you test your idea? (*Note:* Intragenic recombination is a *meiotic* function that *must* occur in a diploid cell; so even though this is a haploid organism, dominance and recessiveness could be involved in this question.)

10. Suppose that you are "reverting" mutant *nic-2* alleles in *Neurospora* with nitrous acid. You treat cells and plate them on medium without nicotinamide and look for prototrophic colonies. The results are as follows:

a. *nic-2* allele 1: no prototrophs at all.

b. *nic-2* allele 2: three prototrophic colonies picked off and each crossed with wild type. From the cross prototroph A $\times$ wild, 100 progeny were isolated and all of them were prototrophic. From the cross prototroph B $\times$ wild, 100 progeny were isolated. Of these 78 were prototrophic and 22 were nicotinamide requiring. From the cross prototroph C $\times$ wild, 1,000 progeny were isolated. Of these, 996 were prototrophic and 4 were nicotinamide requiring.

Explain these results at the molecular level, and indicate how you would test your ideas.

11. Which of the following statements is true:

a. Two multinucleotide deletions at one locus can never complement.

b. Two different single nucleotide deletions at one locus always complement.

c. Two different base-pair substitution mutants affecting the same locus always complement.

d. A single nucleotide deletion and a single nucleotide insertion both at one locus will complement.

e. If two mutants affecting one polypeptide complement, it is likely that the final gene product is a multimeric protein.

12. Electrophoresis is a technique of applying a high voltage across a gel slab. This separates proteins that have different charges because their movement in the electric field is determined by their electric charge. Two yeast strains (haploid) were examined by this technique. One was isolated in Florida and the other in Alaska. A diploid was made from these strains. All three strains were ground up and tested on gels. The gels were then stained to reveal a certain enzyme Q. The results were as follows:

```
Florida strain   |
Alaska strain                 |
Diploid strain   | |  |  | |
```

a. What are the intermediate bands and what do they tell you about enzyme Q?

b. Why are there three intermediate bands?

c. What relative quantity of each would you expect? (Give as ratios.)

d. Why should enzyme Q be different in Florida and Alaska?

13. Among revertants of a given mutant, a large proportion is temperature sensitive; that is, wild type at one temperature but still mutant at another. Why is this so?

14. G. Leblon and J.-L. Rossignol have shown the following in Ascobolus:

a. Single nucleotide pair insertion or deletion mutations show gene conversions of the 6:2 or 2:6 type, rarely 5:3, 3:5 or 3:1:1:3.

b. Base-pair transition mutations show gene conversion of the 3:5, 5:3 or 3:1:1:3 type, rarely 6:2 or 2:6.

Can you think of a reason at the hybrid DNA level why this should be so?

c. Furthermore, they have shown that, for insertions, $6:2 \ll 2:6$ and, for deletions, $6:2 \gg 2:6$. (Here the ratios are $+ :m$). What might this mean? Think once again of pairing at the hybrid DNA level, and perhaps also think about thymine dimer excision.

d. Finally, they have shown that when a frameshift mutation is combined in a meiosis with a transition mutation at the same locus in a *cis* configuration, the asci showing *joint* conversion are all 6:2 or 2:6 for *both* sites; that is, the frameshift conversion pattern seems to have "imposed its will" on the transition site. Propose a mechanism for this.

15. At the *grey* locus in the Ascomycete fungus *Sordaria,* the cross $+ \times g_1$ is made. In this cross, hybrid DNA sometimes extends across the site of heterozygosity and two hybrid DNA molecules are formed, as we have discussed in this chapter. However, correction of hybrid DNA is not 100% efficient. In fact 30% of all hybrid molecules are not corrected at all, whereas 50% are corrected to $+$, and 20% are corrected to g_1. What proportion of aberrant ratio asci will be

    a. 6:2?

    b. 2:6?

    c. 3:1:1:3?

    d. 5:3?

    e. 3:5?

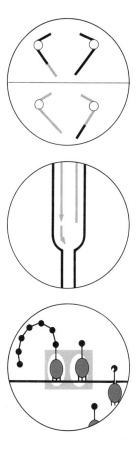

13 / Developmental Genetics

(Or: How do you make an organism out of a fertilized egg?)

A fundamental phenomenon that has long fascinated biologists is how a single fertilized egg is transformed through cell division and differentiation into a multicellular organism comprising perhaps 60 trillion cells (in an average human adult) that perform many different, coordinated functions. In other words, how can an egg give rise to skin cells, nerve cells, muscle cells, bone cells, and so forth; what regulates the orderly sequence of embryological events giving rise to them; and how are different cells and tissues integrated into a single functional unit, the whole animal or plant? These are some of the questions we will pursue in this chapter.

Genes and Cytoplasm

It is clear the genes are actively involved in the developmental process because we know that mutation can alter development in as trivial a way as changing the position of hairs on human knuckles or so severely as to lead to extreme abnormalities and death. However, during the embryological development of most organisms, all cells appear to receive an equal set of genes by mitosis; so it is difficult at the outset to visualize how genes could be responsible for cellular differentiation. On the other hand, the distribution of cytoplasm in dividing cells

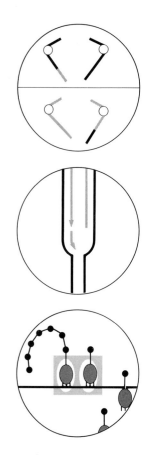

FIGURE 13-1

varies both quantitatively and qualitatively. Thus, mitochondria and yolk are not randomly distributed in a newly fertilized egg and early cleavages partition these factors unevenly. This in turn affects both the rate and position of the cellular division shown in Figure 13-1 so that asymmetries are built up early. This led embryologists to stress that changes in the cytoplasm were the primary factors affecting differentiation by controlling genetic activity. Geneticists, on the other hand, defended the importance of genes by pointing out that the architecture of the egg cytoplasm was in fact under the control of the mother's genotype. This cytoplasm-versus-nucleus controversy resembles the "which-came-first-the-chicken-or-the-egg" conundrum because it is clear that neither the cytoplasm nor the nucleus can exist without the other and the question of primary control becomes more semantic than real. Nevertheless, it kept embryologists and geneticists embroiled in argument for a long time.

We can give you an example of the intricate relationship between maternal genotype and egg cytoplasm "phenotype," which was studied in snails by Sturtevant (who studied crossing over in Drosophila!). He showed that there are two strains of *Limnaea* that differ in the direction of coiling of the shell. Looking into the opening of the shell, it can be seen that one group coils to the right (dextral), whereas the other coils to the left (sinistral). In the cross dextral female $\times$ sinistral male, all the progeny have dextral coils, implying that dextral is dominant over sinistral. However, in the $F_1 \times F_1$ cross, all the F_2 snails are also dextral. The reciprocal cross, sinistral female $\times$ dextral male, produces progeny that are all *left* coilers! In this case, the $F_1 \times F_1$ cross also yields only dextral coils. This certainly seems to contradict the principles we have learned so far. What's going on?

If the F_2 females of either cross are mated with males of any genotype, on the average three females give right coilers for every one that produces left coilers. Here we see the $3:1$ ratio expected in the F_2, but it appears in the F_3! Moreover, it seems to be independent of the paternal genotype. If F_2 males are mated with homozygous right-coiling females, all of their offspring are right coilers but, if mated with homozygous left-coiling females, only left coilers are produced. Clearly, then, the paternal genotype is not crucial in determining the phenotype of their offspring. The $3:1$ ratio in the F_3 gives us a clue that we are dealing with some strange monohybrid cross. It becomes clear if we assume that the gene for left coiling is recessive (s) to the right-coiling pattern (s^+) and that *expression of the phenotype of an individual is determined by the genotype of its mother*. Let's go back now and look at a diagram of the crosses (Figure 13-2). Now you can see that we can easily determine the genotype of a female by looking at the phenotype of her progeny. The explanation for this phenomenon lies in the early

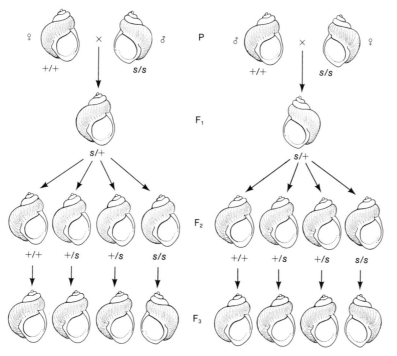

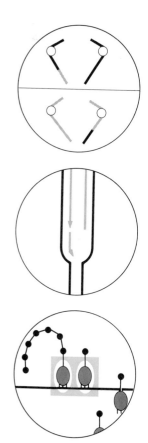

FIGURE **13-2**

cleavages in the fertilized egg. The maternal genotype affects the organization of egg cytoplasm, so that the orientation of the first cleavage plane is affected. If it is tilted to the left, then successive cleavages will produce a spiral to the left, and, if tilted to the right, a dextral pattern will follow. This is a classic illustration of the interdependence of cytoplasm and genotype.

MESSAGE

Cytoplasm may itself be organized by genes acting in the reproductive cells of the mother.

In recent years, the cytoplasm has taken on new significance. Whereas previously the cytoplasm was known to be important as a storage depot of components that were coded for in the nucleus (as in the snail example), it is now certain that there are actually genes in the cytoplasm (often called *plasmagenes*). Their discovery illustrates well the type of deduction that can be made from genetic observation. The best-known cytoplasmic genes are found in two organelles, the mitochondrion and the chloroplast.

One of the first classic demonstrations of *extranuclear inheritance* of plasmagenes came from studies on *Neurospora*. A mutant strain having very slow growth, called *poky,* was crossed as a "female" (i.e., contributor of most of the cytoplasm) parent with a normal strain acting as "male" parent. All of the progeny were poky. In the reciprocal cross, poky male × normal female, all the progeny were normal. This *non-Mendelian* mode of inheritance suggested that the cytoplasm of the maternal parent was important because the only difference between the reciprocal crosses was in the contribution of cytoplasm. Male "gametes" in *Neurospora* contribute very little cytoplasm just as in animals or higher plants. So it was probable that the "poky-ness" resided somewhere in the cytoplasm. The situation differs from the snail example in that segregation of poky from normal is never observed; you can cross poky "females" with wild-type "males" until you're blue in the face and the progeny are always poky. In other words, the nuclear genotype has no effect on phenotype (Figure 13-3).

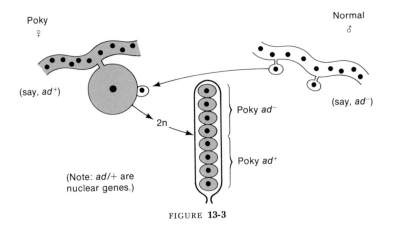

FIGURE **13-3**

Many cytoplasmic genotypes were soon discovered in fungi and because many of them were very bad growers (seemed to lack energy), or had considerably altered mitochondrial cytochrome contents, the mitochondrion was considered to be the likely site of these mutations. This was reinforced when circular DNA about the size of some phage DNAs was discovered in some mitochondria.

It should be stressed that, although there is good evidence for some cytoplasmic organelle-specific genes, their number is very small in comparison with the nuclear genome.

What are the gene products of this mitochondrial DNA? Hybridization experiments show that ribosomal RNA and some transfer RNAs found in the mitochondrion are transcribed from the mitochondrial DNA. They therefore have elements of their translation apparatus that are autonomous and independent of the nucleus. Furthermore, if mitochondrial DNA from *Neurospora* is used in an in vitro transcription and translation system from *E. coli*, two proteins called P1 and P2 are synthesized. P1 and P2 are also seen in the *Neurospora* mitochondria in vivo. What are these proteins? No function has been proved. Possibly they participate in ensuring the precise spatial arrangement that is known to be a necessary feature of mitochondrial enzyme batteries. Thus they

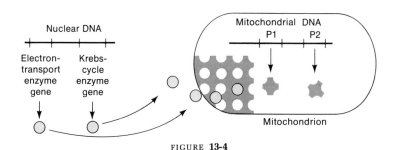

FIGURE **13-4**

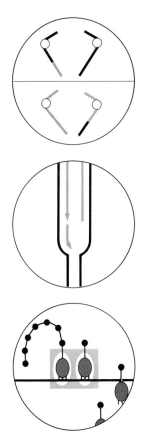

may be part of a scaffold on which Krebs cycle and electron-transport enzymes (the major systems found in mitochondria) are located. Electron-transport enzymes are known to be coded by nuclear genes. The scheme is perhaps that shown in Figure 13-4. In any case, mitochondrial DNA codes for both RNA and proteins. In yeast, one nuclear genome is about 10^{10} daltons of DNA, whereas the mitochondrial DNA genome is only 1/200 as big, or 5×10^7 daltons;* 5×10^7 daltons is about 75,000 nucleotide pairs. About 1/5 of this is probably taken up by ribosomal RNA and transfer RNA genes, and about 1/5 is not transcribed, leaving 45,000 nucleotide pairs. This is equivalent to 15,000 amino acid codons, enough to make about 75 proteins each 200 amino acids long. Animal mitochondrial DNA is much smaller, with enough information for only 25 such proteins. Nevertheless, it is obvious that the P1 and P2 type proteins are by no means the whole story.

Transmission of a hereditary trait such as poky through only one parent (*uniparental transmission*) is one criterion for recognizing cytoplasmic inheritance in fungi. Another is the *heterokaryon test* in which a heterokaryon is made between the prospective cytoplasmic mutant and a strain carrying a known nuclear mutation. If the phenotype of the nuclear mutant can be recovered together with the phenotype of the mutant being tested, then the latter phenotype is probably caused by a cytoplasmic mutation because here there is no genetic exchange between nuclei (Figure 13-5). The slow-growing phenotype has been acquired solely by cytoplasmic contact.

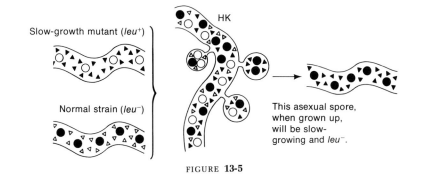

FIGURE **13-5**

*A dalton is a unit of molecular weight; an atom of hydrogen is one dalton.

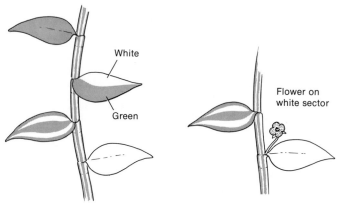

FIGURE **13-6**

In plants, one of the known causes of green and white sectoring in leaves is the presence or absence of mutant chloroplasts. These mutants, for some reason, do not allow the green of the chlorophyll to express itself in the chloroplast. For example, a common houseplant, wandering Jew (*Zebrina*), often shows sectoring due to mutant chloroplasts.

In such plants, a flower growing out of a white sector (Figure 13-6) will produce seeds that grow into plants having only white leaves; the plants die soon after exhausting the food supply in their cotyledons. Thus, the non-Mendelian inheritance is similar to poky. The beautiful stripes of *Zebrina* are produced because most of the mitotic cleavage planes in leaf development are along the long axis of the leaf (Figure 13-7). Chloroplasts also have their own DNA and it is assumed the cytoplasmic mutants are alterations in chloroplast DNA.

As you might expect, geneticists have tried to get recombination between various mutant alleles in chloroplasts or mitochondria. These attempts have been particularly successful in the unicellular alga *Chlamydomonas*. In this plant,

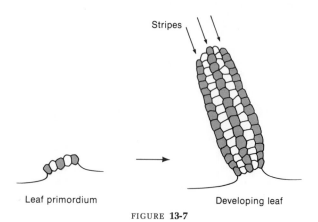

FIGURE **13-7**

many inherited chloroplast markers in addition to the white kind found in *Zebrina* are available. Surprisingly, many of them are drug resistance alleles. Ruth Sager has succeeded in demonstrating recombination between them in what we might call *heteroplasmons* (cells with mixed types of cytoplasm) and has found a circular linkage map! The meaning of this you can deduce for yourselves.

MESSAGE

Some genes are not carried on nuclear chromosomes, but in cytoplasmic organelles. They are recognized by uniparental (non-Mendelian) transmission patterns, or by contact transmission.

The cytoplasm, then, contains its own information codes. This has led some people to speculate that mitochondria evolved from symbiotic microorganisms that gradually lost the ability to exist independently. There is much evidence in support of this. But we digress. We have just established the importance of the cytoplasm in development.

Gene Activation or Mutation?

Although nuclear mutations can produce severe developmental defects, environmental agents such as ether, insulin, temperature, and thalidomide can mimic known genetic defects. The result produced by such an agent is called a *phenocopy*. Suppose that, as cells proliferate in an embryo, changes that vary from cell to cell occur in an ordered sequence to direct the developmental fate of cells. Then mutations or phenocopy-inducing agents can be assumed to deflect or alter the sequence of events, thereby leading to a defect.

If all nuclei have the same gene content and if a differentiated cell reflects the genetic activity occurring within it, how does differentiation take place? We could suppose that mutations occur in a regular manner during development so that, in fact, the genotype of differentiated cells differs from cell type to cell type. On the other hand, the genotype of all cells could remain constant, whereas the activity of specific genetic regions could vary from cell to cell. We have, then, two possible explanations for differentiation:

1. Orderly mutation in somatic cells.
2. Differential activity of genetic loci.

These two hypotheses lead to very different predictions: if differentiation occurs by means of controlled changes in gene activity, then all somatic cells should retain the same set of genetic instructions as the newly fertilized egg, whereas the mutation theory predicts the opposite.

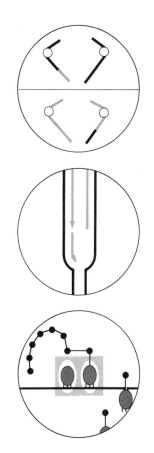

We know from the spontaneous occurrence of multiple births of "identical" individuals derived from a single fertilized human egg that the genetic information is faithfully reproduced through at least the first three cleavages after fertilization (the Dionne quintuplets were genetically identical). But what of the nuclei in differentiated cells? Do they remain *totipotent;* that is, potentially capable of producing an entire organism?

We know that many highly differentiated organisms are capable of regenerating new organs and tissues (e.g., the "arms" of starfish, the tails of reptiles, and the livers of humans). But tissues capable of doing that remain limited both in number and in type. Frederick C. Steward demonstrated that individual highly differentiated phloem cells in the root of a carrot plant are capable of giving rise to an entire carrot plant and therefore are totipotent. Thus, he has succeeded in "cloning" a mature plant (Figure 13-8). An important implication of this

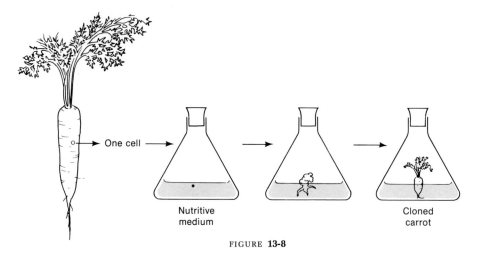

One cell

Nutritive medium

Cloned carrot

FIGURE **13-8**

observation is the economic potential of cloning particularly outstanding phenotypes. One of the problems with any organism that reproduces sexually is that recombination shuffles genes about and can disrupt highly useful gene combinations. Cloning retains genetic combinations intact and is now being used commercially with trees. Nonsexual plant reproduction has, however, long been used through grafting branches of one genotype onto another or generating new plants from roots or rooting stems.

Cells in tissue cultures can be treated and analyzed like microorganisms. Mutations can be induced and selected from cells in culture for such qualities such as the ability to fix nitrogen (thereby simplifying fertilizers) or for higher nutritional value (such as content of certain amino acids). If such mutant cells can be stimulated to differentiate into mature plants, then plant breeding is tremendously simplified and speeded up. But this is a digression. It could be argued that plants must be relatively uncommitted in their state of differentiation because many plants normally reproduce vegetatively.

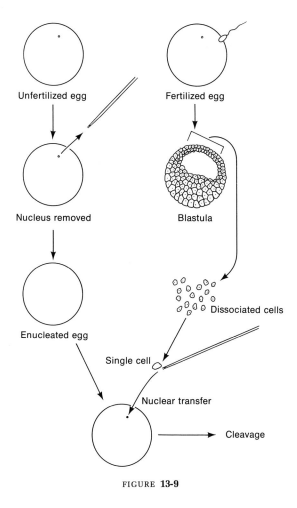

FIGURE **13-9**

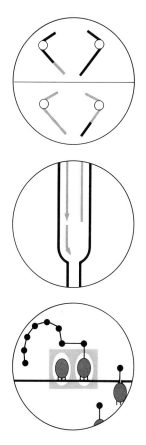

A test of totipotency in more complex higher organisms was made possible when Robert Briggs and Thomas King developed techniques for manipulating nuclei in amphibian cells in the early 1950s. They created *enucleated* frog eggs (lacking nuclei) by inserting fine glass pipettes into unfertilized eggs and sucking the nuclei out. They then sucked nuclei out of differentiated (blastula) tadpole cells and injected them into the enucleated eggs (Figure 13-9). Thus, the developmental capacity of the nucleus of a differentiated cell could be tested in the cytoplasmic environment of an unfertilized egg. Briggs and King found that enucleated eggs would divide when provided with a somatic nucleus. Furthermore, donor nuclei taken from embryos before gastrulation were totipotent, but donor nuclei removed from postgastrulation embryos could not support normal development. So it seems that, at gastrulation, a process of nuclear differentiation begins that cannot be reversed by sticking the differentiated nucleus into an enucleated egg.

It is possible that, once an orderly sequence occurs of genes being turned on and off, it would take a long time for the sequence to revert to its original state (i.e., of a newly fertilized egg) or that, even though the DNA sequences have not changed, the sequence of their activity cannot be reversed. The latter possibility would give results that are operationally similar to mutation. To test the possibility that the trauma of nuclear transplantation might retard the ability of the nuclei to return to a state of genetic activity that would permit complete development, Briggs and King tried another line of experiments. They kept nuclei in the cytoplasmic environment that perpetuates totipotency (before gastrulation) by serially transplanting differentiated nuclei from gastrula embryos into enucleated eggs (Figure 13-10). Even after many serial trans-

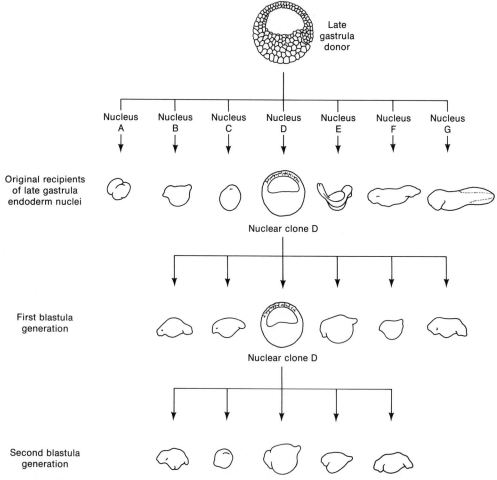

FIGURE 13-10

Serial transplants of gastrula nuclei in *Rana*. All eggs but D are allowed to proceed to the end of their development. The D clone is allowed to proceed to blastula and nuclei are again transplanted and the same procedure repeated.

plantations, the nuclei remained incapable of supporting complete development.

These experiments show that somatic nuclei of *Rana* differentiate irreversibly as defined by the nuclear-transplantation experiments. However, John Gurdon repeated the experiments with a different amphibian, the African clawed toad *Xenopus laevis*. He found that nuclei from highly differentiated cells of the gut of a tadpole (long past gastrulation) *were totipotent* and, when injected into an enucleated egg, allowed development into an adult toad (Figure 13-11). This

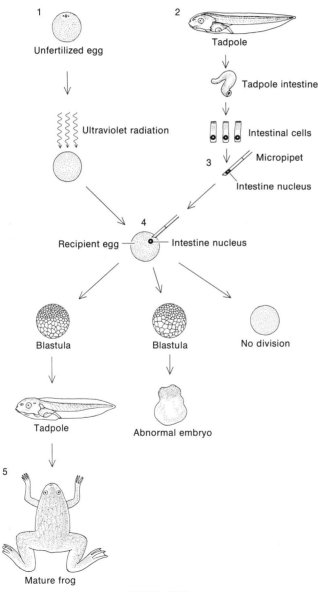

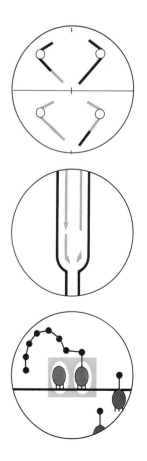

FIGURE **13-11**

result clearly differs from those with frogs (*Rana*) but points out that, in at least one highly differentiated organism, a toad, the genetic material does not appear to have changed irreversibly in its basic informational content. The reason for the different results obtained for the two amphibians remains a mystery. The implications of cloning vertebrates, particularly humans, have received very extensive popular discussion. Huxley's *Brave New World* remains a literary classic in developing futuristic scenarios with human cloning.

MESSAGE

It appears that in many organisms development proceeds by changes in gene activity.

Can we experimentally show these changes in gene function during development? The giant polytene chromosomes of Diptera take center stage once again. Certain specific stains show that these chromosomes contain regions that are rich in RNA, which hints that they might by cytologically visible sites of RNA synthesis. This suggestion can be verified by incubating salivary gland chromosomes in ^{3}H-uracil and preparing autoradiographs of the nuclei. Such studies show that certain parts of the chromosome do indeed incorporate the RNA precursor, especially in the swollen or bloated *puffs* and *Balbiani rings* (Figure 13-12). These chromosomes are so big that it is possible to pull them out of cells with forceps and cut out specific parts. By clipping off Balbiani rings in salivary gland chromosomes of the midge, *Chironomus*, it has been shown that the base ratios of RNA in different Balbiani rings are different from each other but similar to the base ratios of the DNA in the corresponding chromosome segment.

MESSAGE

It would seem that the RNA present in polytene chromosomes represents transcripts of specific gene sequences.

CYTOLOGICAL EVIDENCE FOR GENE ACTIVATION

Remember, we mentioned that polytene chromosomes are found in several different tissues of flies. If puffs really do reflect genetic activity and if differentiation occurs by differential gene activity, would we not expect to see different patterns of puffing in different cells? Wolfgang Beermann answered this by looking at the pattern of puffing that appeared in polytene chromosomes of different tissues of the same chironomid fly. He could identify each chromosome and its bands and found that the spectrum of puffs present in different cells was tissue-specific; some puffs appear in all tissues and others are unique to one or two tissues but the total pattern of puffing is tissue-specific (Figure 13-13). If the size of a puff or ring reflects the amount of gene activity, then changes in RNA

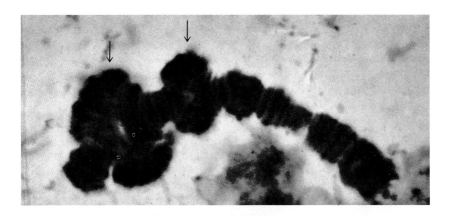

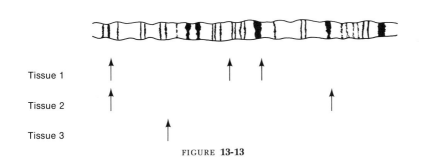

FIGURE **13-12**
Balbiani rings in *Chironomus* salivary gland chromosomes:
(top) arrows point to Balbiani rings;
(bottom) autoradiograph of ³H-uridine incorporation showing heavy
incorporation in Balbiani ring.

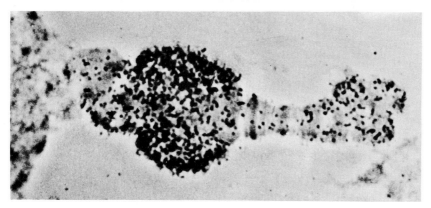

Tissue 1

Tissue 2

Tissue 3

FIGURE **13-13**

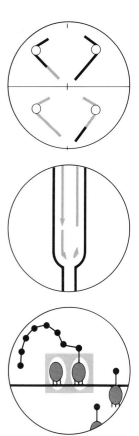

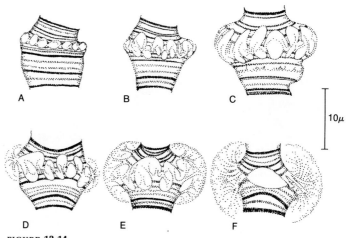

FIGURE 13-14
Sequence of formation of a Balbiani ring in progressively later stages
of larval development.

production are reflected in the growth and regression in the puffed regions
(Figure 13-14). Of course, this does not indicate whether the different patterns
of puff formation are the cause or the result of tissue differentiation, but these
studies provide a very nice cytological model of genes being turned on and off.

Another fly, *Rhynchosciara*, lays batches of eggs, which then develop syn-
chronously. This is handy because puffing patterns can be studied at different
successive stages of development by looking at larvae of different ages from the
same egg batch. Crowaldo Pavan was able to show in *Rhynchosciara* that, in a
given tissue, puffing patterns do change in a regular and ordered sequence
during development. These observations were also verified in *Drosophila*
salivary gland chromosomes. Puffing patterns are not only specific to the
developmental stage but also *reversible*, as evidenced by transplanting salivary
glands from a third instar larva into a younger larva. (Figure 13-15). Puffs
characteristic of the host's stage were observed in the transplanted glands.

The primary stimulus for major changes in puffing at pupation is the
increased level of the molting hormone ecdysone. This can be shown if the
larvae are pinched in two with a fine thread so that the salivary glands are
included on both sides of the constriction (Figure 13-16). Only the part anterior
to the ligature pupates in response to the hormone that is secreted by the ring
glands near the head. Hans Becker showed that the chromosomes in the anterior
part then puff in a pupal fashion, whereas those in the larval part retain the larval
puff pattern. This has been corroborated by the premature induction of pupal
puff patterns on injecting young larvae with ecdysone.

The tissue and stage specificity of puffing patterns in polytene chromosomes
seems to be a visible reflection of differential gene activity. But is there any
evidence that a puff is directly involved in cell differentiation? In the salivary
gland of *Chironomus pallidivittatus*, four cells situated near the opening of the

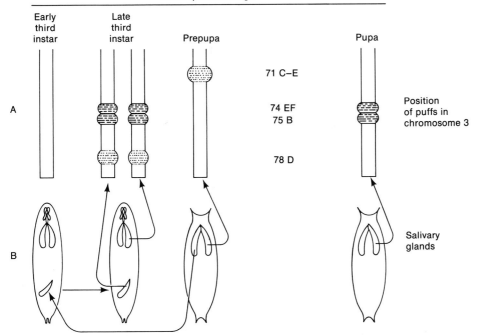

Developmental stage

Early third instar	Late third instar	Prepupa	Pupa	
				71 C–E
				74 EF
				75 B
				78 D

A   Position of puffs in chromosome 3

B   Salivary glands

FIGURE **13-15**
Reversal of puffing patterns in transplanted salivary chromosomes: (A) the position of the puffs present in glands at the different stages shown in part B (early third, late third instar larva, prepupa, pupa, from left to right); (B) the transplant of a gland from a prepupa to an early third larval instar, which is allowed to proceed to a late third instar before puff analysis.

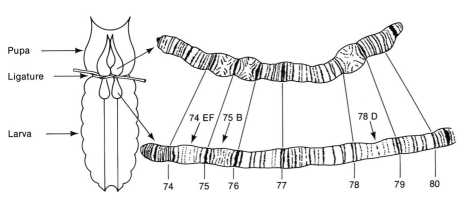

Pupa

Ligature

Larva

74 EF   75 B   78 D

74   75   76   77   78   79   80

FIGURE **13-16**
Ligation of a larva that pupates anteriorly but remains larval posteriorly is shown at the left. The puffs in chromosome 3 in the pupa (top) and larva (bottom) are shown at the right.

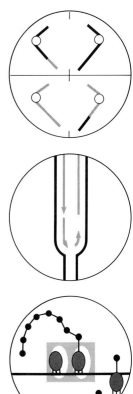

gland have dark cytoplasmic granules, whereas all the other cells of the gland do not. A Balbiani ring at the end of chromosome 4 is found only in those cells, whereas all other puffs are identical with the patterns in the other cells of the gland. In a related species, *Chironomus tentans,* there are no such cytoplasmic granules, nor is there a puff at the end of chromosome 4. Hybrids between the two species exhibit the chromosome-4 ring in the *C. pallidivittatus* homolog and have dark granules in the four cells (i.e., the gene(s) for the granules is dominant). Crossing over can take place between the two fourth chromosomes and Beermann showed that only those progeny carrying the *C. pallidivittatus* chromosome-4 tip exhibited both the puff and the granules. Thus, the granules are a direct reflection of the presence of the chromosome-4 terminal Balbiani ring, which may indeed be responsible for their production (Figure 13-17).

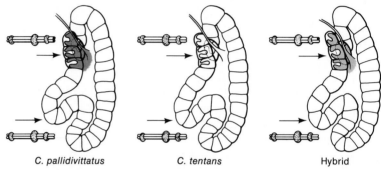

C. pallidivittatus   C. tentans   Hybrid

FIGURE **13-17**
Puffs present in the synapsed chromosomes 4 of *Chironomus* in different cells of the salivary glands. The puff at the tip occurs in *C. pallidivittatus* in cells with granules (left) and in hybrids with *C. tentans* (right) but only in the *C. pallidivittatus* homolog.

MESSAGE

Puffing patterns are tissue and developmental stage specific. The patterns are reversible and can be altered by the molting hormone ecdysome. There is some evidence that puffed regions affect cell phenotype.

Biochemical Evidence for Gene Activity

Puffing changes reflect altered gene activity during development, but the fact that different RNAs are made in different cells can be verified directly by DNA:RNA hybridization tests. Denatured DNA can be fixed to filters of nitrocellulose. When radioactive RNA is then passed through the filters, any RNA complementary to the DNA can "anneal," by forming hydrogen bonds in a DNA:RNA duplex. RNA hybridized to DNA is resistant to ribonuclease (RNase), an enzyme that specifically degrades single-stranded RNA. Thus, after a period of incubation sufficient for maximal duplex formation, RNase digestion will leave only hybridized RNA on the filter. This can be measured as counts per minute after all small nucleotide fragments are washed off. In this way, the proportion of the genome coding for RNA found in a specific tissue can be measured (Figure 13-18).

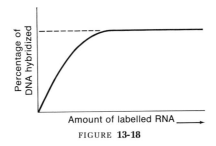

FIGURE **13-18**

Can we determine how much of the RNA present in two different tissues of the same organism is similar and how much is unique to each tissue? This can be measured in "competition" studies in which constant amounts of single-stranded DNA and radioactive RNA are incubated together with varying amounts of unlabelled RNA from specified sources. Let's take an example by asking how much RNA found in liver cells is also found in lungs and how much is found only in liver? By the hybridization of liver RNA to DNA, we can find out what the conditions are for maximum DNA:RNA formation. Now let's fix a known amount of DNA to a filter and enough labelled liver RNA to give maximum hybridization and, to different samples, add increasing amounts of cold competitive RNA from either lung or liver. If DNA:RNA hybrids are formed by the chance alignment of complementary sequences, then increasing the number of cold RNA molecules will mean that a cold RNA molecule will bind to a DNA region more often than a comparable hot RNA. So, as cold competitors increase, the binding of hot RNA will decrease *except for those RNA sequences in the hot sample that are not present in the cold* (Figure 13-19). The result shows that a

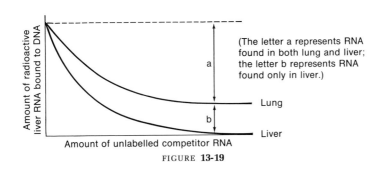

(The letter a represents RNA found in both lung and liver; the letter b represents RNA found only in liver.)

FIGURE **13-19**

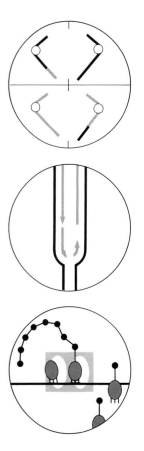

certain proportion of the RNA present in liver cells is also found in lung cells; hence the addition of cold lung RNA decreases the binding of labelled liver RNA. However, even a great excess of cold lung RNA cannot totally eliminate the binding of the labelled RNA. That which is still bound therefore represents RNA sequences not present in the lung, and, with respect to the two tissues, is liver-specific RNA. Studies of this kind clearly show that:

MESSAGE

Genetic activity varies from tissue to tissue although, as for puffs in polytene chromosomes, it cannot be stated whether this is the cause or the result of differentiation.

The Operon Model for the Regulation of Genes

Obviously, a mechanism or mechanisms for regulating gene activity must be found. As we have mentioned, during early embryogenesis in a frog, for example, differences in yolk and mitochondrial distribution in the egg result in asymmetries in cell size, division rate, and cytoplasmic content. There may be differences in internal and external cells in terms of temperature, oxygen and carbon dioxide concentrations, and so forth, as development ensues, but are there more subtle genetic controls? The following experiments indicate that there may be.

If we grow wild-type bacteria on a medium that contains glucose, we detect little or no β-galactosidase, an enzyme responsible for breaking down the sugar lactose into glucose and galactose. However, on transfer of the bacteria to a lactose-containing medium, β-galactosidase activity increases with time (Figure 13-20). We might think that, by depriving cells of glucose while providing

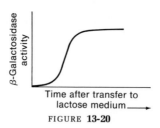

FIGURE **13-20**

lactose, we have simply selected for mutants that can make β-galactosidase and the increase in enzyme follows their multiplication. We can rule this out by growing wild-type cells in a lactose-containing medium without a nitrogen source so that cell division is inhibited. Under these conditions, galactosidase again appears in the same characteristic fashion. Thus, the enzyme begins to appear (we say the enzyme is "induced") in response to the very molecule that is the substrate for the enzyme. This is a very nice regulatory system: does it require turning the gene for the enzyme on and off? A way to probe the system is to induce a number of mutants in the locus whose genetic properties might provide clues to its regulation.

François Jacob and Jacques Monod studied the *lac* locus that determines β-galactosidase production and were able to select missense mutants that produced a protein lacking β-galactosidase catalytic activity but still recognizable as a form of the enzyme by its cross reaction with anti-β-galactosidase serum. They mapped such mutations within the *z* cistron, lying between the *ara* (arabinose) and *trp* (tryptophan) loci. In addition, other mutants were detected that produced wild-type β-galactosidase in the presence of lactose but the *amount* of the enzyme produced was directly proportional to the external concentration of lactose, the "inducer" molecule. These mutants were found to lack *galactoside permease,* a membrane pump protein whose function is to actively concentrate lactose by transporting the molecules into the cell. They were found to map adjacent to *z* in the *y* cistron. Finally, mutants affecting a third enzyme involved in lactose metabolism, *thiogalactoside transacetylase,* were found to map adjacent to *y* in the *a* cistron. Thus, the gene order is *ara-z-y-a-trp.* So we have a "clustering" of cistrons specifying proteins that participate in some aspect of lactose metabolism. What is most interesting is that when the inducer, lactose, is added to the medium of wild-type cells *all three proteins appear simultaneously;* that is, they are "coinduced" by the same molecule. In sexduced (F can insert near this region) merodiploids, $z^+ y^+ a^+ / z^- y^- a^-$ heterozygotes are wild type in phenotype; that is, inducible. The genetic clustering and the coinduction of related cistrons demands further investigation. How can we determine the mechanism that turns on and off the three tightly linked cistrons?

Jacob and Monod found two other kinds of mutations in the vicinity of *z*, *y*, and *a*. They were *regulatory* mutations, which affected not the structures of β-galactosidase, permease, and acetylase, but how much of each enzyme was made in the presence or absence of lactose. One kind was called o^c, or operator constitutive, and maps between *ara* and *z* very close to *z*. Operator constitutives have the effect of causing the three enzymes to be produced even in the absence of lactose, and such enzyme production is said to be constitutive. Most interestingly, the o^c mutations affected only the *z*, *y*, and *a* cistrons that were *adjacent* to them on the same chromosome. Let's illustrate this with an example: because *y* and *a* behave like *z*, we'll consider only *z*. In a merodiploid $o^c z^+ / o^+ z^-$, the cells produce β-galactosidase constitutively; hence o^c is *apparently* dominant over o^+. But in a different arrangement of the same genes, $o^c z^- / o^+ z^+$, the enzyme is produced only in response to lactose (it is *inducible*). Hence o^c can cause constitutive synthesis only in cistrons that are in *cis* arrangement, and it is therefore called *cis-dominant.* Operator mutations, then, define a region that affects the function of the cistrons physically linked to it. The *o-z-y-a* region, referred to as the *lac* locus, was defined as a coordinated unit of function called an *operon.*

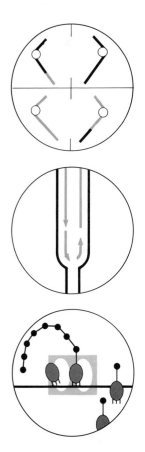

The second class of control mutants, which also fall between *ara* and *o* but are distinct from *o*, belongs to the repressor (*i*) locus, and the mutants in this class are operationally distinguishable from the *o* mutants in that they do not act only in the *cis* arrangement with *z*, *y*, and *a*. Thus, one class of *i* mutants (i^S), is dominant and exhibits *superrepression;* that is, $i^S o^+ z^-/i^+ o^+ z^+$ cannot be induced. On the other hand, another *i* allele, i^c, which causes constitutive enzyme production in $i^c o^+ z^+$, is recessive to i^+, so that $i^c o^+ z^+/i^+ o^+ z^+$ cells are inducible. Jacob and Monod brilliantly recognized that the *i* and *o* loci interact in the control of the "structural" cistrons for the *lac* proteins, *z*, *y*, and *a*. The operator region would act only in *cis* if it determines the transcription of the cistrons adjacent to it. The repressor gene, on the other hand, must act by means of the cytoplasm because certain alleles are transdominant. The repressor gene therefore must produce a product that controls the initiation of transcription at *o*. Jacob and Monod postulated that in a noninduced state, the *i* locus specifies the production of a *repressor*, which interacts physically with the operator to prevent transcription. On addition of an inducer (e.g., lactose), a complex between the repressor and inducer alters the repressor in such a way as to prevent its interaction with the operator site, thereby permitting transcription of the structural genes (Figure 13-21).

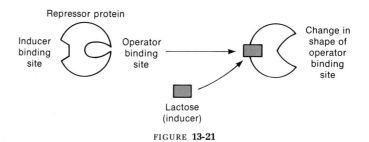

FIGURE **13-21**

Some deletions of the *o* region that extend from *z* to different positions to the left of *o* do not express y^+ and a^+ activity even though *there is no longer a site for repressor binding*. This suggests that there is another site, now called the *promoter* (*p*), that must be present if the *lac* genes are to be expressed. Indeed, mutants mapping between *i* and *o* that coordinately reduce *lac* gene products but are still under *i* gene control appear to be *p* mutants. It is now fairly certain that the *p* represents the site to which RNA polymerase, the enzyme for transcription, binds to the DNA, and starts transcribing on release of the repressor from *o*. Thus, the entire map of the lac region is *ara-i-p-o-z-y-a-trp*.

The control of the lac operon is now described, as shown in Figure 13-22.

What is wrong with the various regulatory mutants? o^c mutants have altered DNA that does not permit repressor binding. i^S mutants produce altered repressor protein that can no longer bind to the inducer, and consequently cannot be released from the operator site. i^c mutants produce repressor protein

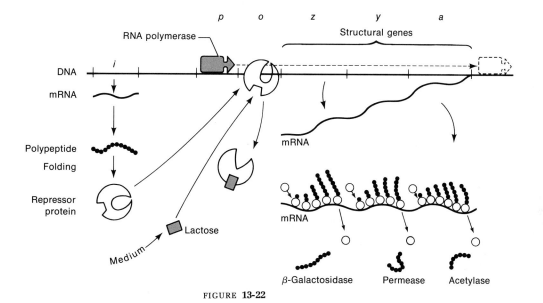

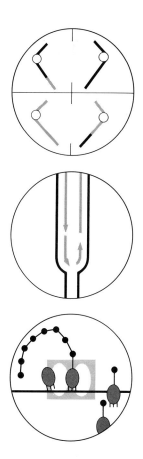

FIGURE **13-22**

that is altered at another position, that which enables it to bind to the operator, and hence the operon can never be shut off.

Polar mutants were recovered within the structural genes that reduce all wild-type activity distal to that mutation. For example, a polar mutation in z results in the loss of z, y, and a products, whereas a polar defect in y affects only y and a. This is easily explained if transcription or translation begins at the operator end of the operon. It has been shown that the polar mutations are suppressible by nonsense suppressors. But if an *induced* nonsense mutation can cause such damage further down the line, why don't the *normal* stop signals that separate the cistrons prevent further transcription beyond? To cut a long story (and a lot of experiments) short, it appears that the "dead space" on the messenger, between the induced stop mutation and the beginning of the next cistron, is never occupied by ribosomes and this makes it susceptible to degradation by various enzymes (Figure 13-23). So the further away a nonsense triplet is from the beginning of the next cistron, the longer the "dead space," the greater the chance of degradation of the messenger, and therefore the more highly polar it is.

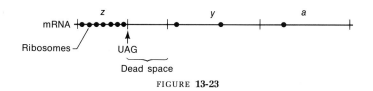

FIGURE **13-23**

The operon unit of regulated gene activity beautifully allows the repression of loci whose products have no available substrate, their induction by substrate, and their subsequent repression on exhaustion of the substrate. Furthermore, the linking of cistrons with related functions to a common operator places them under the same control device. From our standpoint, it is important to emphasize that Jacob and Monod were able to deduce the operon and its mode of regulation by using genetics alone. The biochemical studies confirmed their model years later. Operons consisting of as many as nine cistrons have now been described.

MESSAGE

In bacteria, control of gene action is achieved through operons. *An operon is a linked group of functionally related genes whose separate polypeptide products are translated from a single polycistronic messenger whose formation is under the control of a linked operator site.*

The *lac* operon control system discussed above is a kind of *negative* control; that is, the function of the regulatory protein is to turn off the system. Examples of *positive* control are also known in which the regulatory protein is needed to actively turn on the operon. The operon model can also be invoked to explain the inert state of a prophage (repressed) while it is active (induced) in the lytic cycle.

It is a simple matter to suggest that during development products of an operon might act as repressors or activators of other operons in a sequence, which would result in a "cascade" effect. Whether this would allow the specificity and variety necessary to explain, say, the degree of developmental control that determines the differentiation of pores right next to hairs on a piece of skin remains to be seen.

No examples of operons have been conclusively demonstrated to date in eukaryotes. Most attempts at finding them have been made on lower eukaryotes, especially fungi. Such studies can be summarized very briefly.

First, in some cases, functionally related genes are found clustered next to each other. For example, the *ad-3A* and *ad-3B* loci, which are the structural genes for two successive steps in purine synthesis in *Neurospora*, are found next to each other in the genome. Also, the five genes that control five sequential steps in the synthesis of aromatic amino acids in *Neurospora* are found next to each other in the *arom* cluster and are called *arom-2* through *arom-6*. These two examples seem at first glance to be prospective operons. But read on.

Second, by far the most common situation is one in which functionally related genes are *not* adjacent. In direct contrast to the bacterium *Salmonella*, in which the nine genes that control nine adjacent steps in histidine synthesis are hooked up in an operon, those same genes are scattered throughout the linkage groups of *Neurospora*.

Third, even if genes are grouped, the requirements for an operon may not be fulfilled. For one thing, coordinate control by means of an operator site has not been demonstrated. However, many examples of what look like bacterial polar mutations have been observed (for example in the *arom* cluster), but even these observations are subject to an alternative explanation (given in the next paragraph). In some cases (e.g., *ad-3A* and *ad-3B*), there is *no* evidence of polar mutants, which might indicate transcriptional unity.

Fourth, the demonstrated occurrence of large superproteins coded by one gene that has more than one enzyme function further complicates matters. This occurrence can mimic the properties of an operon because a point mutation (a stop mutant or a missense mutant causing a major conformational distortion) in one function can effect the other functions. Thus, even if only one cistron is involved, it can be mistaken for a one-messenger–several-polypeptides situation. For example, it is still not known whether the *arom* cluster is an operon or one giant gene: all the enzyme activities are in fact found in one large aggregate, but either alternative could be invoked.

Perhaps the *arom* cluster simply represents an *alternative* way of ensuring coordinate protein induction, or perhaps it is a clever way of channelling metabolic intermediates from one enzyme to the next like a bucket brigade. Or perhaps it *is* an operon (Figure 13-24).

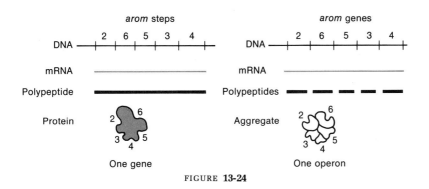

FIGURE **13-24**

Fifth, there are good examples of regulatory genes that can be separated by recombination from their structural loci. However, by far the most common situation is that of positive control; that is the mutated regulatory gene can no longer turn its structural gene (or genes) on.

MESSAGE

No systems completely analogous to the bacterial operon have been found in eukaryotes.

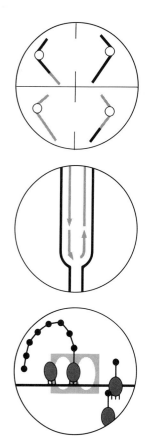

Chromosomal Proteins

Another point that has to be raised concerning the regulation of gene function in eukaryotes is the role of the chromosomal components in masking genes, or perhaps binding like repressors. Various candidates for such binding have been suggested: histone proteins (basic proteins found only in chromosomes), non-histone, acidic chromosomal proteins, and chromosomal RNA. The removal or addition of these components can alter the pattern of gene expression. However, although there is a lot of work going on in this area, at present there is no precise hypothesis involving these components that would explain how the meticulous specificity of gene activity is attained. There are, for example, only a handful of histone types, and it is hard to see how they can specifically control the thousands of genes in a genome. Histones have been primary candidates for the role of nonspecific gene maskers because they are bound to the DNA double helix. They are now believed to be responsible for the supercoiling of DNA. If histones do not seem to have the necessary diversity to explain differential gene regulation, the nonhistone proteins do. In fact in a chromatin (chromosome) reconstruction experiment on mouse tissue, it has been demonstrated that nonhistone protein is responsible for some tissue-specific RNA transcription (Figure 13-25).

Maybe the nonhistone protein is responsible for the removal and replacement of histones on the DNA. But these molecules lie outside the scope of genetic analysis at present.

Sequential Self-assembly in Development

Another aspect of differentiation that has been revealed in microorganisms is the *ability of gene products to form complex structures spontaneously.* It has been known for a long time that tobacco mosaic virus can be separated into its RNA genetic material and protein coat. If these two components are mixed in vitro, infectious virus particles form spontaneously. The process of phage infection, assembly, and lysis is under the control of about 100 genes in T4 and in λ. A remarkable discovery was that infective phages can be produced by mixing lysates of bacteria infected by two different phage mutants neither of which can complete lysis alone. Thus, a mutant blocked in the formation of a component of the tail fibers and another blocked in a head component can complement in vitro by supplying the missing components. By mixing lysates of different mutants, William Wood and his coworkers have been able to show that phage components have the capability of self-assembly and that their temporal sequence in the assembly process can thereby be determined (Figure 13-26).

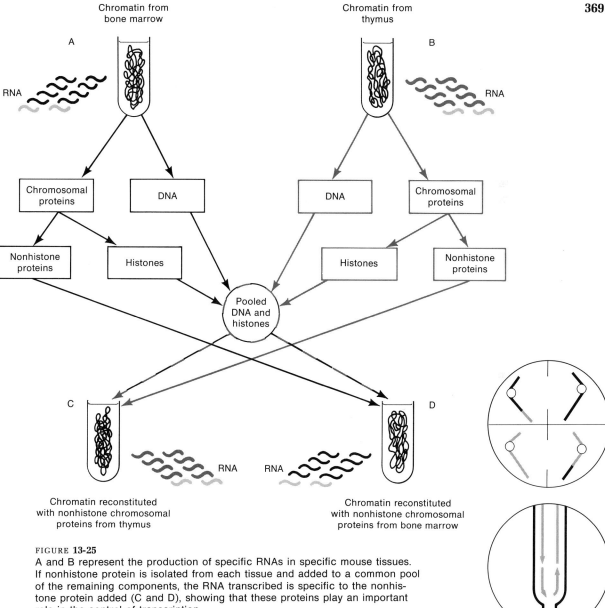

FIGURE **13-25**
A and B represent the production of specific RNAs in specific mouse tissues. If nonhistone protein is isolated from each tissue and added to a common pool of the remaining components, the RNA transcribed is specific to the nonhistone protein added (C and D), showing that these proteins play an important role in the control of transcription.

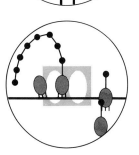

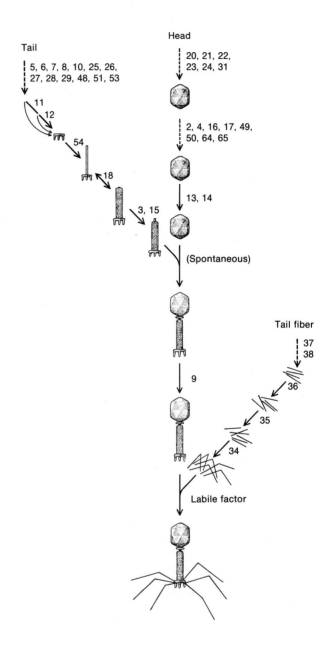

FIGURE **13-26**
The sequence of events in the formation of a mature phage inferred from mutants. Each number represents a specific gene in which a mutation blocks any further development: for example, in mutant 54, separate end plates, heads, and tail fibers are seen; in mutant 20, separate tails and tail fibers; and in mutant 37, whole phages minus tail fibers. If two mutant lysates are mixed, whole phages form by spontaneous assembly.

In a similar way, Masayasu Nomura has demonstrated that completely dissociated protein and RNA components of ribosomal subunits can reassemble spontaneously in a test tube to form a functional element. Moreover, by obtaining mutants for different proteins of the ribosome that block assembly, it can be determined how far ribosome assembly proceeds before stopping. Thus, the order in which RNA and proteins hook up in the ribosome assembly process has been determined. Cell membranes also spontaneously self-assemble.

MESSAGE

Many complex organelles comprising different macromolecules form in the cytoplasm because of the inherent properties of the aggregating molecules.

A method for determining the temporal sequence of gene activity during development was developed by Jonathan Jarvik and David Botstein. They obtained both heat- and cold-sensitive lethal mutations in different cistrons concerned with the maturation of T4 phages and constructed double mutants carrying both a heat-sensitive (hs) and a cold-sensitive (cs) mutation. The defects could act in phage assembly in one of three possible ways:

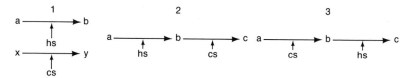

By infecting bacteria at either low or high temperatures and then shifting the cultures to either high or low temperatures, respectively, they could distinguish the relationship between any pair of genes by the pattern of phage production (Figure 13-27).

	1	2	3
High → Low	+	0	0	+
Low → High	+	0	+	0

FIGURE **13-27**

How do we interpret the results if a type-2 relationship exists? In this case, phages grown initially at a high temperature are blocked in a → b, and so a accumulates. The shift to a low temperature then permits the a → b step but blocks the b → c step, and so no viruses are expected. However, if the infection is started at a low temperature, the a → b reaction can occur, thereby allowing the accumulation of b. On shifting to a high temperature, the b already formed can proceed to c, which completes phage maturation. This kind of analysis permits the determination of temporal and catalytic sequences of genes in prokaryotes even in the absence of any information about the molecular nature of the genes' activities.

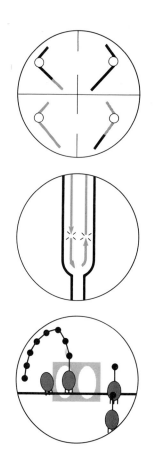

Dosage Compensation

In those organisms in which sex is determined by X and Y chromosomes that differ considerably from each other in their genetic content (such as in *Drosophila* and humans), Y chromosome functions seem to be concerned primarily with male determination or fertility. The X chromosome carries a large number of loci vital to both sexes, which means that X/X females carry twice as many X-linked genes as X/Y males. Remarkably, if one assays for the amount of an X-linked gene product such as glucose 6-phosphate dehydrogenase (G-6-PD), one will find little difference between males and females. This suggests that there is some mechanism to compensate for the difference in gene dosage in the two sexes.

A clue to the mode of "dosage compensation" in mice was obtained by Liane Russell in studies on mutations after radiation. She irradiated wild-type male mice and crossed them with females homozygous for several recessive autosomal coat-color mutations. Any F_1 individual exhibiting a mutant coat color would be presumed to carry a radiation-induced mutant allele of one of the loci. Among the F_1 progeny, several females were recovered that exhibited a variegated phenotype of patches of mutant and wild-type fur. On testcrossing these variegated females, two types of male progeny were recovered, completely mutant or completely wild type, and two types of females were recovered, completely mutant or variegated. On testcrossing the wild-type males, she obtained completely mutant males and variegated females, as shown in Figure 13-28 (in which an asterisk indicates radiated chromosomes). The last cross

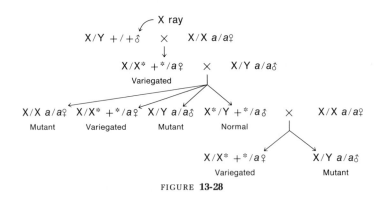

FIGURE **13-28**

shows sex linkage of the phenotype with the alteration expressing a normal phenotype in males. What is going on? Russell and Jean Bangham found that variegation resulted from a translocation between the chromosome carrying a wild-type coat color allele and the X. This suggested that, in females heterozygous for the translocation, in some cells the autosome did not function in the production of pigment in fur (therefore the cells were mutant), whereas in other cells the X-autosome part did produce wild-type gene product. The transloca-

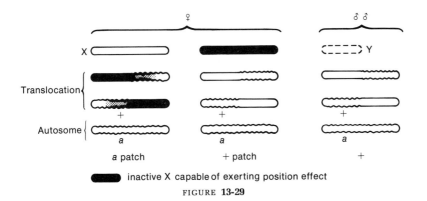

♀ ♂♂

X

Translocation

Autosome

a a a

a patch + patch +

■■■■ inactive X capable of exerting position effect

FIGURE **13-29**

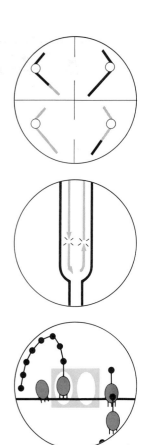

tion in males, on the other hand, always functioned to produce normal fur color in all cells (Figure 13-29).

Russell and Mary Lyon independently noted that many sex-linked mutations in mice and humans exhibit a variegated phenotype in heterozygous females. They suggested therefore that, in mammals, dosage compensation may occur in females by the *inactivation* of one of the two X's, thereby producing a functional equivalence of X's in males and females. This can be tested in humans using females heterozygous for two alleles of the *G-6-PD* locus, which produce electrophoretically distinct enzymes (F and S). If isolated cells from a skin biopsy of heterozygotes are cloned, each clone contains either the F or the S forms of G-6-PD, but never both. This tells us a number of things:

1. The *G-6-PD* locus and, by inference, most or all of the loci on *one* of the X chromosomes in females are inactive.

2. The X chromosome received from either the mother or the father can be inactive.

3. Because skin samples taken from a small area contain cells of both phenotypes, the inactivation must occur fairly late in development or otherwise very large homogeneous patches would have resulted.

4. Once the inactivation occurs, that state is inherited somatically because a clone of cells is uniform in the expression of the same allele.

The abnormal appearance of X/0 (sterile females) individuals shows either that not all of the X is shut off in normal X/X females (or else they would functionally be equivalent to X/0 females) or that at some critical early stage in development some or all of the loci of both X chromosomes must function.

The inactive X hypothesis inferred from genetic data was corroborated cytologically by the observation that one of the X chromosomes in X/X females forms a functionally inert heterochromatic body (often referred to as a "Barr body" after its discoverer, Murray Barr). All individuals bearing more than two X chromosomes have only a single active X; the other X chromosomes form heterochromatic bodies. This is a gene dosage effect and not a response to sexual differences because X/X/Y males have a Barr body. The mechanism regulating the inactivation of extra X chromosomes remains to be solved.

Tortoise-shell (calico) cats owe their appearance to inactivation of X chromosomes in females that are heterozygous for the sex-linked alleles for black and orange. If you come across a male tortoise-shell cat, either you can't tell a male from a female or it is X/X/Y. Another very dramatic demonstration of X inactivation is seen in human females who are heterozygous for X-linked anhidrotic dysplasia (Figure 13-30 shows the phenotypic effects in three generations of women). The mutant allele causes the absence of sweat glands, and mutant sectors can be detected by altered electrical resistance of the skin, or by various sprays.

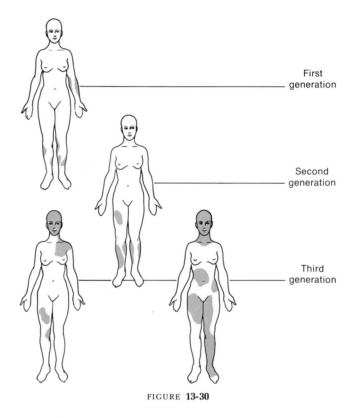

First generation

Second generation

Third generation

FIGURE **13-30**

MESSAGE

Compensation for the doubled gene dosage on the X chromosomes of mammalian females is accomplished by inactivation of either of the X chromosomes at some time during development.

The differentiation of organisms of a species into two different sexes is a remarkable example of the role of genes in development and merits special mention. An extreme example of the importance of genes in this process is the detection in humans of XY individuals who appear to be well-developed females. The phenotype is called testicular feminization and appears to be produced by a mutation on the X chromosome that acts only in XY zygotes. In most higher organisms, the potential sex of a fertilized cell is determined at the time of gametic fusion, which establishes a combination of sex chromosomes or sex-determining genes. Nevertheless, the actual differentiation of distinct sexual characteristics may take place much later in development and may even be altered from the sex determined.

In many species of plants and animals, a single individual may function as both male and female. Such individuals are *hermaphroditic,* if animals, and *monoecious,* if plants. Most monoecious plants have both sexual parts together in the flower. However, you are probably all familiar with corn plants, which are monoecious and which have spatially separated eggs (ear of the plant) and pollen (tassels). Worms and snails, which have both testes and ovaries, are examples of hermaphrodites in which the two kinds of sexual organs develop from the same genotype; clearly, some developmental mechanism must regulate the expression of specific sets of genes in the two regions of the organism.

In other organisms, a single genotype can produce one or the other sex. In such organisms, environmental and cytoplasmic elements play an important role in determining which sex develops. A few striking examples illustrate this kind of *phenotypic sex determination*. The marine annelid *Ophryotrocha* differentiates into a sperm-producing male as a young animal and changes into an egg-laying female when it gets older. If part of an older female is amputated, the worm reverts to a male, thus showing that size rather than age is the important element controlling sex. Sex in the marine archiannelid *Dinophilus*, on the other hand, appears to depend solely on the size of the eggs produced by females. Thus, small eggs always produce males whereas eggs twenty-seven times as large develop into females. The question of how the nucleus recognizes the environmental agents to turn on or off the proper set of genes to develop sexual organs is obviously the same as the problem of how a zygote differentiates.

In many other organisms, separate sexes develop in response to genetic differences between them. The X-Y and W-Z sex-determining mechanisms were clearly established by the 1920s. In these cases, the heterogametic sexes produce two types of gametes, which determine sex on fertilization. Among the organisms in which sex is determined in this way, individuals have been detected that have cells of both sex types, some parts of the organism being male and other parts female. In *Drosophila*, for example, a cross between a *y w m f/y w m f* female and a + + + +/Y male can produce a sexual mosaic, called a *gynandromorph,* that is phenotypically y w m f and male on one side and

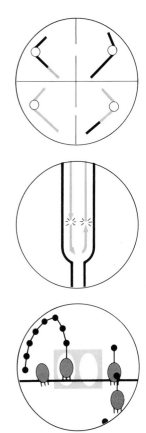

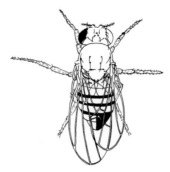

FIGURE **13-31**
A bilateral gynandromorph. The left
half of the fly is wild-type and female;
the right half is white-eyed, miniature-
winged, and male.

$+ + + +$ and female on the other (Figure 13-31). Somatic mutation is ruled
out in this case by the simultaneous appearance of several recessive markers;
rather, the gynandromorphism is explained by the loss of the wild-type X
chromosome in some cells to produce a mosaic of $y\,w\,m\,f/+ + + +$ female
and $y\,w\,m\,f/0$ male cells. We will return to gynandromorphs later: they can be
put to useful functions.

The ability of hormones to cause embryos for which the sex has already been
determined chromosomally to develop into the other sex has been strikingly
demonstrated in fish and frogs. Medaka fish have an X-Y sex determining
mechanism, yet exposure of young XY fish (which can be distinguished from
females by genetic markers) to female hormones (estrogens) results in their
development into XY "females"; these XY "females" can be mated with XY
males to produce YY males. The YY males on exposure to estrogens can also be
changed to females. Thus, a cross of YY "females" with YY males produces
only male progeny. Obviously the X and Y chromosomes must be very similar
in their genetic content or else the YY males would lack a number of genes
carried by the X chromosome. Analogous results have been obtained in frogs,
which have a W-Z sex-determining mechanism. If male hormones (androgens)
are fed to WZ females, WW females are obtained, and they in turn can be
modified by hormones to become "males".

In Hymenoptera (bees, wasps, etc.), sex is usually determined by ploidy,
which is controlled by the female. Thus, the queen of honey bees, whose diploid
number is 32, can lay two types of eggs. By controlling the sphincter of her
sperm receptacle, she can allow a sperm to fertilize an egg (producing a zygote
that will have 32 chromosomes and will be a female) or lay an unfertilized egg
(which will have 16 chromosomes and will develop as a male). The diet of the
diploid zygote will determine whether it will differentiate into a worker or a
queen. This is a striking illustration of sex determination by chromosomal

constitution and the effect of environmental factors on subsequent differentiation.

Remember that Bridges, in his studies on nondisjunction in *Drosophila,* demonstrated that XXY flies are normal females and X0 flies are sterile males. Obviously, the Y chromosome is necessary for male fertility but *not* for determining the male sex. What determines sex in *Drosophila?* Triploid flies carrying three X's and three sets of autosomes (we'll refer to a complete set of autosomes as A) are normal females. On crossing triploid females with normal males, offspring having different combinations of X's and autosomes can survive. Bridges found that flies carrying two X's and three sets of autosomes (2X:3A) were *intersexes,* possessing phenotypic characteristics intermediate between the two sexes. He concluded that there are *sex-determining genes* both on the X and on the autosomes and that the *balance* between them determines sex. Thus, if you suppose that male-determining genes are on the autosomes and that female-determining genes are on the X, then two sets of A's to one X "tip the scale" to maleness, whereas the female-determining genes on two X's outweigh the male-determining genes on two sets of A's. This can be illustrated by taking the ratio of X's to sets of autosomes: 2X:2A = 1.0 (female) and 1X:2A = 0.5 (male), but 2X:3A flies have an intermediate ratio of 0.67 and do indeed appear to be intersex. Other progeny with 1X:3A (0.33) and 3X:2A (1.5) ratios are weak, extreme males and extreme females, respectively. This so-called balance theory of sex determination seems to explain the observations in *Drosophila* very well.

MESSAGE

In Drosophila, *sex is determined by the balance between male-determining genes on the autosomes and female-determining genes on the X chromosome.*

Until the late 1950s, it was assumed that sex determination in mammals was similar to that in *Drosophila* (i.e., a balance between male- and female-determining genes). In 1959, William Welshons and William Russell reported experiments in mice suggesting that this was incorrect. They had a dominant mutation, Tabby (*Ta*), on the X chromosome that produces a dark coat color. *Ta/Ta* females have the same dark-furred phenotype as *Ta*/Y males (called the "Ta" phenotype) but different from *Ta*/+ females, which have patches of dark and light fur. Wild-type +/+ females and +/Y males are a uniform light color. On crossing *Ta*/+ females × +/Y males, they recovered females that were phenotypically similar to *Ta/Ta* females. (You can see that, if disjunction is normal, all females should be *Ta*/+ and +/+.) If sex determination is like that in *Drosophila,* these exceptions are expected to be *Ta/Ta*/Y nondisjunctional females. Welshons and Russell then crossed these "Ta" exceptions with +/Y males and recovered females that were phenotypically *Ta*/+ and + and *Ta*/Y

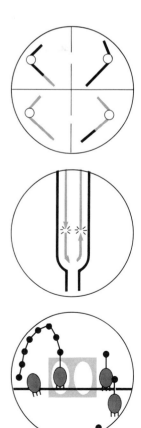

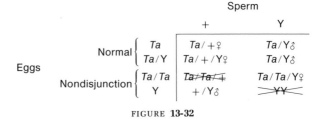

FIGURE 13-32

males in a 1:1:1 ratio. If the females had indeed been $Ta/Ta/Y$, then we would have expected the progeny results given in Figure 13-32.

Obviously, the recovery of + females and no Ta/Ta females shows that this is not correct. Welshons recognized that the female could have been a $Ta/0$ *female* in which case she would produce Ta and 0 eggs and give rise to $Ta/+$ and $+/0$ female offspring. He concluded that, in mice, the Y chromosome determines maleness and its absence results in a female. This was corroborated cytologically by the finding that the "Ta" females had only one X chromosome and genetically by the discovery of X/X/Y males. Both X/0 "females" and X/X/Y "males" have, of course, also been demonstrated in humans.

MESSAGE

In mammals, sex is determined by the Y chromosome and the extra X-chromosome dosage in females is compensated by the inactivation of one of them.

Sex determination and differentiation illustrate the range of differential genetic activity and its response to environmental and intrinsic factors.

Mosaics for sex chromosomes are common in humans. Many turn up in the sex tests administered to participants in the Olympic games, which examine chromosomes. One person examined was a "female" athlete who turned out to be an XYY/X0 mosaic. She (he?) probably started life as a male XY zygote, and very early, perhaps at the first mitotic division, nondisjunction of the Y chromosome produced the two cell types.

Presumably the X0 tissues were enough to account for the female phenotype.

Gene Amplification

We have now seen that regulation of genetic activity can occur through the organization of operons and by dosage compensation. Another method of regulation is performed by genes whose products are needed in large amounts at specific stages of development, or must be present in excess of all other gene products in certain cells. This is exemplified by the genes that code for ribosomal components. Ribosomes are vital to the functioning of any cell, but, in a highly specialized cell such as an egg, the rate of cell division immediately following fertilization far exceeds the nuclear capacity to synthesize cytoplasmic components. Consequently, the cytoplasm contains a large *excess* of ribosomes prior to fertilization. Two of the ways in which a cell might be able to fill the need for ribosomal material would be (1) to increase the number of ribosomal genes in the chromosome and (2) to produce excess copies of these genes, which can be transported to and transcribed in the cytoplasm. In fact, one or the other of these ways is used by different cells.

In most organisms, the nucleolus can be found at a specific region of a chromosome, the nucleolus organizer (NO). Electron microscopic analysis of nucleoli suggests that they are packages of ribosomes or ribosomal precursors. We can ask whether the chromosomal region of nucleolar formation (the NO) contains genes coding for ribosomal components. RNA is one constituent of ribosomes, and this ribosomal RNA can be used to quantify complementary DNA, by hybridization techniques. In *Drosophila*, the NO region is located in proximal heterochromatin of the X and Y chromosomes, and it can be asked, Does the NO contain DNA coding for ribosomal RNA? To answer the question, Ferruccio Ritossa and Sol Spiegelman used inversions to construct chromosomes carrying different numbers of NO regions. The inversions, scute-8 (sc^8) and scute-4 (sc^4) have left breakpoints close to the same place but right breaks on either side of NO (Figure 13-33).

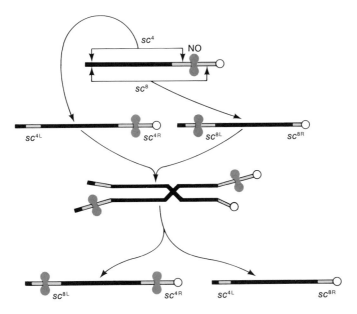

FIGURE **13-33**

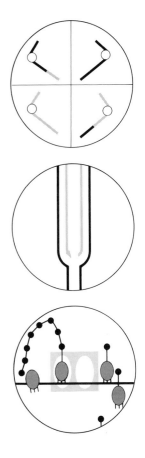

In females heterozygous for sc^8 and sc^4, you can see that a crossover will generate a chromosome called $sc^{4L}sc^{8R}$, which carries the left part of sc^4 and the right part of sc^8 and lacks an NO region. The reciprocal product, $sc^{8L}sc^{4R}$ carries two NO regions (remember, we discussed the production of duplication and deficiency chromosomes in Chapter 7). Thus, different doses of NOs can be established with specific combinations of X and Y chromosomes (Table 13-1).

TABLE **13-1**

Number of NOs	Genotype	
1	$sc^{4L}\ sc^{8R}/Y\ ♂$	$sc^{4L}\ sc^{8R}/X\ ♀$
2	X/Y ♂	X/X ♀
3	$sc^{8L}\ sc^{4R}/X\ ♀$	
4	$sc^{8L}\ sc^{4R}/sc^{8L}\ sc^{4R}\ ♀$	

By annealing radioactive ribosomal RNA to known amounts of DNA, Ritossa and Spiegelman could measure the amount of that DNA that would hybridize to ribosomal RNA. They found a linear relationship between the number of NOs and amount of RNA hybridized (Figure 13-34). This shows that some of the

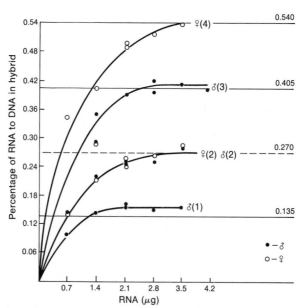

FIGURE **13-34**
Amount of ribosomal RNA hybridized to a constant amount of DNA. The plateau is reached when all DNA complementary to the RNA is hybridized. Each of the curves represents DNA samples isolated from males and/or females with different numbers of NO regions.

genes for ribosomal RNA are located in the NO region. Furthermore, because the size of the RNA molecules is known, as well as the percentage of the total DNA that hybridizes, an estimate of the number of genes coding for the ribosomal RNA can be computed. This was found to be about 130 copies per chromosome. Obviously, this "redundancy" of ribosomal RNA genes is one method of guaranteeing that a large amount of product of one type can be made at a time. The mechanism whereby the duplicate copies remain identical with one another rather than changing with random mutations remains a complete enigma.

The other method for increasing products of specific loci was to replicate an excess of DNA of that region and then transcribe such "amplified" or extra-chromosomal DNA elsewhere. In amphibian oocytes, such *specific gene amplification* has been identified as the basis for the large number of nucleoli present in the cells. In fact, 1,000 to 1,500 extra copies of DNA specifying ribosomal RNA are amplified from the already redundant NO region and packaged into the nucleoli. This was shown using a deletion of the NO region called *anucleolate (an)*, which lacks the corresponding ribosomal RNA cistrons. (How do you think this was demonstrated?) Yet from the cross $an/+ \times an/+$, the an/an homozygotes proceed through development to the tadpole stage (with a twitching tail) before death occurs, even though *no new ribosomal RNA is made by the embryo*. This shows that the $+$ allele in the female parent produced enough ribosomal RNA before meiosis to carry an embryo through a considerable part of development. The way in which this is accomplished seems to be by excessive production of DNA coding for an^+. Obviously such amplification must occur *before* segregation of the an^+ locus in meiosis. The amplified an^+ DNA molecules can be isolated by means of sucrose gradient centrifugation and shown to be tandemly repeated for the ribosomal RNA cistrons. Hybridization of ribosomal RNA with the amplified DNA demonstrates their base complementarity.

The redundance of ribosomal RNA cistrons in the nucleolar organizer region and in the cytoplasm makes a certain teleologic sense. It is now known, however, that the genome of higher organisms abounds with repeated DNA sequences, some of the multiples being astronomically high, whose function is a mystery. This can be deduced from DNA annealing rates as follows: When DNA is sheared and melted to separate the polynucleotide strands, annealing requires that the separated complementary strands "find each other." DNA sequences that are repeated will anneal much more quickly than others because there are many more complementary pairing partners for any given repeated sequence. Such rapidly annealing DNA can be readily demonstrated and measured in a variety of organisms and some results are shown in Figure 13-35. It can be seen that there is a correlation of the presence of various repetitive sequences with evolutionary complexity. Why do you think this should be so? Your ideas will be as valid as any. Many of the repeated sequences have no detectable RNA transcript! The function of such sequences is also the subject of much entertaining speculation, which you can also try.

Repeated sequences have also been shown by electron microscopic studies of individual DNA molecules. After partial alkaline denaturation of amplified an^+

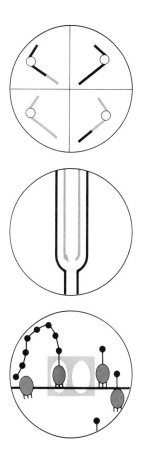

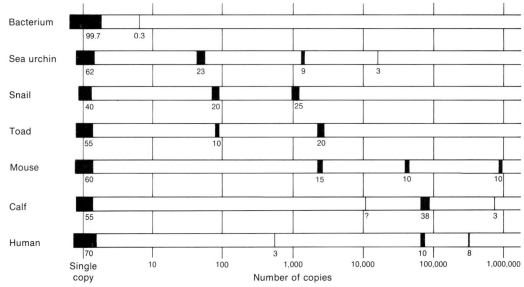

FIGURE **13-35**
Abundance of repeated sequences in the DNA of seven organisms. The width of each band represents the percentage (given below the band) of the total DNA that has that degree of repetition.

DNA, for example, electron microscopic inspection clearly reveals repeated patterns of DNA melts, thus confirming the repetitive sequences in this material. Although specific gene amplification is an important process in development, the mechanism of its control remains to be determined.

MESSAGE

Some gene sequences appear to be tandemly repeated and some can be amplified as extrachromosomal copies. This allows a higher production of transcripts from specific loci. The function of other repeated sequences is less clear.

Cell Lineage

If you have ever watched time-lapse movies of embryogenesis, you know that cell and tissue movements are extremely complex. In the dynamic process of embryogenesis, cells, by folding or migrating, may end up in areas of the embryo a long way from their original positions in the blastula. Is it possible to trace this cell lineage through embryogenesis? The classical approach was to mechanically "mark" specific cells of amphibian embryos with carbon particles so that the movement of the marked cells could be traced through development. Such marking makes it possible to construct "fate" maps that show the destinies of cells through embryogenesis.

An alternative method of tracing cells is to mark them genetically so that they and their descendants can be distinguished phenotypically. This is possible if we have genetic mosaics, individuals that comprise several genetically distinct populations of cells. They can occur naturally. For example, the exchange of blood cells in the placentas of twin cattle can produce *chimeras* carrying two blood types. Also, somatic nondisjunction can produce a person whose cells carry different numbers of X and Y chromosomes, as discussed in the section on sex determination. Furthermore, every female mammal is a mosaic of her two X chromosomes. Sexual mosaics are called gynandromorphs: we have already talked about these.

We will see several ways of artificially inducing genetic mosaics before the end of this chapter. Lyon has artificially produced mosaic mice by injecting cells of one type into embryos of a different type. Beatrice Mintz has developed a very elegant technique for fusing the developing embryos of two different mouse genotypes (Figure 13-36). When implanted in a host female, such "tetra-parental" mice develop as single animals.

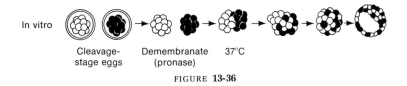

In vitro

Cleavage-stage eggs   Demembranate (pronase)   37°C

FIGURE **13-36**

Because the sex of the embryos being fused is not known, half of the tetraparental mice will be male-female chimeras. Under the influence of male hormones, most of these sexual chimeras differentiate as males even though the XX cells can be demonstrated. Because all daughter cells of the cells introduced at the time of embryo fusion represent a genetic clone, the recovery of geno-typically identical clusters of cells points to their common origin through division. Mintz fused embryos from a strain having wild-type fur color to embryos from a white-fur stock. Assuming that cells from either genotype can become precursors of skin, the pattern of fur color would reflect the clonal origin of these cells. She found that the pattern of fur of such mice was always in bands of black and white tissue circling the body, producing a dorsal-to-ventral stripe on each side (Figure 13-37). This shows that the cells of the skin originate in pairs along the midline of the embryo and then, by division, form clonal sheets that meet on the other side. So you can see how useful genetic markers can be in allowing us to trace cell lineage during development. This kind of genetic dissection of development has been honed to a fine edge in *Drosophila,* as we shall see in the next section.

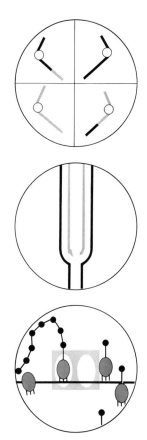

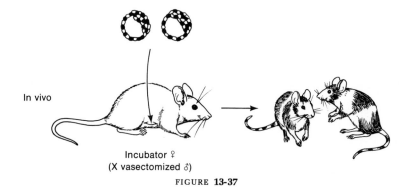

In vivo

Incubator ♀
(X vasectomized ♂)

FIGURE 13-37

Drosophila as a Model for Developmental Studies

Drosophila melanogaster, the tiny beast that played such a key role in elucidating the basic mechanics of chromosome behavior, has once again become the organism of choice for studies of development and behavior. In large part, the resurgence of interest in *Drosophila* is due to the extensive array of genetic manipulations that have been perfected by generations of drosophilists. Up to now we have regarded flies as convenient vehicles for the study of chromosomes; so let's take another look at them as complex aggregates of cells and tissues that are somehow coordinated to function as whole animals.

DEVELOPMENTAL SEQUENCE

Following fertilization, the zygotic nucleus undergoes rapid synchronous divisions without separation into cells. Thus, a newly fertilized egg develops as a *syncitium* (single multinucleated cell). In the posterior part of the egg is a region characterized by cytoplasmic occlusions called "polar granules," which are maternally deposited elements that determine differentiation of gonadal tissue. (How could you prove that such differentiation is maternally determined?) About 2.5 hours after fertilization, a few nuclei migrate to the region of polar granules to be pinched off for future development as gonadal cells. At 3 hours, when there are about 4,000 nuclei, the nuclei migrate to the cortical area of the egg and are cellularized to form a hollow sphere, the blastoderm (Figure 13-38).

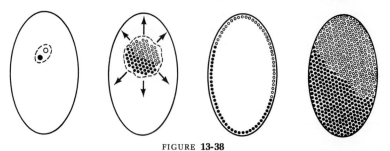

FIGURE 13-38

An important feature of blastoderm formation is that nuclear migration is not a helter skelter affair—those nuclei most recently related by nuclear division remain closer together in the peripheral sheet of cells that form than do more distantly related nuclei.

The complex embryological processes leading to larval formation then follow. Clearly, the larva is a highly developed organism with, for example, a nervous system to coordinate movement and responses to environmental cues of temperature, light, chemicals, and so forth. But that worm must also give rise as a pupa to a totally different animal, a fly, which has legs, eyes, wings, and genitalia. This remarkable transformation is anticipated in the larva by the presence of packets of cells, programmed to differentiate into adult tissue on exposure to the proper hormonal cues. These packets are called *imaginal discs* (often just discs) and can be recognized by their size, shape, and location in the larva (Figure 13-39).

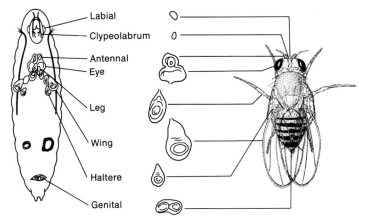

FIGURE **13-39**
Imaginal discs of *Drosophila:* (left) third instar larva (disproportion-ately larger than the adult) showing the position of the discs; (center) name and morphology of each disc; (right) the part of the adult that is formed by each disc.

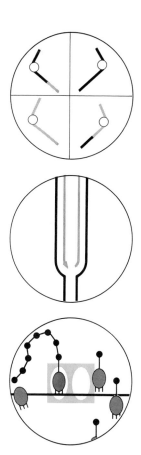

How do we know that the discs are already programmed or determined for an adult fate? If the growing, undifferentiated larval discs are transplanted into the abdomen of an adult, they continue to proliferate without differentiating. However, if an eye disc, for example, is implanted in another larval host that then pupates and emerges as an adult, the adult will have within it extra adult eye structures derived from the implanted disc! And that can be shown for each of the disc types. The stimulus for the differentiation of the already determined disc is the molting hormone ecdysone released in the late third larval instar. Again, this can be shown by exposing isolated discs from late third instar larvae to ecdysone and observing their differentiation.

On pupation, then, the larval carcass begins to break down and acts as a thick

medium nourishing and imbedding each of the discs, which can now differentiate into their proper adult structures. Each part of the adult is thus formed as
a separate element, which then fuses with its correct neighbors to form a
complete adult. The method whereby each disc forms an adult part is a beautiful
story that would take too many pages to tell. This incredibly intricate and
precise program is acted out each time a fly is "born" and you can see that
mutations blocking any of the developmental processes could tell us about their
genetic control.

EARLY DETERMINATION

Now let's go back and ask a few questions. When during development do the
determinative events for disc cells occur? Lillian Chan and Walter Gehring
developed a test to answer this question. They took blastoderms of a genotype
distinct from wild type and bisected them at the equator. The cells were then
dissociated and treated as indicated in Figure 13-40. When they implanted the
cells in wild-type larval hosts, which then hatched as flies, they found that
anterior adult structures (head and thorax) formed from the anterior part of the
egg and posterior adult structures (thorax and abdomen) formed from the

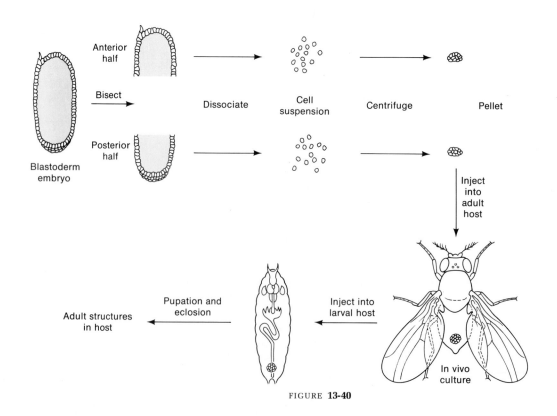

FIGURE **13-40**

posterior embryonic cells. Thus, the determinative events appear to occur at the time of blastoderm formation and it is reasonable to suggest that factors in the egg cortex are the important triggering elements. As in shell coiling in snails, mutants that act through the female parent to affect the development of the progeny would be expected in *Drosophila* if this were so. It now appears that such "maternal mutants" are not rare and therefore they will provide information about this process.

That the larva is a fully functional organism is shown by the recovery of a series of mutants that lack all discs or specific discs. Larvae lacking discs are completely viable and die only after pupation.

DISC DETERMINATION

What is the nature of the determined state in discs? The answer to that is far from clear, but two interesting observations are known. One is the existence of a class of mutations, called *homeotic*, that change the fate of a disc from one adult structure to another. For example, *ophthalmoptera* produces wing structure in an eye disc and *aristapedia* transforms the feathery arista to a leg (Figure 13-41).

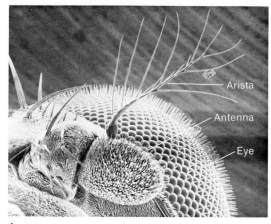

A.

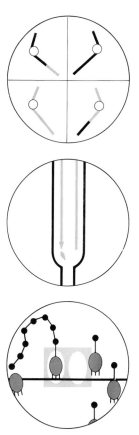

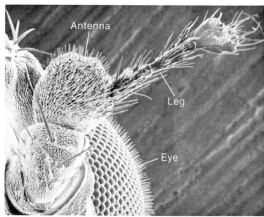

FIGURE **13-41**
A. Wild type.
B. Aristapedia.

B.

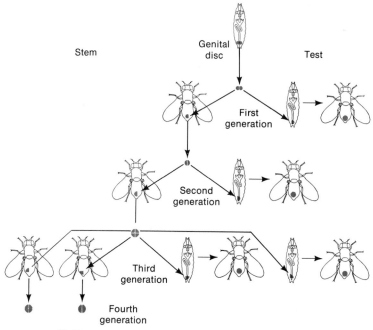

Stem   Genital
                     disc   Test

First
generation

Second
generation

Third
generation

Fourth
generation

FIGURE **13-42**
Serial transfers of a genital disc through adults. At each generation,
the disc is subdivided, some part(s) being implanted in an adult for
further growth, and the other part(s) implanted into a larva. The im-
plant in the adult proliferates without differentiation, whereas, when
the larval host undergoes metamorphosis, the different structures developed from the disc are revealed.

In the 1940s, it was found that many of the homeotic mutants are temperature-
sensitive and temperature-shift experiments were performed to find out when,
during development, the homeotic transitions occur. Almost all of these studies
reveal that the temperature-sensitive periods are in the third larval instar. In
other words, until that time, the disc can go one way or the other.

Another observation was made by Ernst Hadorn in the 1960s while growing
disc cells in adult hosts. As we've said, the cells divide in adults but do not
differentiate so serial transplantation of a disc through successive hosts is a way
of getting a lot of tissue from one original disc (Figure 13-42). By periodically
injecting a part of the cultured disc into a larva, its developmental potential can
be checked. Hadorn made an astonishing observation: on successive transplants
through adult hosts, the determined state of the disc changes. In other words,
what began as a genital disc, for example, could start to give rise to leg
structures! Hadorn called this change of state "transdetermination." Transdetermination is not randomly directed, but follows well-defined paths. Thus, a
genital disc may go to leg but never directly to an eye (Figure 13-43).

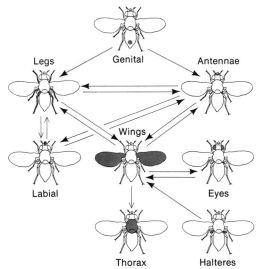

FIGURE **13-43**
Types of transdetermination produced from an adult culture of different imaginal discs. The shaded areas show the fly parts that develop on testing serially transplanted disc material. The arrows indicate the changes observed in the tests.

Interestingly, homeotic mutants cause alterations only in the same directions as the known transdetermination pathways. What all of these observations mean at the molecular level remains to be found, but there is no question that, as a developmental system, imaginal discs are fertile grounds for study.

BEHAVIOR

At this point, we have to mention the adult and its behavior. As a functional entity, the adult must control the movement of its legs and wings to walk and fly; it responds to sexual stimuli by courting and mating; it detects light and gravity; and it senses odor and movement. All this activity is coordinated through its central nervous system, which can therefore be probed by the selection of mutants that alter a behavioral phenotype. Of course, this is tricky because the observation that a vestigial wing mutant doesn't fly or that an eyeless mutant doesn't see isn't really telling us anything about the nervous system. Nevertheless, a number of studies of mutant behavioral phenotypes such as hyperactivity, paralysis, blindness, flightlessness, and shock sensitivity show that the neurobiology of *Drosophila* can be studied. Interestingly, many of the mutations selected as behavioral aberrations have been highly useful in studying development.

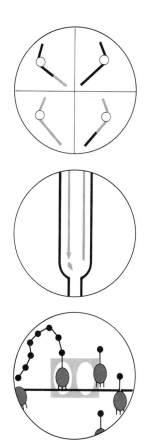

As we mentioned earlier, genetic mosaicism is a useful analytic tool. However, the chance detection or tedious experimental production of mosaics makes systematic study difficult in many organisms. In *Drosophila,* a variety of methods allows the relatively simple and frequent production of mosaics. Two general mechanisms can be used to make mosaics: mitotic crossing over and chromosome loss.

As we saw in Chapter 5, crossing over between homologous chromosomes that are heterozygous during mitosis will result in homozygous daughter cells half of the time. Stern showed, if you remember, that, when both homologs are marked, somatic twin spots of mutant tissue on a wild-type background are seen. Such mitotic crossovers can be induced at specific times by irradiating developing larvae; the later the time of irradiation, the smaller the twin spots will be. Because such twin spots are rare (usually less than one per fly even when increased by radiation), the two mutant patches can be assumed to be reciprocal products of a single event. All mutant cells of the same genotype represent a clone from the crossover cell and their distribution provides a graphic map of the distribution of that cell's daughter cells in the structure examined. This is often called the *cell lineage* of a structure. Comparing the size of mutant patches resulting from mitotic crossing over at different times in development provides information on the kinetics of cell division. Thus, the occurrence of cell divisions, the relative rates of division of different cell types, and their termination in development can be measured from spot size.

Becker used larvae heterozygous for the eye-color mutant alleles w^{co} and w^a. The eye color of all three genotypic classes for these alleles is different: w^{co}/w^a produces dark red; w^{co}/w^{co}, red; and w^a/w^a, lemon. By inducing mitotic crossovers in third instar larvae and examining the distribution of mutant spots in the adult eye, he was able to deduce the cell lineage for the bottom half of the eye, which comes from eight original cells (Figure 13-44).

Howard Schneidermann and his associates have used the same method to show the distribution of cells along the longitudinal axes of different organs (Figure 13-45).

Antonio Garcia-Bellido and John Merriam used somatic crossing over to find out whether the end of gene activity could be determined. They induced crossovers at different developmental stages in females that were $y\,Hw\,+/$ $+\,+\,sn^3$ (*Hw* is a dominant, closely linked to *y*, and causes the growth of extra bristles along wing veins; sn^3 produces short, gnarled bristles). Induced crossovers produce twin spots (Figure 13-46). They found that yellow, *Hw* mutant patches could be induced throughout the third instar and first 24 hours of the pupa. We wouldn't expect sn^3 spots with the *Hw* (extra wing vein bristles) phenotype because those cells have lost the *Hw* allele by the segregation of the crossover product. However, they found sn^3 spots with the *Hw* phenotype by irradiating during the last 12 hours of the third instar and the first 24 hours

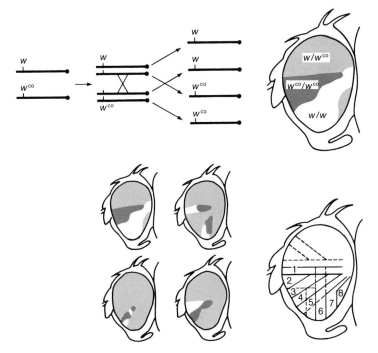

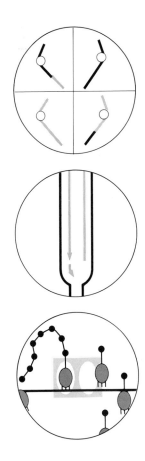

FIGURE **13-44**
Cell lineage patterns in the eye. By mitotic crossing over, twin spots are generated in the eye (top) that indicate the clonal derivatives of each cell after crossing over. Various twin spot patterns of the type shown at the lower left may be used to deduce that eight original cells produce the lower half of the eye as shown at the lower right.

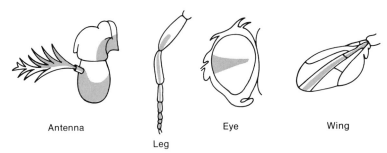

Antenna

Leg

Eye

Wing

FIGURE **13-45**
Mutant clones (shaded area) derived from mitotic crossing over in different imaginal discs.

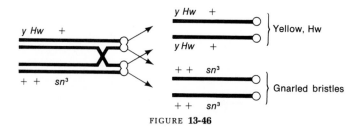

FIGURE **13-46**

of the pupa! They concluded that the *Hw* allele had acted to irreversibly imprint its phenotype on the wing cells before the last 12 hours of larval life.

Sturtevant observed in 1929 that eggs from females homozygous for an autosomal recessive mutation in *Drosophila simulans, ca*nd (claret-nondisjunction), frequently lost a chromosome during the first or second cleavage after fertilization. If the chromosome lost was an X, then a mosaic of XX (♀) and X0 (♂) cells was formed. If the original fertilized egg was heterozygous ($+/m$), then loss of the $+$-bearing chromosome allowed expression of the recessive phenotype in the hemizygous tissue (Figure 13-47). The fact that the flies

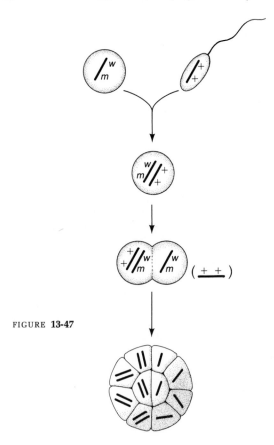

FIGURE **13-47**

produced did not have a "salt and pepper" phenotype of male and female cells all mixed up shows that nuclei more closely related by cell division tend to stay together. Sturtevant recognized that the distribution of mutant and nonmutant tissue in a mosaic could provide information about the spatial relationships *in the embryo* of prospective adult nuclei.

Since Sturtevant's time, several different mutants that cause chromosome loss have been studied. But the most useful aberration was described by Claude Hinton in 1955. It is a ring X chromosome called $In(1)w^{vC}$ (we'll call it w^{vC} for short) that has the completely mysterious property of great instability in the newly formed zygote nucleus. Thus, w^{vC} is lost at a very high frequency and, if the egg was initially w^{vC}/m, gynandromorphs of w^{vC}/m (wild-type ♀) and $m/0$ (mutant ♂) cells are created. This is an extremely useful aberration.

Using data originally obtained by Sturtevant in the 1930s, Garcia-Bellido and Merriam in 1969 set out to determine whether the pattern of mosaicism in adult flies could be used to infer relationships between cells in the embryo. So long as we assume that related nuclei distribute in close proximity in the blastoderm, such an analysis is possible (Figure 13-48). It is probable that the orientation of

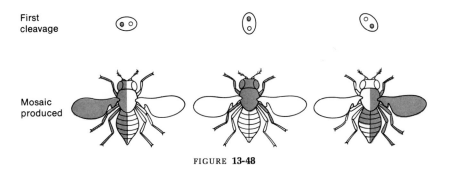

First
cleavage

Mosaic
produced

FIGURE **13-48**

the first cleavage plane in any given fertilized egg is randomly tilted through a 360° arc within a sphere because all kinds of mosaic patterns are formed in the adult. We can also assume that prospective cells for imaginal discs and for larval structures are distributed throughout the surface of the blastoderm.

If cells destined to form two imaginal discs are located close to each other within the blastoderm sphere, the chance that they will be genetically *different* in a mosaic is small. This is because the dividing line between mutant and nonmutant cells determined at the first cleavage is randomly distributed on the blastoderm. The further the cells are apart in the blastoderm, the greater the likelihood that the dividing line will fall between them and so the adult structures they develop into will be genetically different. Thus, the "distance" between blastoderm cells determined for different adult structures can be quantified as the percentage of mosaics in which the two adult structures are

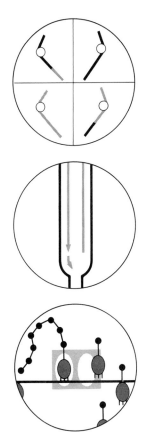

genotypically different; that is,

$$\frac{\text{Number of mosaics in which structures are different}}{\text{Total number of mosaics scored}} \times 100$$

We have no idea what this value means in real physical terms, but neither did we in constructing linkage maps. In this case, we develop a two-dimensional "map" because the spatial distribution is over the surface of a hollow spheroid. We can standardize the surface as that seen when the spheroid is cut along the axis of

FIGURE **13-49**

symmetry (Figure 13-49). If we map structures "a" and "b" as 10 units apart, we can introduce a second dimension by measuring their distances from a third structure, "c." If "a" is 8 units from "c" and "b" is 4 units from "c", we get

So we now have another unit, a distance between embryonic "foci" for adult structures. Garcia-Bellido and Merriam were able to construct an embryonic "fate map" like that shown in Figure 13-50. You should see that, if your eye is good enough to spot differences between structures such as bristles formed by specific imaginal discs, a fate map can be constructed for each disc. We assume, of course, that this map will have some kind of congruence with the actual spatial distribution of cells in the blastoderm.

In 1972, Yasuo Hotta and Seymour Benzer took this analysis one step further by analyzing the foci pertinent to mutant *behavioral* phenotype. If we have a mutant that causes the legs of a fly to twitch when anesthetized, the cells responsible for the twitch can be mapped relative to the external markers. Let's call the behavioral mutant *kic* (for kicker) and make a cross to generate $y\,kic/+\,+$ flies. A female that has no yellow tissue but exhibits leg licking must have internal cells that are $y\,kic/0$. We can calculate the distance of the focus for *kic* from the right front leg, for example, as

$$\frac{\text{Number of } y \text{ nonkicking legs} + \text{Number of } y^+ \text{ kicking legs}}{\text{Total mosaics}} \times 100$$

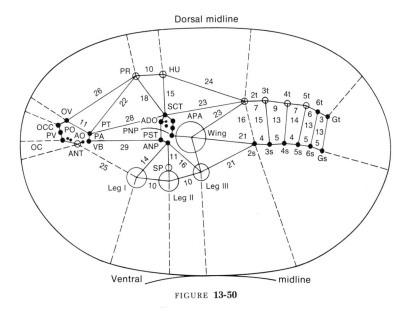

FIGURE **13-50**

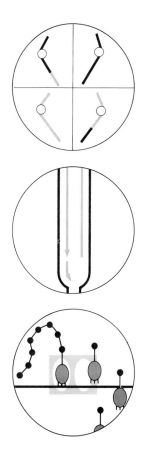

Again, by triangulating with a third structure such as a thoracic bristle, the focus for kicking can be mapped.

Interestingly, mutants with presumed muscle and nerve defects mapped below each other and the leg foci of the fate map. Their positions corresponded very well with cytological studies of neural and muscular tissues coming from the blastoderm. This allows us to align the fate map with known reference points in the embryo and now we can construct a fanciful map of the adult cells in the embryo (Figure 13-51). Many neurological mutants act through defects in

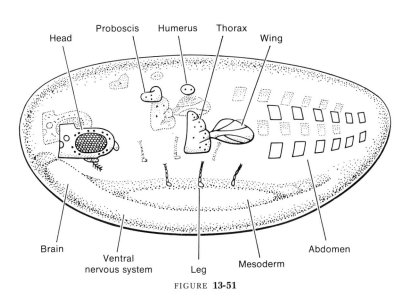

FIGURE **13-51**

several anatomical positions and here the analysis becomes much more complex. But for our purposes,

---

MESSAGE

The location of cells destined to form adult structures can be "mapped" relative to each other in the embryo by the use of mosaics.

The complexity and intricate beauty of developmental events can only be appreciated by a study of embryology. But modern developmental biology is a classic example of how the tools of genetic dissection can be combined with biochemical and cytological methods to further our understanding of the regulatory mechanisms of an intricate biological process.

Problems

1. The "copper" color variant is very common in trees (e.g., copper beeches, Japanese maples). If a copper branch arises spontaneously on a tree, how would you test to see if it is a nuclear or a cytoplasmic mutation? How would you propagate it in either case?

2. In *Aspergillus,* a "red" mycelium arises in a haploid strain. You make a heterokaryon with a nonred, paba-requiring haploid and from this recover some paba-requiring, *red* cultures, along with several other phenotypes. What does this tell you about *red?*

3. Assume that diploid plant A has a different cytoplasm from plant B. You wish to obtain a plant with the cytoplasm of plant A and the nuclear genome predominantly of plant B to examine nuclear-cytoplasmic relations. How would you go about producing such a plant?

4. During the lytic cycle after infection of *E. coli* by the phage λ, it is found that λ messenger RNA made immediately after infection is transcribed from the center of the chromosome to the right. Later in the cycle, the RNA is transcribed from the center to the left-hand side. What does this tell you?

5. Suppose that you have a translocation in mice in which the closely linked autosomal genes a^+, b^+, and c^+, which affect the fur, are attached to the X chromosome in the order a^+-b^+-c^+-X. Each female heterozygous for the translocation (a normal X chromosome and an autosome carrying the recessive mutations a, b, and c) has the following phenotype:

> Most of the fur is $a^+ b^+ c^+$
>
> Some spots are $a^+ b^+ c$
>
> A few spots are $a^+ b c$

What does this suggest?

6. You have four different temperature-sensitive phage mutants that are each defective in the assembly of the tail proteins. Two are alleles in cistron x, one dying at high temperature (call it hs for heat-sensitive), the other dying at low temperature (call it cs for cold sensitive). Let these mutants be x^{hs} and x^{cs}. The third mutant, in cistron $y(y^{hs})$, dies at high temperature and the fourth mutant, in cistron z (z^{cs}), dies at low temperature. You construct the following double mutant strains: $x^{hs} z^{cs}$, $x^{cs} y^{hs}$, and $z^{cs} y^{hs\cdot}$. You then infect bacteria with each of these strains and divide each group into A and B samples. Sample A is grown for 10 minutes at a low temperature and then for 10 minutes at a high temperature. Sample B is grown for 10 minutes at a high temperature and then for 10 minutes at a low temperature. You then determine whether any mature phages are produced and obtain:

	Phage recovery in each sample	
Phage strain	Sample A	Sample B
$x^{hs}z^{cs}$	0	0
$x^{cs}y^{hs}$	0	0
$z^{cs}y^{hs}$	+	0

Zero means no mature phages recovered; plus means phages are released. How do you interpret these results?

7. The lactose system of *E. coli* contains the following five important genes: i, the regulator; p, the promotor; o, the operator; z, the structural gene for β-galactosidase; and y, the structural gene for permease. Their map positions are as follows:

i		p	o	z	y

The promoter region is thought to be the site of initiation of transcription of the operon, and could represent the initial binding site for the RNA-polymerase

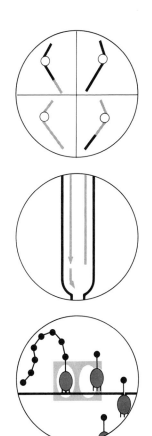

molecule before actual mRNA production. Mutants of the promoter (p^-) apparently cannot bind the RNA-polymerase molecule. Certain predictions can be made on the effect of p^- mutants: use *your* predictions, and your knowledge of the lactose system to complete the following table. Insert a plus where enzyme is produced, and a minus where none is produced.

	β-galactosidase		Permease	
Genotype of partial diploid	No lactose	Lactose	No lactose	Lactose
Example $i^+p^+o^+z^+y^+/i^+p^+o^+z^+y^+$	−	+	−	+
a. $\quad i^C p^- o^C z^+ y^+ / i^+ p^+ o^+ z^- y^-$				
b. $\quad i^+ p^- o^+ z^+ y^+ / i^C p^+ o^+ z^+ y^-$				
c. $\quad i^+ p^+ o^0 z^- y^+ / i^+ p^- o^+ z^+ y^-$				
d. $\quad i^S p^+ o^+ z^+ y^+ / i^C p^+ o^+ z^+ y^+$				
e. $\quad i^C p^+ o^0 z^+ y^+ / i^C p^+ o^+ z^- y^+$				
f. $\quad i^C p^- o^+ z^+ y^+ / i^C p^+ o^C z^+ y^-$				
g. $\quad i^+ p^+ o^+ z^- y^+ / i^C p^+ o^+ z^+ y^-$				

Note: o^0 is a polar mutation now known to be in z, close to the o region.

8. The table below is from the *E. coli lac* operon system, but the symbols a, b, and c have been used to represent the repressor (i) gene, the operator (o) region, and the structural gene for β-galactosidase (z), although not necessarily in that order. Furthermore, the order in which the symbols are given in the table is not necessarily the real *lac* system order. From the information in the table:

   a. State which of the symbols stand for the repressor, operator, and β-galactosidase genes.
   b. In the table, a minus sign merely indicates "mutant," but as you are aware in this system there are some special mutant behaviors, which are given special mutant designations. Redraw the genotype part of the table using the conventional gene symbols of the lac system.[1]

Genotype	Inducer absent	Inducer present
$a^- b^+ c^+$	+	+
$a^+ b^+ c^-$	+	+
$a^+ b^- c^-$	−	−
$a^+ b^- c^+ / a^- b^+ c^-$	+	+
$a^+ b^+ c^+ / a^- b^- c^-$	−	+
$a^+ b^+ c^- / a^- b^- c^+$	−	+
$a^- b^+ c^+ / a^+ b^- c^-$	+	+

Note: A plus means that β-galactosidase is made. A minus means that no β-galactosidase is made.

---

[1] Problem 8 is from *Genetics: Questions and Problems* by J. Kuspira and G. W. Walker. McGraw-Hill, 1973.

9. In *Neurospora,* all mutants affecting the enzymes carbamyl phosphate synthetase or aspartate transcarbamylase map at the *pyr-3* locus. If you induce *pyr-3* mutations with ICR-170, you find that either both enzyme functions are lacking or only the transcarbamylase function is lacking, but the synthetase function is never lacking alone. Interpret these results in terms of a possible operon.

10. In *Drosophila,* an autosomal locus, *Adh⁺,* specifies the production of alcohol dehydrogenase, an enzyme that converts alcohols into aldehydes. Wild-type flies, when fed pentenol, convert the alcohol into a toxic compound and die, whereas *Adh⁻* mutants survive. Conversely, ethanol-fed flies die if they're *Adh⁻* but survive if they're *Adh⁺.* Describe methods for the selection of mutants regulating the production and the amino acid sequence of alcohol dehydrogenase. How would you distinguish various mutant types operationally?

11. Thomas Kaufmann and Burke Judd concentrated on a section of the X chromosome of *Drosophila* consisting of fifteen bands in the salivary gland chromosome. They induced hundreds of mutations in this region and found that almost all of them were lethal or semilethal. By using overlapping deletions and duplications, they were able to localize the mutants to specific bands. They were able to show by complementation studies that there are fifteen "functional" groups and each group was localized to a single band. This has become known as the one-gene–one-band observation. If it is generally applicable to the entire genome, it suggests that there are about 5,000 "genes" in *Drosophila.* However, if *Drosophila* genes are the same size as *E. coli* genes, DNA studies suggest that *Drosophila* cells have 100,000 genes. How could you reconcile these two estimates of gene number (give several alternatives) and design experimental tests to distinguish between the possibilities.

12. There is a sex-linked gene in mice that causes muscular dystrophy. Alan Peterson fused embryos from the muscular dystrophic strain with wild-type embryos. Because the two strains also differed in enzyme patterns, he could determine the parental origin of cells in the tetraparental mice. He found that animals with muscle cells from the wild type ennervated by nerves from the dystrophic strain had muscular dystrophy. However, in parabiosis studies (in which two animals are surgically bound together), switching nerves from dystrophic animals to the muscles of normal mice did not induce dystrophy. What do these experiments signify?

13. David Suzuki and his associates have recovered a number of sex-linked recessive temperature-sensitive (*ts*) paralysis mutations in *Drosophila.* At 29°C, the mutants are completely immobilized, whereas, at 22°C, they move about like wild-type flies. What kinds of experiments would you perform to determine the cellular bases for the defects?

14. Let us assign numbers to each leg of *Drosophila* such that legs 1, 2, and 3 are the front, mid, and hind legs on the left side, respectively, and legs 4, 5, and 6 are the front, mid, and hind legs on the right side. Normally, a fly walks by moving legs 1, 3, and 5, and then moving legs 2, 4, and 6. A mutation, *wobbly,* causes the fly to get its mid legs tangled with its front or its hind legs. What could be wrong and how would you study it?

15. Suppose that you observe that during larval development of *Drosophila* an enzyme specifically found in the salivary glands appears, increases, and disappears in exactly the same manner as a specific puff on the X chromosome.

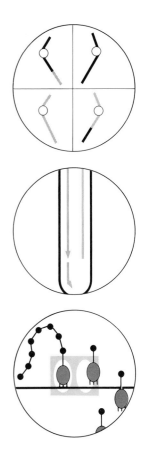

a. Does this prove that the enzyme is specified by the puff? Why?

b. Design experiments to prove whether the puff specifies the enzyme.

16. Joseph Gall and Mary Lou Pardue recognized that polytene chromosomes are like columns of DNA laid out in proper sequence. They developed a method for denaturing the DNA in these chromosomes without disrupting the spatial organization of the chromosomes. By incubating radioactive RNA on the denatured chromosomes and then removing the unbound RNA, they were able to localize the RNA-DNA hybrids by autoradiography. This is called *in situ* hybridization.

a. What parameters would affect the resolving power of this technique?

b. How would you demonstrate that such RNA binding does indeed reflect complementary sequences of DNA?

c. What do you infer when a certain RNA binds to a certain locus?

17. Highly redundant DNA sequences (distinct from sequences coding for ribosomal RNA) can be demonstrated and separated from unique sequences by their rapidity in reannealing. Very few, if any, of these sequences form duplexes with RNA. In situ hybridization with redundant DNA in *Drosophila* suggests that most of the sequences are located in heterochromatin around all of the centromeres. Yet these very regions have a disproportionately (in terms of DNA content) small number of loci detectable genetically. For that reason, Painter and Müller suggested that heterochromatin is "inert." In addition, very little crossing over occurs in heterochromatin. What "function" might these redundant sequences have that would explain the absence of detectable loci and their nonrandom distribution? (This is a wide open question in that no one knows the answer, although there are lots of hypotheses.)

18. One of the theories of aging or senescence is that, with time, randomly occurring mutations accumulate in different cells within an individual, gradually disrupting the normal cellular processes necessary for life. This explanation is sufficiently specific to allow direct tests to be made. Design experiments to test the hypothesis. (*Note:* Choose your organism and the kinds of tests carefully.)

19. Of what significance are phenocopies in medical genetics? How could phenocopies be used to study gene action? How can an investigator determine whether an altered phenotype is due to a phenocopy or a mutation?

20. (This problem concerns the ascomycete fungus *Saccharomyces cerevisiae* (yeast), which has a very simple life cycle: two haploid cells of opposite sex fuse to form a diploid; this diploid may divide mitotically, or it may be stimulated to go through meiosis and produce haploid products in an ascus.) When a haploid culture of yeast is plated, most cells form large colonies, but a few (about 0.2%) form small colonies, which are called "petite." You isolate several petites and find they fall into three types:

Type 1: When crossed with wild-type (grande) cells, the haploid products of meiosis are all grande. On examination, petites of this kind are found to contain no mitochondrial DNA.

Type 2: When crossed with wild type, the haploid products of meiosis are all petite. The mitochondrial DNA of this type of petite has an altered base ratio.

Type 3: When crossed with wild type, a 1:1 ratio of petites to grandes is obtained. The mitochondrial DNA is normal in petites of this type.

a. Propose hypotheses to account for the production of these three types of petites, and make sure your hypothesis accounts for the special inheritance patterns.

Suppose that you have made a yeast diploid between haploid cells differing in their antibiotic resistances. One is resistant to erythromycin and to spiramycin, and the other is resistant to paramomycin. Thus the diploid is $E^R \, S^R \, P^S + E^S \, S^S \, P^R$. You allow the diploid to divide *mitotically* for a while and then plate out the cells. You select 119 diploid clones at random and score them for their antibiotic resistance phenotypes. The results are as follows:

| Antibiotic | | | Number |
E	S	P	of clones
S	S	R	49
R	R	S	25
S	S	S	14
R	R	R	23
S	R	S	3
S	R	R	4
R	S	S	1

Each of the 119 diploid clones was then allowed to undergo meiosis. In each case the haploid products of meiosis were phenotypically identical with the clone from which they were derived (e.g., an $E^S \, S^R \, P^S$ diploid clone gave all $E^S \, S^R \, P^S$ haploid progeny).

b. Propose a hypothesis to account for these results. Account for the inheritance patterns of each resistance type individually, and explain how all the seven clone phenotypes might be produced, and at the frequencies detected. (Mitotic recombination won't work. Explain why.)

A haploid $E^R \, S^R \, P^R$ strain is treated with ethidium bromide, a compound that enhances the production of large numbers of petites. Suppose that the petites are of two kinds only, $E^R \, S^R \, P^R$ and $E^S \, S^S \, P^R$.

c. What does this suggest?

d. Is the result compatible with your hypothesis above? (If not, derive a hypothesis with which it is compatible!)

e. Would you predict that the $E^S \, S^S \, P^R$ petites would be of type 1 or type 2 or type 3?

21. In a haploid eukaryotic organism you are studying two enzymes that perform sequential conversions on a nutrient A supplied in the medium: thus

$$A \longrightarrow B \longrightarrow C$$
$$E_1 \qquad E_2$$

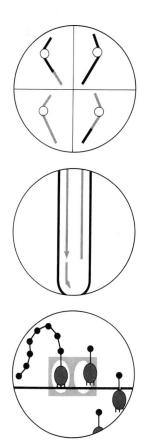

Treatment of cells with mutagen produces several mutant types with respect to these functions:

Type 1: Loss of E_1 function. All map to one locus on linkage group II.

Type 2: Loss of E_2 function. All map to one locus on linkage group VIII.

Type 3: Loss of E_1 *and* E_2 function. All map to one locus on linkage group I.

Compare this system with the *E. coli lac* system and point out the similarities and differences. (Be sure to account for each mutant type at the molecular level.) If you were to intensify the mutant hunt, would there be any other types you might predict that you could find on the basis of your model?

14 / Population Genetics

(Or: How the biosphere becomes more ordered.)

On October 2, 1836, a small, battered sailing ship turned out of the English Channel into Falmouth Harbor. The ship was the H.M.S. *Beagle,* which had just returned from a voyage around the world charting coastlines for the British Admiralty. On board was a passenger who had formed on the trip the seed of an idea that was to revolutionize not only biology, but man's view of himself and his creation. The passenger was Charles Darwin and the idea was evolution through natural selection.

The concept of evolution was later described by Darwin in these words:

> As many more individuals of each species are born than can possibly survive, and as consequently there is a frequently recurring struggle for existence, it follows that any being, if it *vary* however slightly in any manner profitable to itself . . . will have a better chance of surviving and thus be naturally selected. . . . This preservation of favourable individual differences and *variations,* and the destruction of those which are injurious, I have called Natural Selection, or the Survival of the Fittest. (Italics added.)

By this simple statement, Darwin suggested that the diversity of life forms now present on Earth had evolved from common ancestors rather than appearing by a single, divine creative act. Darwin thus saw evolution as a process in which selection and variation interact to produce a population of individuals who are more fit for their environment. This was, of course, pre-Mendel and,

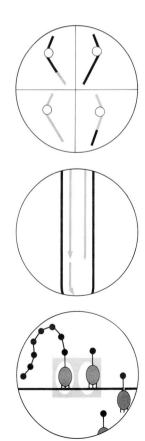

although Darwin could see how selection could act on variants, he had no idea of how these variants were arising. More specifically, he didn't know that these variants were produced by mutation and that more variation resulted from recombination. The central goal of population genetics in this century has been to apply the findings of genetics to evolution, in order to see whether selection of different genotypes could in fact explain evolution as Darwin theorized.

This work has been of paramount importance to biology. Evolution provides perhaps the most powerful unifying theme in biology and so it is essential that it be thoroughly understood. In fact, we now have a good understanding of how mutation, selection, and a few other important phenomena can make evolution occur and that is what we shall discuss in this chapter. The analysis is, in the true tradition of genetics, highly quantitative; the precision of quantitative analysis is essential if we are to make accurate predictions about real situations.

One spin-off of population genetics theory is that it has had major practical applications in human affairs. Examples are seen in medicine, agriculture, and politics. In medicine, population genetics provides a way, as we shall see, of estimating the prevalence of recessive genes that produce heritable defects in our human population. Such information is useful not only in counseling prospective parents about the risks of having affected children, but also in apportioning funds available for medical research into appropriate channels. In agriculture, population genetics is essential in controlling pest populations, either in the introduction of new, resistant varieties of crops, or in the addition of male sterile pests to control reproduction. Finally, in politics, intelligent discussions about the biological consequences of atomic warfare, compulsory sterilization, directed breeding (eugenics), or the relationship between I.Q. and race are dependent on informed opinions of the genetic population parameters involved.

MESSAGE

Population genetics attempts to understand the process of evolution. It is also very useful in practical matters.

Populations

What is a *population?* In general, people use the term very loosely. In the loose sense, expressions like "the pet population of Vancouver" have meaning. But in population genetics, the use is more narrow; to a geneticist, a population is a group of *interbreeding* individuals. Gene exchange, or *gene flow*, within the group, is thus the important defining factor. The largest possible population, therefore, is a species, but local subdivisions are also thought of as populations. By this definition, mankind is a rather loose population divided into tighter local populations by virtue of such factors as language, geography, religion, nationality, and so on. By this definition also, the blue-eyed people, fat people, dwarfs, or criminals in your city are all part of one population. Note, however, that selfing and asexual organisms are not easily accommodated by this definition.

As a species (*Homo sapiens*), we are characterized by many traits: most of us are tall, naked, pigmented, five-fingered, and so on. However, there are people we still consider human who do not possess some of these traits: some are short, some hairy, some have six fingers, some are albino, and so on. The reason for this is that some gene alleles—for example, those causing five fingers—are relatively more common or *frequent* compared with their allelic counterparts —for example, those causing six fingers or more.

A preliminary step in analyzing the genetics of a population, then, is to measure *allele frequencies*. The allele frequencies of all gene loci would define a species very precisely and a knowledge of how these frequencies change is tantamount to a knowledge of how evolution can occur. Theoretically, it is very simple to measure allele frequencies: one simply looks at a population, assumes that each person is just one cell (perhaps the zygote from which he developed), and counts how many alleles of each kind there are for any particular locus. In some practical cases, too, it is relatively simple. Take, for example, the locus controlling the MN blood groups. People are either blood group M, MN, or N: all three genotypes, $L^M L^M$, $L^M L^N$, and $L^N L^N$, at the L locus thus can be identified directly, and allele frequencies come easily, as follows. In a sample of 100 people, let's assume that there are 49 of M blood group, 42 of MN, and 9 of N. Table 14-1 shows how the frequencies of the L^M and the L^N alleles are calculated. Then the proportion of the 200 L alleles that are L^M (= frequency

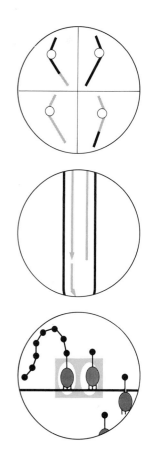

TABLE **14-1**

Blood group	Genotype	Number of people	Number of L^M alleles	Number of L^N alleles
M	$L^M L^M$	49	98	
MN	$L^M L^N$	42	42	42
N	$L^N L^N$	9		18
		100	140	60

of L^M allele) = 140/200 = 0.7 (or 70%) and the proportion of L alleles that are L^N = 60/200 = 0.3 (or 30%). If there are only two alleles at one particular locus, their frequencies are generally called "p" and "q." In this example, we can arbitrarily call the L^M frequency p and the L^N frequency q. If one allele is dominant over the other, p is usually given to the dominant allele, and q to the recessive. In our example:

$$p = 0.7$$
$$q = \underline{0.3}$$

and

$$p \text{ plus } q = 1.0$$

Obviously, if there are only two allelic states, p plus q, by definition must be 1.

One important point to notice here is that the ratios associated with families just don't show up in populations. The relative frequencies of the phenotypes M, MN, and N depend on the allele frequencies. We should *not* necessarily expect Mendelian ratios in a population: a 1:2:1 ratio of M:MN:N would come *only* from matings of MN × MN, and of course all possible matings occur in populations, M × MN, M × M, M × N, N × N, N × MN, and MN × MN.

Equilibrium

Having described the population in terms of genotype and allele frequencies, we now have to investigate the possibility of any spontaneous tendency for these frequencies to change. This is directly relevant to the issue of evolution because evolution surely involves the alteration of allele frequencies and we are trying to discover the basis for this change.

Let's take our population and allow it to form another generation. The simplest assumption we can make is that mating is random (that is, irrespective of genotype). If you think about it, you will see that what this random mating amounts to is that each person contributes an equal number of gametes into a "bucket." These gametes will fuse pair by pair at random and the resultant zygotes will represent the next generation (Figure 14-1). It should be clear that

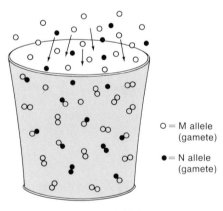

O = M allele
(gamete)

● = N allele
(gamete)

FIGURE **14-1**

the frequencies of gametes in the bucket should be a direct reflection of the allele frequencies in the contributing population; that is, the ratio of L^M-bearing gametes to L^N-bearing gametes in the bucket should be 0.7:0.3. Because we have made no distinction between males and females, the eggs should show 0.7 and 0.3, as well as the sperm. The formation of the next generation can then be visualized as shown in Figure 14-2. The total genotypes of the next generation

Eggs

	L^M-bearing (0.7) = p	L^N-bearing (0.3) + q	
Sperm L^M-bearing (0.7) = p	L^ML^M zygote (0.49) = p^2	L^ML^N (0.21) = pq	⎫ The next
L^N-bearing (0.3) = q	L^ML^N (0.21) = pq	L^NL^N (0.09) = q^2	⎭ generation

FIGURE **14-2**

will therefore be 0.49 blood group M, 0.21 + 0.21 = 0.42 group MN, and 0.09 group N. In other words, there will have been no change in either the allele or the genotype frequencies from one generation to the next: this is an *equilibrium* situation. A special name is given to this equilibrium: it is called a *Hardy-Weinberg equilibrium* (HWE).

MESSAGE

HWE is defined by a set of genotype frequencies of p^2 AA to 2 pq Aa to q^2 aa, in which A and a represent alleles at any locus and p and q are the frequencies of A and a, respectively. If the ratio of AA:Aa:aa is $p^2:2 pq:q^2$, then the population is in HWE, and the allele and genotype frequencies do not change from one generation to the next.

To show that this is not a trick of numbers, we can use the same allele frequencies for L^M and L^N, but non-HWE genotype frequencies, and see the consequences from one generation to the next. Take, for example, a population of 45 M, 50 MN, and 5 N blood groups (Table 14-2).

By using the bucket trick, you'll see that our next generation of zygotes will be represented by frequencies of p^2 (0.49) L^ML^M (M group), 2 pq (0.42) L^ML^N (MN group), and q^2 (0.09) L^NL^N (N group). So this time we *do* have a change: the 45:50:5 ratio becomes a 49:42:9 ratio in one generation. The original population is obviously *not* in equilibrium.

TABLE **14-2**

Blood group	Genotype	Number of People	Number of L^M alleles	Number of L^N alleles
M	L^ML^M	45	90	—
MN	L^ML^N	50	50	50
N	L^NL^N	5	—	10
		100	140	60

Note: p is once again 140/200 = 0.7; q is once again 60/200 = 0.3.

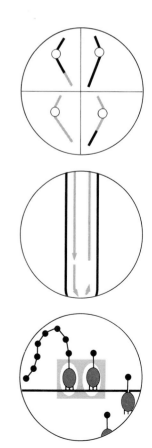

Another important point emerges here, too:

MESSAGE

If a population is not *in HWE, then it takes only one generation of random mating (in our example, the bucket trick) to establish an HWE.*

These are rather surprising conclusions! Not only are there equilibrium genotype frequencies that are potentially persistent through any number of generations, but also any deviation from these frequencies is snapped back into the equilibrium situation in one generation of random mating. Obviously there are no forces here capable of driving evolution, which we define as changes in allele frequencies.

So far, we have been very lax in setting up a generalized model. We will now look at the assumptions inherent in the example we have analyzed. We will then demonstrate the truth of the preceding conclusions algebraically (in other words, for *any* situation). After that, we will relax the assumptions one by one in adapting the model to the world of real populations. When this is completed we will have some clues about how evolution is driven.

Assumptions of the Hardy-Weinberg Model

The following assumptions are inherent in our population analysis:

1. Autosomal loci.
2. No mutation.
3. No selection.
4. Random mating (irrespective of genotype of mate).
5. Very large populations (no sampling errors).
6. No migration.

ALGEBRAIC DEMONSTRATION OF EQUILIBRIUM

Allele Frequencies Don't Change. Instead of the actual numbers used to construct Table 14-2, we use proportions to obtain the frequency of alleles in Table 14-3. All we are saying in this kind of table is that, if *AA*, *Aa*, and *aa* people are present in the ratio $p^2 : 2pq : q^2$, then the gametic contributions of these three genotypes to the next generation bucket will also be in the ratio $p^2 : 2pq : q^2$.

The new frequency of *A* is $p^2 + pq = p(p + q)$: $p + q$ always equals 1, and so the new frequency of *A* is $p \times 1 = p$; in other words, no change in p has occurred. Similarly, q doesn't change either.

TABLE **14-3**

Proportion of people	Genotype	Relative contribution to next generation	
		A	a
p^2	AA	p^2	—
2 pq	Aa	pq	pq
q^2	aa	—	q^2
		$p^2 + pq$	$q^2 + pq$

Genotype Frequencies Don't Change. If we start with p^2 of *AA*, 2 pq of *Aa* and q^2 of *aa*, there are six possible types of mating and these are best represented by Figure 14-3, in which the numbers in the grid indicate the frequencies (or probabilities) of each possible kind of mating, and similar configurations designate genetically identical matings. For example, the mating $Aa \times aa$ will

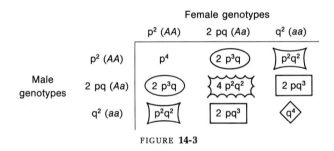

FIGURE **14-3**

occur at a frequency $2\ pq^3 + 2\ pq^3 = 4\ pq^3$. It is all simply the product and sum rules of probability.

We now know the frequencies of all the matings; so the contribution of progeny to the next generation will be a direct reflection of these frequencies. That means that we can examine directly the genotypes of the next generation in Table 14-4.

TABLE **14-4**

Mating	Frequency	Progeny		
		AA	Aa	aa
$AA \times AA$	p^4	p^4	—	—
$AA \times Aa$	$4\ p^3q$	$2\ p^3q$	$2\ p^3q$	—
$AA \times aa$	$2\ p^2q^2$	—	$2\ p^2q^2$	—
$Aa \times Aa$	$4\ p^2q^2$	$1\ p^2q^2$	$2\ p^2q^2$	$1\ p^2q^2$
$Aa \times aa$	$4\ pq^3$	—	$2\ pq^3$	$2\ pq^3$
$aa \times aa$	q^4	—	—	q^4

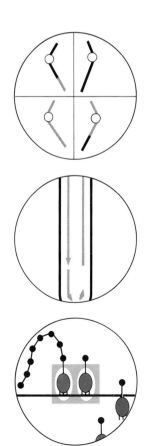

We have simply used Mendelian genetics to derive the progeny columns.

$$\begin{aligned}
\text{The total } AA \text{ progeny} &= p^4 + 2\,p^3q + p^2q^2 \\
&= p^2(p^2 + 2\,pq + q^2) \\
&= p^2(p + q)^2 \\
&= p^2 \times 1^2 \text{ (because } p + q = 1) \\
&= p^2
\end{aligned}$$

$$\begin{aligned}
\text{The total } Aa \text{ progeny} &= 2\,p^3q + 2\,p^2q^2 + 2\,p^2q^2 + 2\,pq^3 \\
&= 2\,pq(p^2 + pq + pq + q^2) \\
&= 2\,pq(p + q)^2 \\
&= 2\,pq
\end{aligned}$$

$$\begin{aligned}
\text{The total } aa \text{ progeny} &= p^2q^2 + 2\,pq^3 + q^4 \\
&= q^2(p^2 + 2\,pq + q^2) \\
&= q^2(p + q)^2 \\
&= q^2
\end{aligned}$$

Thus we have proved that, for any HWE, the next generation of genotype frequencies will be identical with the first. Perhaps you would have believed this on the basis of the single MN group demonstration, but we hope that you are not persuaded by single examples to swallow general laws!

A very useful picture of the HWE model is given in Figure 14-4, which shows all HWE values for all values of p and q. If you remember nothing else from this chapter, try to remember this diagram because it provides a sound jumping-off point for many discussions in population genetics.

Let's try a little applied population genetics. The MN blood group example we discussed earlier was very simple in that we could easily identify each genotype because codominance was involved. However, where there is dom-

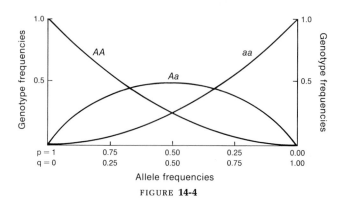

FIGURE **14-4**

inance of one allele over another, genotypes are often identical in phenotype; for example, AA looks the same as Aa. If, however, we assume that an HWE prevails (maybe a big assumption), then we can estimate the number of AA compared with Aa. For example, take the autosomal recessive allele that causes albinism in humans. In human populations, one person in ten thousand is an albino. If our population is in HWE for this locus, then q^2 (the frequency of albinos) must $= 1/10,000$; therefore,

$$q = \sqrt{\frac{1}{10,000}} = \frac{1}{100}$$

$$p = 1 - \frac{1}{100} = \frac{99}{100}$$

and the frequency of Aa genotypes (i.e., "carriers") $= 2\,pq = 2 \times 1/100 \times 99/100$. If we assume that $99/100$ is about 1, then the frequency of carriers is about $2/100$ or $1/50$. Thus, although homozygotes for the disease are rare, the recessive allele is surprisingly common in the population. Similarly, the crippling disease cystic fibrosis results from an autosomal recessive and occurs in $1/2,500$ people; so $q = \sqrt{1/2,500} = 1/50$ and $2\,pq = 2 \times 49/50 \times 1/50$, which is approximately $1/25$.

There are two important practical consequences of this. First, *most* of the deleterious recessive alleles in a population are masked in heterozygous "carriers" (who, although phenotypically normal, will pass the gene on to half of their offspring). We will pursue this point later in the discussion on selection. Second, a known carrier of a recessive deleterious allele can predict the probability of marrying a similar carrier and hence the probability of having a handicapped child. (Try the calculation for yourself). Now we can start relaxing the assumptions underlying our model. The first to go is autosomal linkage.

SEX-LINKED ALLELES

What about alleles whose locus is on the X chromosome? These alleles behave in a slightly more complicated way because a recessive is expressed in a hemizygous male but not in a carrier female. Is an equilibrium still possible? Yes, but, as we shall see, the equilibrium genotype frequencies are different in males and females. Furthermore, the return to equilibrium from nonequilibrium is less fast.

The equilibrium situation for X-linked alleles is described in Table 14-5. You are probably wondering how these genotype frequency distributions are compatible with an overall value of p and q for the entire population. Table 14-6 will convince you of this. In this table we will have to be careful in juggling totals because of the fact that $2/3$ of all X-linked alleles in the population are borne by females, and only $1/3$ by males!

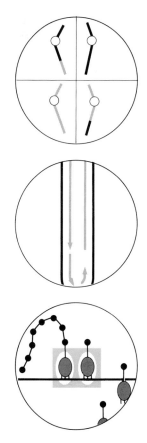

TABLE **14-5**

Sex	Genotype	Frequency in that sex	Chromosomes
Female	AA	p^2	$X^A X^A$
	Aa	2 pq	$X^A X^a$
	aa	q^2	$X^a X^a$
Male	A	p	$X^A Y$
	a	q	$X^a Y$

TABLE **14-6**

Genotype	Relative proportion or frequency of genotype	Relative proportion of alleles	Relative proportion of A and a	
$X^A X^A$	p^2	$2 p^2$	$2 p^2$	—
$X^A X^a$	2 pq	4 pq	2 pq	2 pq
$X^a X^a$	q^2	$2 q^2$	—	$2 q^2$
$X^A Y$	p	p	p	—
$X^a Y$	q	q	—	q
	2	3	3	

$$\text{The total } A \text{ alleles} = \frac{2 p^2 + 2 pq + p}{3}$$

$$= \frac{p}{3}(2 p + 2 q + 1)$$

$$= \frac{p}{3}[2(p + q) + 1]$$

$$= \frac{p}{3} \times 3$$

$$= p$$

The same kind of calculation shows that the frequency of $a = q$. In other words, the equilibrium genotypes stated *do* render an overall mean of p and q for the population frequencies as a whole.

To show that such genotypes do represent equilibrium populations, it is necessary to do only one checkerboard representing a generation of random mating (Figure 14-5).

We see that the same genotype frequencies in each sex are generated in the next generation and, thus, the population *is* in overall equilibrium if these genotype frequencies prevail. The equilibrium genotype frequencies for X-linked alleles demonstrate very graphically why it is that a far higher propor-tion of men suffer from X-linked recessive diseases than do women.

	Eggs	
	X^A (p)	X^a (q)
X-bearing sperm $\;\begin{cases} X^A \text{ (p)} \\ X^a \text{ (q)} \end{cases}$	AA (p^2) female	Aa (pq) female
	Aa (pq) female	aa (q^2) female
Y-bearing sperm $\quad$ Y	A (p) male	a (q) male

FIGURE **14-5**

What if the population is not in equilibrium? The approach to equilibrium on random mating is an interesting one for X-linked loci. In this situation, the allele frequency of, say, A in the female population (p_f) is not necessarily equal to that in the male population (p_m) and neither is necessarily equal to the overall population frequency ($\bar{p}$). Because females carry twice as many X-linked alleles as males, the three terms are related by the expression $\bar{p} = \frac{2}{3}p_f + \frac{1}{3}p_m$. (This is a way of weighting the contribution of each sex to the overall frequency.)

Some other interesting points emerge that can be verified by the inspection

		Eggs	
		A (p_f)	a (q_f)
Sperm	A (p_m)	$p_m p_f$ AA	$p_m q_f$ Aa
	a (q_m)	$p_f q_m$ Aa	$q_m q_f$ aa
	Y	p_f $\;$ A	q_f $\;$ a

FIGURE **14-6**

of a simple checkerboard (Figure 14-6). These points are:

1. Because males get their X-linked alleles only from their mothers, $p_m = p'_f$, in which p' represents the frequency in the previous generation.
2. Because females get X-linked alleles equally from both parents, the p_f will equal the simple mean of the p_m and p_f of the previous generation, or $p_f = \frac{1}{2}(p'_m + p'_f)$.
3. The difference between the female and the male allele frequencies is equal to minus one-half the difference in the previous generation.

$$p_f - p_m = \tfrac{1}{2}(p'_m + p'_f) - p'_f$$
$$= \tfrac{1}{2}p'_m + \tfrac{1}{2}p'_f - p'_f$$
$$= -\tfrac{1}{2}p'_f + \tfrac{1}{2}p'_m$$
$$= -\tfrac{1}{2}(p'_f - p'_m)$$

If we plot this difference we get a visualization of the approach to equilibrium (Figure 14-7). Of course the zero line is never reached in theory, but, for all practical purposes, it soon is.

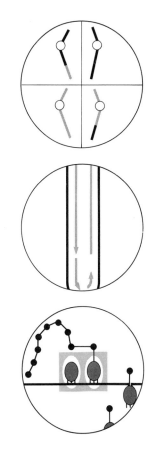

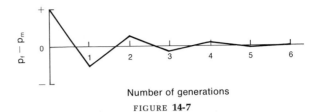

Number of generations

FIGURE **14-7**

It is interesting to note that $\bar{p}$ remains constant throughout because

$$\bar{p}' = \tfrac{2}{3}p'_f + \tfrac{1}{3}p'_m$$

and

$$\begin{aligned}
\bar{p} &= \tfrac{2}{3}p_f + \tfrac{1}{3}p_m \\
&= \tfrac{2}{3}[\tfrac{1}{2}(p'_m + p'_f)] + \tfrac{1}{3}p'_f \\
&= \tfrac{1}{3}p'_m + \tfrac{1}{3}p'_f + \tfrac{1}{3}p'_f \\
&= \tfrac{2}{3}p'_f + \tfrac{1}{3}p'_m = \bar{p}'
\end{aligned}$$

Therefore

$$\bar{p} = \bar{p}'$$

So even in the more complicated case of X-linked alleles, equilibrium, or absence of change, is again the order of the day. The second assumption to be relaxed is mutation.

MUTATION

Alleles change from one to another by various processes of mutation. Because forward and reverse mutations do not usually occur at the same frequency, we can assume that mutation from A to a occurs at rate u, and the reverse, a to A, at rate v. The rates are expressed in proportions of alleles mutating per generation. In any one generation, the actual proportion of total alleles changing from A to a will be pu, and from a to A will be qv. The situation will reach equilibrium (no change of allele frequency) when pu = qv in any generation. This we will call *mutational equilibrium*.

$$pu = qv$$

Therefore

$$(1 - q)u = qv$$

Therefore

$$u - uq = qv$$

Therefore

$$u = uq + vq = q(u + v)$$

Therefore

$$q = \frac{u}{u + v}$$

$$p = 1 - \frac{u}{u + v}$$

$$= \frac{u + v}{u + v} - \frac{u}{u + v} = \frac{v}{u + v}$$

Hence the u and v values will determine the p and q values in a mutational equilibrium. Note that, in this situation, if a mutagen affects u and v *proportionally* the same, p and q are not changed! This is of interest in relation to the potential hazards of environmental mutagens in producing deleterious alleles in human populations, a subject on which we will have more to say later.

In this analysis, we see for the first time how mutant alleles can get into a population. The term used to describe this is *mutation pressure*. For example, if u is greater than v (as it usually is), new *a* mutant alleles are pumped into the population by mutation pressure. If u is 10^{-5} and v is 10^{-6}, theoretically a very high value of q will pertain at mutational equilibrium because

$$q = \frac{10^{-5}}{10^{-5} + 10^{-6}} = \frac{10}{11} \approx 0.9$$

This calculation demonstrates the *potential* of mutation pressure in changing allele frequency, and hence in causing evolution (review Chapter 12).

Now it is appropriate to relax our third assumption in the simple model.

SELECTION

The environment imposes rigid rules and restrictions on what genotypes are successful. And success in evolution is measured by the number of fertile progeny you leave in relation to other individuals in your population. Another word used by geneticists to describe success of this kind is *fitness* (w), and it is a direct index or measurement of the contribution of one genotype to the next generation, in relation to others in the population. Fitness is conversely related to selection. We can measure selection more precisely as the *selection coefficient* (s), which is the *reduction* in contribution to the next generation compared with others in the population. Hence,

$$w = 1 - s$$

When we go to a store to select a new coat or a new hat, we usually have a pretty good idea of what we want and therefore our sense of the role of selection is very positive. The environment, however, is unmindful. It has no idea of what it wants. It simply sets up barriers that impede the basically divergent process of evolution in certain areas, as shown in Figure 14-8.

It is obvious that, if there is no mutation and we start selecting against a particular genotype, then allele frequencies will change. This genotype selected

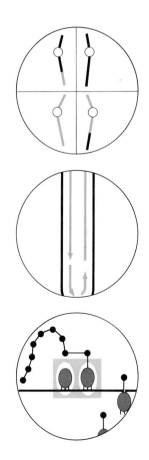

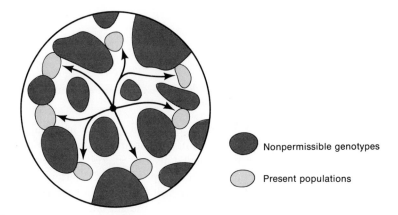

FIGURE **14-8**
The interaction of mutation and selection in evolution. Starting with one genotype at the center, the circle represents all possible genotypes that can be derived from this genotype by sequential mutation. Random mutation generates a diverse array of genotypes (an inherently divergent process) of which only some are compatible with life in the contemporary environment. Thus the divergent process is funnelled between the nonpermissible genotypes to produce today's array.

against can be either the homozygous recessive *aa* or the genotypes containing dominant alleles *AA* and *Aa*. This kind of selection, which pushes p and q values across the board, is called *directional selection,* for obvious reasons. There are several different kinds of selection possible theoretically. Another one that we will say more about later is *stabilizing selection.* In this type of selection, the two homozygous genotypes, *AA* and *aa*, are selected against. But first let's discuss directional selection.

The algebra of directional selection will be illustrated first by selection against the homozygous recessive, *aa.* We will try to be as thorough as possible and juggle both selection and mutation. Table 14-7 shows the way that s is used algebraically.

The total is now $p^2 + 2pq + q^2 - sq^2 = 1 - sq^2$. Hence a fraction sq^2 of all gametes (or alleles, effectively) is eliminated by selection against *aa*. If v is so small compared with u as to be ignored, and if q is small anyway so that v has very few *a* alleles to act on, then the fraction representing *a* alleles entering the population is pu, and sq^2 leaving is equal to pu entering; if q is small, $p \rightarrow 1$,

TABLE **14-7**

Genotype	Initial frequencies	Relative contribution to next generation	Actual contribution to next generation
AA	p^2	1	$1 \times p^2$
Aa	2 pq	1	$1 \times 2\,pq$
aa	q^2	$1 - s$	$(1 - s)q^2$

and

$$q = \sqrt{\frac{u}{s}}$$

(In this chapter, you will learn that the art of population genetics is the art of knowing when to make simplifying assumptions.)

Equilibrium is reached if q and, of course, p are held constant and determined by the ratio of u/s. Again, this is a simplification because reversion does occur. We are not taking reversion into consideration, however, because the algebra becomes very complex and the picture doesn't change much except that a small extra drain (v) is introduced into the equilibrium.

From the standpoint of evolution, it is interesting to consider the efficiency of selection against homozygous recessives in terms of how long it takes to change allele frequencies appreciably. Even if one imposes the maximum selection possible (an s of 1, a fitness of zero) on the *aa* phenotype, the decline of q is slow. In a simple algebraic representation

$$
\begin{array}{ccc}
AA & Aa & aa \\
p^2 & 2\,pq & \text{sterilized or wiped} \\
 & & \text{out } (s = 1,\ w = 0)
\end{array}
$$

The new heterozygote frequency is

$$\frac{2\,pq}{p^2 + 2\,pq}$$

Now

$$\frac{2\,pq}{p^2 + 2\,pq} = \frac{2\,q}{p + 2\,q} = \frac{2\,q}{p + q + q} = \frac{2\,q}{1 + q}$$

The probability of two heterozygotes mating and having an *aa* baby is

$$\frac{2\,q}{1 + q} \times \frac{2\,q}{1 + q} \times 1/4 \ (\text{product rule}) = \frac{q^2}{(1 + q)^2}$$

This is effectively the new q^2 or $q_1{}^2$

$$q_1{}^2 = \frac{q^2}{(1 + q)^2}$$

and so

$$q_1 = \frac{q}{1 + q}$$

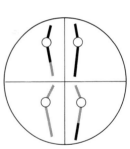

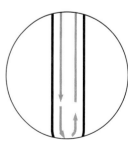

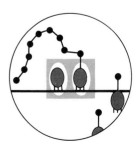

The new generation's q is derived in the same way so that

$$q_2 = \frac{q_1}{1 + q_1}$$

or

$$q_2 = \frac{q/(1 + q)}{1 + [q/(1 + q)]} = \frac{q}{1 + q} \times \frac{1 + q}{1 + q + q} = \frac{q}{1 + 2q}$$

After n generations,

$$q_n = \frac{q}{1 + nq}$$

Therefore

$$q_n + nqq_n = q$$

$$n = \frac{q - q_n}{qq_n} = \frac{1}{q_n} - \frac{1}{q}$$

To plug in some figures, a modest halving of the present allele frequency for cystic fibrosis of $1/50$ to $1/100$ by this most severe form of selection would take $1/(1/100) - 1/(1/50) = 100 - 50 = 50$ generations.

Fifty generations in man would require many centuries; so we see that:

MESSAGE

Any kind of eugenics program proposing to rid a population of a deleterious recessive allele by eliminating homozygotes for it is practically impossible.

The efficiency of selection against a recessive allele is obviously low. As one might expect, the selection coefficient in nature is not often as severe as the value of 1 that we have just discussed. More often, s values are closer to zero; it is these small differences that, over long periods, funnel evolution. Directional selection against dominant alleles is inherently much quicker because there is no heterozygote sheltering effect. Ideally, one could rid a population of all deleterious dominant alleles in one generation, provided that they all reveal themselves in the phenotype. Phenomena of epistasis often prevent this complete penetrance however. Furthermore, new alleles constantly arise through mutation.

The algebra of directional selection against dominant alleles is shown in Table 14-8. Of the alleles lost, the proportion that are A is $sp^2 + pqs$. At equilibrium,

TABLE **14-8**

Genotype	Frequency	Relative contribution to next generation	Gametic contribution
AA	p²	1 − s	p² − sp²
Aa	2 pq	1 − s	2 pq − 2 pqs
aa	q²	1	q²

this must equal the *A* alleles gained by mutation, v × q. If *A* is rare and q → 1, then sp² + pqs = v, or p² + pq = v/s. Because q → 1, p must be very small and p² negligible; so

$$pq = \frac{v}{s}$$

at equilibrium.

Also of interest is directional selection against the allele itself. This is particularly important in haploid organisms, in the haploid gametophyte stages of 2n/n life cycles, and in diploids if the alleles are codominant. In all such cases the selective forces can be thought of as acting on the allele, and the situation can be represented as shown in Table 14-9. Note that *A'* and *A''* are used to

TABLE **14-9**

Genotype	Frequency	Relative contribution to next generation	Gamete contribution
A'	p	1	p
A''	q	1 − s	q − qs

represent alleles instead of *A* and *a*, which imply dominance and recessiveness—a relationship not applicable in this case.

The loss of *A''* alleles, qs, is balanced at equilibrium by forward mutation, *A'* → *A''*, which will be u × p. Again, for a rare *A''* allele, qs = u or

$$q = \frac{u}{s}$$

This concept of selection against an allele is useful because it enables us to think in relatively simple terms about the "lifespan" of a given allele in a population. The selection coefficient against an allele can be thought of as the probability of that allele's surviving to the next generation. Let's assume that

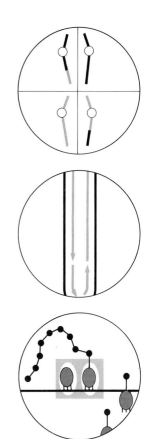

s = 1/10; then only one out of ten alleles of that kind gets eliminated. Doing a further piece of "hard thinking" should enable you to see that what this also means is that, on an average, one of these alleles should last ten generations.

In other words, the lifespan of a given allele is determined by the expression 1/s. If s is 1/100, then each allele will last an average of 1/(1/100) or 100 generations. This is important because it tells us that each new deleterious mutation will ultimately undergo a *genetic death,* which is basically the death of a genotype in a population. This may occur through the death of individuals or through their failure to reproduce. A deleterious allele may have to wait hundreds of generations, but it will ultimately cause that genetic death. The presence of deleterious mutations that undergo genetic death is called the "mutational load" of a population.

Many of the parameters of population genetics—selection, mutation, allele frequency, and the lifespan of alleles—are united graphically in the following analogy based on the model of selection against an allele (the simplest to explain). A trucker buys two new trucks per year (u = 2), and 1/3 of his fleet wears out every year (s = 1/3). His equilibrium size fleet of trucks (q) will be 6 [or 2/(1/3)] and each truck will last three years [1/(1/3)]. The head of each arrow in Figure 14-9 represents a genetic death. Analogous relations hold for other forms of directional selection.

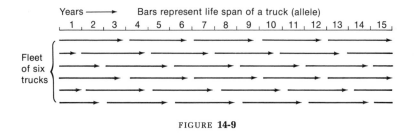

FIGURE **14-9**

At this point it is worth pausing to recapitulate the situations we have discussed. We have seen that either mutation or directional selection can drive p or q values across the board and could theoretically eliminate an allele from a population. These are forces that are obviously important in evolution. However, we have also seen that, when these forces are working in opposite *directions,* allele frequencies can reach an equilibrium at which they will remain constant.

MESSAGE

Evolution can occur only in a nonequilibrium situation.

Stabilizing selection can lead to an equilibrium without the help of mutation. Stabilizing selection usually acts against both homozygotes, AA and aa, and hence the heterozygote is preferred. The two different homozygote types are generally selected against by two different environmental agents, which can act either simultaneously or at different times.

The best understood example of stabilizing selection in humans is that of the hemoglobin locus. A mutant allele, H^S, produces hemoglobin in which the normal three-dimensional properties of hemoglobin are altered, and the $H^S H^S$ genotypes show an anemic phenotype called *sickle-cell anemia*. However, the H^S allele confers a resistance to malaria when in the heterozygote $H^A H^S$. Although heterozygotes suffer from a mild form of anemia, their genotype is superior in a malarial environment because malaria selects against $H^A H^A$ with a selection coefficient of s_1 and the anemic phenotype of $H^S H^S$ selects against its genotype with a selection coefficient of s_2. Such a situation is called *heterozygote advantage*.

$$\text{Fitness of } H^A H^S = 1$$
$$\text{Fitness of } H^A H^A = 1 - s_1$$
$$\text{Fitness of } H^S H^S = 1 - s_2$$

A selection coefficient of s_1 will eliminate a proportion of alleles, $s_1 p^2 / p$ (all of which will be H^A), and one of s_2 will eliminate a proportion, $s_2 q^2 / q$ (all of which will be H^S). These proportions are equal at equilibrium and $s_2 q^2 / q = s_1 p^2 / p$; therefore

$$s_2 q = s_1 p$$

and

$$s_2(1 - p) = s_1 p$$
$$s_2 - s_2 p = s_1 p$$
$$s_1 p + s_2 p = s_2$$
$$p = \frac{s_2}{s_1 + s_2}$$

Similarly,

$$q = \frac{s_1}{s_1 + s_2}$$

Thus a stable situation is set up with p and q values effectively determined by s_1 and s_2. This is a striking example of the critical role of the environment in evolution. The *same* genotypes in a nonmalarial environment have completely different selection coefficients!

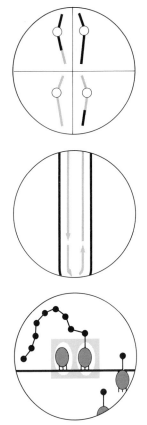

The situation in which there are several common forms or phenotypes associated with one locus is called a *polymorphism*. The phenomenon of heterozygote advantage (discussed above) is one way in which polymorphisms can be maintained in a population. Polymorphisms with severely deleterious homozygous phenotypes are quite common—cystic fibrosis and Tay-Sachs disease are examples—and perhaps the heterozygotes do have some advantage that we don't understand. Other well-known polymorphisms are those associated with blood groups, ABO, MN, and Rh. How are these polymorphisms maintained? Once again, perhaps heterozygote advantage is responsible, either through selection against both homozygotes or, as we shall see later, because of a *molecular* superiority of heterozygotes.

The whole question of polymorphism has received considerable attention in the last several years because it was discovered that as much as 66% of *loci* show polymorphism in a population. These polymorphic allele forms are often detectable only by electrophoresis apparatus, which detects minute differences between proteins that are the products of alleles of a specific gene. (Figure 14-10 shows an electrophoretic gel stained so that only one species of protein is visible. Three forms of the protein are detectable in this example, plus one hybrid band.)

The pressing question is, What are these protein polymorphisms doing in a population? There are two main ways of viewing polymorphism in an attempt to find the answer: First, they might represent insignificant changes at the functional level of the protein. In other words, all of them fulfill the wild-type function. (In our *A/a* symbolism *all* would be considered *A*.) This way of looking at the situation underlines the fact that changes within a population can be nonadaptive—in other words, *selectively neutral*. But, at the same time, this *is* a kind of evolution. We have viewed evolution as a divergent process involving an increase in diversity and changes in allele frequencies. The allele forms in a polymorphism *do* represent diversity and change even though they may be

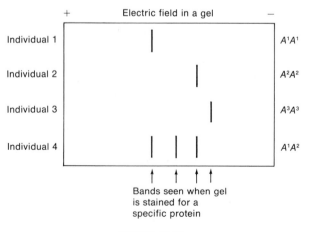

FIGURE **14-10**

nonadaptive. So one would have to conclude that some evolution is non-Darwinian and does not depend on selection. (We have considered this point before in Chapter 12.) The second way of viewing protein polymorphism is that there is heterozygote advantage acting at these loci. In other words, the variation is adaptive, or potentially adaptive. The trouble with this view is that if most loci are polymorphic through heterozygote advantage then there is a very high probability that a zygote will be homozygous for *either* locus 1, *or* 2, *or* locus 3, . . . , *or* locus *n*, and a lot of less fit zygotes might be expected (the expression of this type of weakness is called *segregational load*). However, each of these loci might be expected to contribute individually very little to the overall fitness of an organism and the heterozygote advantage gained from loci 1 and 2 (for example) might more than outweigh the disadvantages from homozygosity at loci 3 and 4.

Genetic polymorphism seems to be the rule in nature rather than the exception. This is supported not only by protein studies but also by careful observation of natural populations. Take a good look at the organisms in your area: most flowers, berries, snails, insects, and even weeds reveal some kind of polymorphism for any characteristic if carefully studied. Are they neutral? Is there heterozygote advantage? One fascinating idea is that heterozygote advantage has actually been selected for in evolution in the following way: Any allele A_2 usually has some favorable effects and some unfavorable ones: evolution acting through modifying genes may change the chemical milieu of the cell so that the favorable effects are dominant and the unfavorable ones recessive. Then A_1A_2 heterozygotes will have only advantages but A_1A_1 and A_2A_2 homozygotes will have both advantages and disadvantages. Alternatively, some polymorphisms may be transient—that is, a polymorphism is analogous to a frame from a movie depicting the gradual replacement of one allele by another. Other polymorphisms may represent range overlaps. The whole subject is exciting and the object of intense debate and study.

An Aside on Comparative Protein Sequences. Although no answers to these questions are available at present, protein evolution is currently the greatest obsession of population geneticists. Another particularly exciting series of studies on protein evolution has also raised more interesting questions than it has answered; this is in the area of comparative protein sequences. Certain proteins are found doing the same job in a wide range of organisms. One such protein is cytochrome *c*, an electron-transport protein found in membranes specializing in oxidation (for example, in mitochondria). Cytochrome *c* from many organisms has been analyzed and the amino acid sequences have been elucidated. Some of them are shown in Figure 14-11. As you can see, some amino acid sites are remarkably constant; others (e.g., site 60) are highly variable. The constant or *conservative* sites probably represent regions of the protein that are functionally crucial either as active sites or as sites involved in precise folding or bonding to adjacent membrane proteins in the electron-transport battery. The variable sites could be simply "filler" or "spacer" regions.

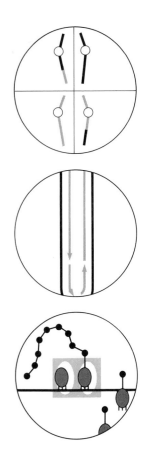

Amino acid sequences of cytochrome c (positions −9 to 23 and 49 to 80).

Positions −9 to 23

Group	Species	−9	−8	−7	−6	−5	−4	−3	−2	−1	1	2	3	4	5	6	7	8	9	10	11	12	13	14	15	16	17	18	19	20	21	22	23
Mammals	Man, chimpanzee									a	G	D	V	E	K	G	K	K	I	F	I	M	K	C	S	Q	C	H	T	V	E	K	G
Mammals	Rhesus monkey									a	G	D	V	E	K	G	K	K	I	F	I	M	K	C	S	Q	C	H	T	V	E	K	G
Mammals	Horse									a	G	D	V	E	K	G	K	K	I	F	V	Q	K	C	A	Q	C	H	T	V	E	K	G
Mammals	Donkey									a	G	D	V	E	K	G	K	K	I	F	V	Q	K	C	A	Q	C	H	T	V	E	K	G
Mammals	Cow, pig, sheep									a	G	D	V	E	K	G	K	K	I	F	V	Q	K	C	A	Q	C	H	T	V	E	K	G
Mammals	Dog									a	G	D	V	E	K	G	K	K	I	F	V	Q	K	C	A	Q	C	H	T	V	E	K	G
Mammals	Rabbit									a	G	D	V	E	K	G	K	K	I	F	V	Q	K	C	A	Q	C	H	T	V	E	K	G
Mammals	California gray whale									a	G	D	V	E	K	G	K	K	I	F	V	Q	K	C	A	Q	C	H	T	V	E	K	G
Mammals	Great gray kangaroo									a	G	D	V	E	K	G	K	K	I	F	V	Q	K	C	A	Q	C	H	T	V	E	K	
Other vertebrates	Chicken, turkey									a	G	D	I	E	K	G	K	K	I	F	V	Q	K	C	S	Q	C	H	T	V	E	K	G
Other vertebrates	Pigeon									a	G	D	I	E	K	G	K	K	I	F	V	Q	K	C	S	Q	C	H	T	V	E	K	G
Other vertebrates	Pekin duck									a	G	D	V	E	K	G	K	K	I	F	V	Q	K	C	S	Q	C	H	T	V	E	K	G
Other vertebrates	Snapping turtle									a	G	D	V	E	K	G	K	K	I	F	V	Q	K	C	A	Q	C	H	T	V	E	K	G
Other vertebrates	Rattlesnake									a	G	D	V	E	K	G	K	K	I	F	T	M	K	C	S	Q	C	H	T	V	E	K	G
Other vertebrates	Bullfrog									a	G	D	V	E	K	G	K	K	I	F	V	Q	K	C	A	Q	C	H	T	C	E	K	G
Other vertebrates	Tuna									a	G	D	V	A	K	G	K	K	T	F	V	Q	K	C	A	Q	C	H	T	V	E	N	G
Other vertebrates	Dogfish									a	G	D	V	E	K	G	K	K	V	F	V	Q	K	C	A	Q	C	H	T	V	E	N	
Insects	Samia cynthia (a moth)					h	G	V	P	A	G	N	A	E	N	G	K	K	I	F	V	Q	R	C	A	Q	C	H	T	V	E	A	G
Insects	Tobacco hornworm moth					h	G	V	P	A	G	N	A	D	N	G	K	K	I	F	V	Q	R	C	A	Q	C	H	T	V	E	A	G
Insects	Screwworm fly					h	G	V	P	A	G	D	V	E	K	G	K	K	I	F	V	Q	R	C	A	Q	C	H	T	V	E	A	A
Insects	Drosophila (fruit fly)					h	G	V	P	A	G	D	V	E	K	G	K	K	L	F	V	Q	R	C	A	Q	C	H	T	V	E		
Lower plants	Baker's yeast				h	T	E	F	K	A	G	S	A	K	K	G	A	T	L	F	K	T	R	C	E	L	C	H	T	V	E	K	G
Lower plants	Candida krusei (a yeast)			h	P	A	P	F	E	Q	G	S	A	K	K	G	A	T	L	F	K	T	R	C	A	E	C	H	T	I	E	A	G
Lower plants	Neurospora crassa (a mold)					h	G	F	S	A	G	D	S	K	K	G	A	N	L	F	K	T	R	C	A	E	C	H	G	E	G	G	N
Higher plants	Wheat germ	a	A	S	F	S	E	A	P	P	G	N	P	D	A	G	A	K	I	F	K	T	K	C	A	Q	C	H	T	V	D	A	G
Higher plants	Buckwheat seed	a	A	T	F	S	E	A	P	P	G	N	I	K	S	G	E	K	I	F	K	T	K	C	A	Q	C	H	T	V	E	K	G
Higher plants	Sunflower seed	a	A	S	F	A	E	A	P	P	G	D	P	T	T	G	A	K	I	F	K	T	K	C	A	Q	C	H	T	V	E	K	G
Higher plants	Mung bean	a	A	S	F	B	E	A	P	P	G	B	S	K	S	G	E	K	I	F	K	T	K	C	A	Q	C	H	T	V	D	K	G
Higher plants	Cauliflower	a	A	S	F	B	E	A	P	P	G	B	S	K	A	G	E	K	I	F	K	T	K	C	A	Q	C	H	T	V	D	K	G
Higher plants	Pumpkin	a	A	S	F	B	E	A	P	P	G	B	S	K	A	G	E	K	I	F	K	T	K	C	A	Q	C	H	T	V	D	K	G
Higher plants	Sesame seed	a	A	S	F	B	E	A	P	P	G	B	V	K	S	G	E	K	I	F	K	T	K	C	A	Q	C	H	T	V	D	K	G
Higher plants	Castor bean	a	A	S	F	B	E	A	P	P	G	B	V	K	A	G	E	K	I	F	K	T	K	C	A	Q	C	H	T	V	D	K	G
Higher plants	Cottonseed	a	A	S	F	Z	E	A	P	P	G	B	A	K	A	G	E	K	I	F	K	T	K	C	A	Q	C	H	T	V	D	K	G
Higher plants	Abutilon seed	a	A	S	F	Z	E	A	P	P	G	B	A	K	A	G	E	K	I	F	K	T	K	C	A	Q	C	H	T	V	E	K	G
	Number of different amino acids										1	3	5	5	5	1	3	3	4	1	4	3	2	1	3	3	1	1	2	4	3	4	2

Positions 49 to 80

Group	Species	49	50	51	52	53	54	55	56	57	58	59	60	61	62	63	64	65	66	67	68	69	70	71	72	73	74	75	76	77	78	79	80
Mammals	Man, chimpanzee	T	A	A	N	K	N	K	G	I	I	W	G	E	D	T	L	M	E	Y	L	E	N	P	K	K	Y	I	P	G	T	K	M
Mammals	Rhesus monkey	T	A	A	N	K	N	K	G	I	I	W	G	E	D	T	L	M	E	Y	L	E	N	P	K	K	Y	I	P	G	T	K	M
Mammals	Horse	T	D	A	N	K	N	K	G	I	T	W	K	E	E	T	L	M	E	Y	L	E	N	P	K	K	Y	I	P	G	T	K	M
Mammals	Donkey	T	D	A	N	K	N	K	G	I	T	W	K	E	E	T	L	M	E	Y	L	E	N	P	K	K	Y	I	P	G	T	K	M
Mammals	Cow, pig, sheep	T	D	A	N	K	N	K	G	I	T	W	G	E	E	T	L	M	E	Y	L	E	N	P	K	K	Y	I	P	G	T	K	M
Mammals	Dog	T	D	A	N	K	N	K	G	I	T	W	G	E	D	T	L	M	E	Y	L	E	N	P	K	K	Y	I	P	G	T	K	M
Mammals	Rabbit	T	D	A	N	K	N	K	G	I	T	W	G	E	D	T	L	M	E	Y	L	E	N	P	K	K	Y	I	P	G	T	K	M
Mammals	California gray whale	T	D	A	N	K	N	K	G	I	T	W	G	E	E	T	L	M	E	Y	L	E	N	P	K	K	Y	I	P	G	T	K	M
Mammals	Great gray kangaroo	T	D	A	N	K	N	K	G	I	I	W	G	E	D	T	L	M	E	Y	L	E	N	P	K	K	Y	I	P	G	T	K	M
Other vertebrates	Chicken, turkey	T	D	A	N	K	N	K	G	I	T	W	G	E	D	T	L	M	E	Y	L	E	N	P	K	K	Y	I	P	G	T	K	M
Other vertebrates	Pigeon	T	D	A	N	K	N	K	G	I	T	W	G	E	D	T	L	M	E	Y	L	E	N	P	K	K	Y	I	P	G	T	K	M
Other vertebrates	Pekin duck	T	D	A	N	K	N	K	G	I	T	W	G	E	D	T	L	M	E	Y	L	E	N	P	K	K	Y	I	P	G	T	K	M
Other vertebrates	Snapping turtle	T	E	A	N	K	N	K	G	I	T	W	G	E	E	T	L	M	E	Y	L	E	N	P	K	K	Y	I	P	G	T	K	M
Other vertebrates	Rattlesnake	T	A	A	N	K	N	K	G	I	I	W	G	D	D	T	L	M	E	Y	L	E	N	P	K	K	Y	I	P	G	T	K	M
Other vertebrates	Bullfrog	T	D	A	N	K	N	K	G	I	T	W	G	E	D	T	L	M	E	Y	L	E	N	P	K	K	Y	I	P	G	T	K	M
Other vertebrates	Tuna	T	D	A	N	K	S	K	G	I	V	W	N	N	D	T	L	M	E	Y	L	E	N	P	K	K	Y	I	P	G	T	K	M
Other vertebrates	Dogfish	T	D	A	N	K	S	K	G	I	T	W	Q	Q	E	T	L	R	I	Y	L	E	N	P	K	K	Y	I	P	G	T	K	M
Insects	Samia cynthia (a moth)	S	N	A	N	K	A	K	G	I	T	W	G	D	D	T	L	F	E	Y	L	E	N	P	K	K	Y	I	P	G	T	K	M
Insects	Tobacco hornworm moth	S	N	A	N	K	A	K	G	I	T	W	Q	D	D	T	L	F	E	Y	L	E	N	P	K	K	Y	I	P	G	T	K	M
Insects	Screwworm fly	T	N	A	N	K	A	K	G	I	T	W	Q	D	D	T	L	F	E	Y	L	E	N	P	K	K	Y	I	P	G	T	K	M
Insects	Drosophila (fruit fly)	T	N	A	N	K	A	K	G	I	T	W	Q	D	D	T	L	F	E	Y	L	E	N	P	K	K	Y	I	P	G	T	K	M
Lower plants	Baker's yeast	T	D	A	N	I	K	K	N	V	L	W	D	E	N	N	M	S	E	Y	L	T	N	P	K	K	Y	I	P	G	T	K	M
Lower plants	Candida krusei (a yeast)	T	D	A	N	K	R	A	G	V	E	W	A	E	P	T	M	S	D	Y	L	E	N	P	X	K	Y	I	P	G	T	K	M
Lower plants	Neurospora crassa (a mold)	T	D	A	N	K	Q	K	G	I	T	W	D	E	N	T	L	F	E	Y	L	E	N	P	X	K	Y	I	P	G	T	K	M
Higher plants	Wheat germ	S	A	A	N	K	N	K	A	V	E	W	E	E	N	T	L	Y	D	Y	L	L	N	P	X	K	Y	I	P	G	T	K	M
Higher plants	Buckwheat seed	S	A	A	N	K	N	K	A	V	T	W	G	E	D	T	L	Y	E	Y	L	L	N	P	X	K	Y	I	P	G	T	K	M
Higher plants	Sunflower seed	S	A	A	N	K	N	M	A	V	I	W	E	E	N	T	L	Y	D	Y	L	E	N	P	X	K	Y	I	P	G	T	K	M
Higher plants	Mung bean	S	T	A	N	K	N	M	A	V	I	W	E	E	K	T	L	Y	D	Y	L	L	N	P	X	K	Y	I	P	G	T	K	M
Higher plants	Cauliflower	S	A	A	N	K	N	K	A	V	E	W	E	E	K	T	L	Y	D	Y	L	E	N	P	X	K	Y	I	P	G	T	K	M
Higher plants	Pumpkin	S	A	A	N	K	N	R	A	V	I	W	E	E	N	T	L	Y	D	Y	L	E	N	P	X	K	Y	I	P	G	T	K	M
Higher plants	Sesame seed	S	A	A	N	K	N	M	A	V	I	W	E	E	N	T	L	Y	D	Y	L	E	N	P	X	K	Y	I	P	G	T	K	M
Higher plants	Castor bean	S	A	A	N	K	N	M	A	V	Q	W	G	E	N	T	L	Y	D	Y	L	E	N	P	X	K	Y	I	P	G	T	K	M
Higher plants	Cottonseed	S	A	A	N	K	N	M	A	V	Q	W	G	E	N	T	L	Y	D	Y	L	E	N	P	X	K	Y	I	P	G	T	K	M
Higher plants	Abutilon seed	S	A	A	N	K	N	M	A	V	N	W	G	E	N	T	L	Y	D	Y	L	E	N	P	X	K	Y	I	P	G	T	K	M
	Number of different amino acids	2	5	1	1	2	6	4	3	2	7	1	7	4	5	2	2	5	4	1	1	3	1	1	1	1	1	1	1	1	1	1	1

Top block (positions 25–48)

25					30					35					40					45			
K	H	K	T	G	P	N	L	H	G	L	F	G	R	K	T	G	Q	A	P	G	Y	S	Y
K	H	K	T	G	P	N	L	H	G	L	F	G	R	K	T	G	Q	A	P	G	Y	Y	Y
K	H	K	T	G	P	N	L	H	G	L	F	G	R	K	T	G	Q	A	P	G	F	T	Y
K	H	K	T	G	P	N	L	H	G	L	F	G	R	K	T	G	Q	A	P	G	F	S	Y
K	H	K	T	G	P	N	L	H	G	L	F	G	R	K	T	G	Q	A	P	G	F	S	Y
K	H	K	T	G	P	N	L	H	G	L	F	G	R	K	T	G	Q	A	P	G	F	S	Y
K	H	K	T	G	P	N	L	H	G	L	F	G	R	K	T	G	Q	A	V	G	F	S	Y
K	H	K	T	G	P	N	L	H	G	L	F	G	R	K	T	G	Q	A	V	G	F	S	Y
K	H	K	T	G	P	N	L	N	G	I	F	G	R	K	T	G	Q	A	P	G	F	T	Y
K	H	K	T	G	P	N	L	H	G	L	F	G	R	K	T	G	Q	A	E	G	F	S	Y
K	H	K	T	G	P	N	L	H	G	L	F	G	R	K	T	G	Q	A	E	G	F	S	Y
K	H	K	T	G	P	N	L	H	G	L	F	G	R	K	T	G	Q	A	E	G	F	S	Y
K	H	K	T	G	P	N	L	N	G	L	I	G	R	K	T	G	Q	A	E	G	F	S	Y
K	H	K	V	G	P	N	L	Y	G	L	I	G	R	K	T	G	Q	A	A	G	F	S	Y
K	H	K	V	G	P	N	L	W	G	L	F	G	R	K	T	G	Q	A	E	G	Y	S	Y
K	H	K	T	G	P	N	L	S	G	L	F	G	R	K	T	G	Q	A	Q	G	F	S	Y
K	H	K	V	G	P	N	L	H	G	F	Y	G	R	K	T	G	Q	A	P	G	F	S	Y
K	H	K	V	G	P	N	L	H	G	F	F	G	R	K	T	G	Q	A	P	G	F	S	Y
K	H	K	V	G	P	N	L	H	G	L	F	G	R	K	T	G	Q	A	A	G	F	A	Y
K	H	K	V	G	P	N	L	H	G	L	I	G	R	K	T	G	Q	A	A	G	F	A	Y
P	H	K	V	G	P	N	L	H	G	I	F	G	R	H	S	G	Q	A	Q	G	Y	S	Y
P	H	K	V	G	P	N	L	H	G	I	F	S	R	H	S	G	Q	A	Q	G	Y	S	Y
T	Q	K	I	G	P	A	L	H	G	L	F	G	R	K	T	G	S	V	D	G	Y	A	Y
G	H	K	Q	G	P	N	L	H	G	L	F	G	R	Q	S	G	T	T	A	G	Y	S	Y
G	H	K	Q	G	P	N	L	N	G	L	F	G	R	Q	S	G	T	T	A	G	Y	S	Y
G	H	K	Q	G	P	N	L	N	G	L	F	G	R	Q	S	G	T	T	A	G	Y	S	Y
G	H	K	Q	G	P	N	L	N	G	L	F	G	R	Q	S	G	T	T	A	G	Y	S	Y
G	H	K	Q	G	P	N	L	N	G	L	F	G	R	Q	S	G	T	T	P	G	Y	S	Y
G	H	K	Q	G	P	N	L	N	G	L	F	G	R	Q	S	G	T	T	P	G	Y	S	Y
G	H	K	Q	G	P	N	L	N	G	L	F	G	R	Q	S	G	T	T	A	G	Y	S	Y
G	H	K	Q	G	P	N	L	N	G	L	F	G	R	Q	S	G	T	T	P	G	Y	S	Y
3	4	2	1	4	1	1	2	1	5	1	3	3	2	1	3	2	1	3	3	6	1	2	3	1

FIGURE **14-11**
Composition of cytochrome *c* in thirty-eight species. The letter *a* indicates that a methyl group is attached to the molecule at the amino end; the letter *h* indicates that the methyl group is absent. Amino acids that have similar chemical properties (largely because of their R, or side, groups) are shown bracketed in the key.

Bottom block (positions 85–104)

	85					90					95					100					104		
I	F	V	G	I	K	K	K	E	E	R	A	D	L	I	A	Y	L	K	K	A	T	N	E
I	F	V	G	I	K	K	K	E	E	R	A	D	L	I	A	Y	L	K	K	A	A	N	E
I	F	A	G	I	K	K	K	T	E	R	E	D	L	I	A	Y	L	K	K	A	T	N	E
I	F	A	G	I	K	K	K	T	E	R	E	D	L	I	A	Y	L	K	K	A	T	N	E
I	F	A	G	I	K	K	K	G	E	R	E	D	L	I	A	Y	L	K	K	A	T	N	E
I	F	A	G	I	K	K	K	T	G	E	R	A	D	L	I	A	Y	L	K	K	A	T	K	E
I	F	A	G	I	K	K	K	D	E	R	A	D	L	I	A	Y	L	K	K	A	T	N	E
I	F	A	G	I	K	K	K	G	E	R	A	D	L	I	A	Y	L	K	K	A	T	N	E
I	F	A	G	I	K	K	K	G	E	R	A	D	L	I	A	Y	L	K	K	A	T	N	E
I	F	A	G	I	K	K	K	S	E	R	V	D	L	I	A	Y	L	K	D	A	T	S	K
I	F	A	G	I	K	K	K	A	E	R	A	D	L	I	A	Y	L	K	Q	A	T	A	K
I	F	A	G	I	K	K	K	S	E	R	A	D	L	I	A	Y	L	K	D	A	T	A	K
I	F	A	G	I	K	K	K	A	E	R	A	D	L	I	A	Y	L	K	D	A	T	S	K
V	F	T	G	L	S	K	K	K	E	R	T	N	L	I	A	Y	L	K	E	K	T	A	A
I	F	A	G	I	K	K	K	G	E	R	Q	D	L	I	A	Y	L	K	S	A	C	S	K
I	F	A	G	I	K	K	K	G	E	R	Q	D	L	V	A	Y	L	K	S	A	T	S	–
I	F	A	G	L	K	K	K	S	E	R	Q	D	L	I	A	Y	L	K	K	T	A	A	S
V	F	A	G	L	K	K	A	N	E	R	A	D	L	I	A	Y	L	K	E	S	T	K	–
V	F	A	G	L	K	K	A	N	E	R	A	D	L	I	A	Y	L	K	Q	A	T	K	–
I	F	A	G	L	K	K	P	N	E	R	G	D	L	I	A	Y	L	K	S	A	T	K	–
I	F	A	G	L	K	K	P	N	E	R	G	D	L	I	A	Y	L	K	S	A	T	K	–
A	F	G	G	L	K	K	E	K	D	R	N	D	L	I	T	Y	L	K	K	A	C	E	–
A	F	G	G	L	K	K	A	K	D	R	N	D	L	V	T	Y	M	L	E	A	S	K	–
A	F	G	G	L	K	K	D	K	D	R	N	D	I	I	T	F	M	K	E	A	T	A	–
V	F	P	G	L	X	K	P	Q	D	R	A	D	L	I	A	Y	L	K	K	A	T	S	S
V	F	P	G	L	X	K	P	Q	E	R	A	D	L	I	A	Y	L	K	D	S	T	E	–
V	F	P	G	L	X	K	P	Q	E	R	A	D	L	I	A	Y	L	K	T	S	T	A	–
V	F	P	G	L	X	K	P	Q	D	R	A	D	L	I	A	Y	L	K	E	S	T	A	–
V	F	P	G	L	X	K	P	Q	D	R	A	D	L	I	A	Y	L	K	E	A	T	A	–
V	F	P	G	L	X	K	P	Q	D	R	A	D	L	I	A	Y	L	K	E	A	T	A	–
V	F	P	G	L	X	K	P	Q	E	R	A	D	L	I	A	Y	L	K	E	A	T	A	–
V	F	P	G	L	X	K	P	Q	D	R	A	D	L	I	A	Y	L	K	E	A	T	A	–
V	F	P	G	L	X	K	P	Q	D	R	A	D	L	I	A	Y	L	K	E	S	T	A	–
V	F	P	G	L	X	K	P	Q	D	R	A	D	L	I	A	Y	L	K	E	S	T	A	–
3	1	5	1	2	2	1	6	9	2	1	7	2	2	2	2	2	2	6	4	4	5	4

KEY

F = Phe
W = Trp
Y = Tyr

I = Ile
L = Leu
M = Met
V = Val

G = Gly

H = His
K = Lys
R = Arg
X = methylated Lys

D = Asp
E = Glu

A = Ala
B = Asn or Asp
C = Cys
N = Asn
P = Pro
Q = Gln
S = Ser
T = Thr
Z = Gln or Glu

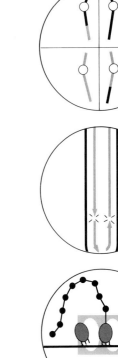

Such studies reveal large sections of identical amino acid sequences and make it extremely likely that the cytochrome *c* genes in different organisms are related to each other evolutionarily. This realization makes it possible to study rates of evolution at the molecular level. Because we know from paleontological studies the approximate eras in which various groups of organisms diverged from each other, we can plot, on one axis of a graph, the time elapsed since divergence between any two groups of organisms and, on the other axis, the average number of amino acid differences per 100 sites along the protein. A plot of this kind for cytochrome *c* and some other ubiquitous proteins is given in Figure 14-12.

The plots, as you can see, are all straight lines! In other words, the rates of protein evolution (the slopes) are constant for each protein, but different proteins evolve at different rates. Why do fibrinopeptides evolve faster than hemoglobins, and hemoglobins faster than cytochrome *c*, and so on? It is thought that this is because fibrinopeptides (for example) have more sites that

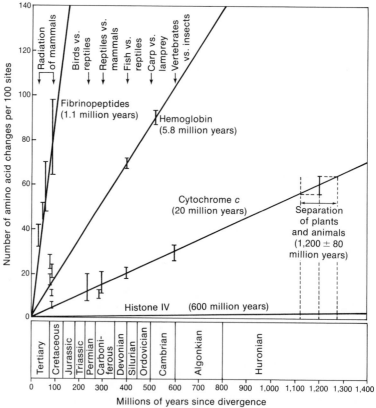

FIGURE 14-12
Evolution of four proteins found in a wide array of organisms: fibrinopeptides, hemoglobin, cytochrome *c*, and histone IV. Slopes represent evolution rates and are standardized as the time required for an amino acid sequence to change by 1%; these values are given in parentheses.

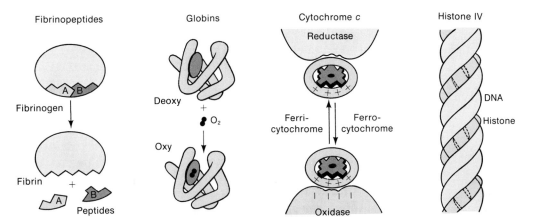

Fibrinopeptides	Globins	Cytochrome c	Histone IV

FIGURE **14-13**
Molecular precision needed in the four proteins: fibrinopeptides simply act as props to prevent fibrin from assuming a clotting conformation; hemoglobin must combine with or release oxygen when appropriate; cytochrome c is part of the electron-transport chain of proteins and must physically fit in with its protein neighbors; and histone IV combines with DNA to promote supercoiling.

can be "tinkered with" without producing a change that is incompatible with function. Histones, at the other extreme, have very few such sites and there is little opportunity for change. This makes sense because fibrinopeptides simply hold blood-clotting proteins apart (presumably there are many ways of doing this), whereas histones have to fit precisely into the grooves and supercoils of the DNA molecule. Some idea of the precision necessary in the four protein types can be seen in Figure 14-13. If this picture is true, it means that the protein evolution we are observing in these cases is at the unimportant sites. Although this is nonadaptive, it once again indicates that some evolution is driven by mutation pressure and is therefore non-Darwinian. One might expect this kind of evolution to follow a straight line as the graphs show.

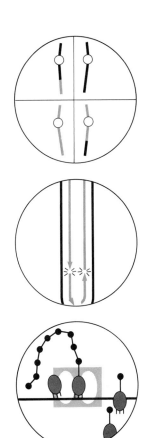

MESSAGE

Mutations can be advantageous, disadvantageous, or neutral at any given time. Selection acts on the first two but not on the last one. Evolution through selection is called Darwinian and evolution by accumulation of neutral mutations is called non-Darwinian.

Biological Selection in Vitro. Many experiments have set up model selective systems, using organisms in the laboratory to test various ideas about how selection works. Not surprisingly, they have shown by and large that selection does work when it is operative at the organism level. After all, the organism is presumably what selection acts on. But organisms cannot devote *all* their energies *directly* to reproductive efficiency: they must eat, move around, photosynthesize, and so forth, as well. What would happen if we could strip an organism of *all* its paraphernalia not directly concerned with reproduction? Could selection still act on the basic essence of the reproductive apparatus, the nucleic acid? After all, the organism is only a vehicle for getting nucleic acid replicated!

A system for investigating this was devised in the laboratory of Spiegelman. A certain phage, $Q\beta$, has RNA as its nucleic acid. The RNA made by the phage's replicase enzyme can be used as a template for further replication, and so on, all in an in vitro system if necessary. Selection was imposed by serially transferring the replicating mixture at decreasing intervals, so that, at each transfer, there would be enrichment for fast-replicating RNA molecules.

It was found that in fact fast-replicating nucleic acids *could* be selected in this manner. Not surprisingly, these turned out to be very short compared with the original. In fact, after seventy-four transfers, 83% of the original genome had been eliminated! Of course this RNA was no longer capable of directing phage synthesis, but it does show that nucleic acids are amenable to selection in a Darwinian manner. This experiment is relevant to population genetics in that it provides some inkling of the Darwinian processes that may have prevailed in the precellular stages of evolution, soon after the origin of life. At that time, selection presumably acted on molecules: we know from the above that such is possible.

Now it is time to relax the fourth assumption.

NONRANDOM MATING

The most common type of nonrandom mating is inbreeding, in which the fertilization of a gamete is by a related partner instead of a partner chosen at random from the population. The most extreme form of inbreeding is selfing, and the results of selfing show many of the features associated with inbreeding in general. The main feature, as we have already examined quantitatively in Chapter 3, is an increase in the proportion of homozygous individuals. Very important here is the bringing to homozygosity of various recessive alleles, most of which are deleterious. The continued inbreeding of a normally outbreeding organism leads, for this reason, to a phenomenon known as the *inbreeding depression* of fitness.

We can illustrate this with an example in which an organism (say, a plant) is heterozygous at five unlinked loci, *Aa Bb Cc Dd Ee*. If this plant and its descendants are continually selfed, the ultimate result will be thirty-two, equally frequent, different genotypes that are pure-breeding: *AA BB CC DD EE*, *AA BB CC DD ee*, *AA BB CC dd EE*, *AA BB CC dd ee*, . . . , *aa bb cc dd ee*. If, on the other hand, random mating is permitted, an HWE will be set up in which $p = 1/2$ and $q = 1/2$ at each locus. In this case, a proportion $(p^2 + q^2 = 1/4 + 1/4 = 1/2)$ of all individuals will be homozygous for one particular locus, but only $(1/2)^5$ or 1/32 will be homozygous at all of the loci. In fact, $(3/4)^5$ or 243/1024 or about 25% will be phenotypically identical with phenotype A- B- C- D- E-. This illustrates a second important point concerning inbreeding in a normally outbreeding population, which is an increase in the variation (or, as it is called, *variance*) between individuals in the population.

Many plants and animals devote a considerable part of their energies to mechanisms (often bizarre) that promote outcrossing. Some plants and animals, however, maintain very high levels of inbreeding. Although there are no easy explanations for precisely why this is so, the answers presumably lie in the vagaries of their ecological niches. Outbreeding populations maintain high levels of potentially adaptive variation in their heterozygous loci and free recombination, and they presumably can quickly call upon this store of variation to meet an evolutionary challenge. Inbreeders are predominantly homozygous and tend to form local pure-breeding races. Some degree of outcrossing, however, can result in enough variant races to maintain what is probably an adequate level of evolutionary flexibility.

We viewed inbreeding depression in terms of bringing deleterious recessive alleles to homozygosity. This is undoubtedly true, but homozygosity apparently has another disadvantage, which is distinguishable from that disadvantage. It seems possible that homozygosity per se is a less favorable state than heterozygosity. We have already talked about heterozygote advantage in terms of selection against homozygotes. We can now postulate an additional mechanism for this kind of advantage, involving interactions at the molecular level. The best clue we have about this is that complementation occurs between the protein gene products of a heterozygous locus, and that the dimer consisting of two different polypeptides is usually more efficient than a dimer consisting of two of either polypeptide. One can visualize this in terms of the mutual "propping up" of various minor faulty areas in the polypeptide. (The diagram for intragenic complementation, which we discussed in Chapter 8, is shown again in Figure 14-14.)

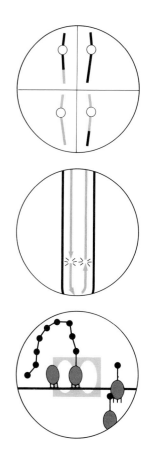

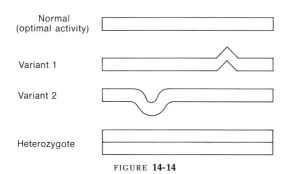

Normal (optimal activity)	
Variant 1	
Variant 2	
Heterozygote	

FIGURE **14-14**

Heterozygote advantage was made use of in developing hybrid corn. Two high-yielding strains can be made into a superefficient hybrid, perhaps by a mechanism of heterozygote advantage at many loci, as shown in Figure 14-14.

High-yielding   High-yielding
pure line (1)   pure line (2)

$$A^1A^1\,B^1B^1\,C^1C^1\,D^1D^1 \times A^2A^2\,B^2B^2\,C^2C^2\,D^2D^2$$
$$\downarrow$$

Very high
yield hybrid

$$A^1A^2\,B^1B^2\,C^1C^2\,D^1D^2$$

This is probably more realistic than the deleterious recessive way of looking at the situation, which would be written:

$$AA\,BB\,cc\,dd \times aa\,bb\,CC\,DD$$
$$\downarrow$$

Hybrid
$Aa\,Bb\,Cc\,Dd$

This is because, if the latter picture were true, one should be able to breed very high yielding pure lines $AA\,BB\,CC\,DD$, and this has been found to be impossible. Both pictures are probably important and, of course, both act in the same direction.

Our present society frowns on inbreeding in human populations. There is a biologically sound basis for this taboo: consider the "risk" in a random marriage compared with the risk in a first-cousin marriage for a "typical" deleterious allele d of frequency $q = 0.01$. The probability of having an affected child in a random marriage is, of course, $q^2 = 0.0001 = 1/10,000$. If a person marries his cousin, however, this risk intrinsically increases *despite* the fact that he doesn't know if the recessive allele is in his family pedigree. Consider the pedigree shown in Figure 14-15. Two first cousins, 1 and 2, decide to marry. What is the

FIGURE **14-15**

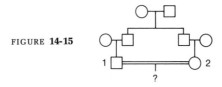

risk of having an affected child dd? The probability of cousin 1 being heterozygous is $2\,pq$, which is $2 \times 99/100 \times 1/100 =$ approximately $2/100$ or $1/50$. If he is heterozygous, then there is a 50% chance that he obtained that allele from his father, who *must* have received it from one of the grandparents. If all this is so, then one of the grandparents was heterozygous; so there is a 50% chance of the prospective bride's father being heterozygous, and, if that is so, a further 50% chance that cousin 2 herself is heterozygous. Multiplying all of these probabilities, we get $1/50 \times 1/2 \times 1/2 \times 1/2$ for the probability that they are both heterozygous and we have to multiply all this by $1/4$ to produce an affected child: $1/50 \times 1/2 \times 1/2 \times 1/2 \times 1/4 = 1/1600$, which is about six times as high as in the random marriage.

In HWE, the assumption of very large populations was intended to get around the problems of *sampling error*. In small populations, the alleles that are destined to become the next generation can be a nonrandom sample of the alleles in the population, simply because of sampling error. For example, in a small village with a constant population of twenty with a p of 0.9 and a q of 0.1, it is possible that the forty alleles that form the next generation will all be *A*. In fact, the probability of this occurring is $(0.9)^{40} = 1.5\%$. You can see that, in such populations, the allele frequencies are prone to large random changes. A q value can drop considerably in one generation, but may recover in later generations by a similar random-sampling process. The allele frequencies are said to *drift*. Genetic drift is most pronounced in smaller populations. When one allele is lost altogether, then the other is said to become *fixed*. Drift is illustrated in small and large populations in Figure 14-16.

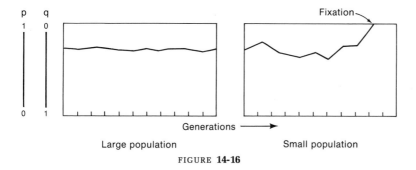

FIGURE **14-16**

A good example of drift in human populations has been observed in the Parma Valley in Italy. Studies of church records in the isolated mountain villages, together with testing blood types of the present inhabitants, have yielded data on blood type allele frequency stretching back many generations. A computer analysis showed that drift was a major factor in the evolution of these village populations.

Drift is probably also important in getting new alleles into a population. Mutation pressure is the ultimate source of evolutionary variability, but it is of interest to wonder how a new allele becomes established. A new recessive mutation in a sperm population, for example, still has a long way to go genetically. It must be the fertilizing sperm, the resultant zygote must mate and have offspring, and there must be inbreeding at some point to bring the allele to homozygosity. It is only at this point that selection can act on the allele; until then, the allele has been effectively drifting into the population, subject to many stages of random sampling for its survival.

Another feature of small populations that we must discuss is the possibility that in small populations many of the advantages of sexual reproduction

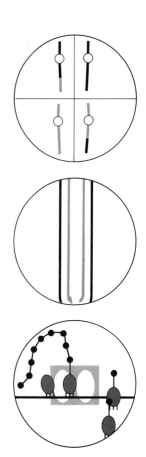

disappear. The advantage of sexual reproduction, of course, is that a new genotype *A B C* is instantly possible through recombination at meiosis, whereas, in an asexual system, each new allele has to arise in sequence to build up *A B C*. (This may have to go through a specific sequence too.) Therefore, in sexual reproduction, mutations are incorporated "in parallel"; in asexual reproduction, "in series" (Figure 14-17). In a small population, however, there is "not enough

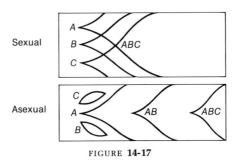

FIGURE **14-17**

room" for mutations to occur in parallel, because numbers are low (Figure 14-18); that is, the difference in speed with which *A B C* appears has vanished.

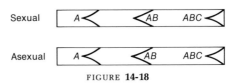

FIGURE **14-18**

The final assumption can now be relaxed; it is a topic that we will not pursue at length.

MIGRATION

Mass immigration and emigration have tremendous potential in changing allele frequencies. The best examples in human populations are seen in newly colonized areas, notably in North and South America. As well as the more recent European and African immigrations, there were several prehistoric waves of immigration from Asia via the Bering Strait. The latter migrations are evident in the blood group allele distribution in Amerindian people, which undoubtedly reflect colonization from the northwestern tip of the continent (Figure 14-19).

Similar migration patterns are visible in Europe, such as that shown in Figure 14-20, which indicates migration in an east to west direction.

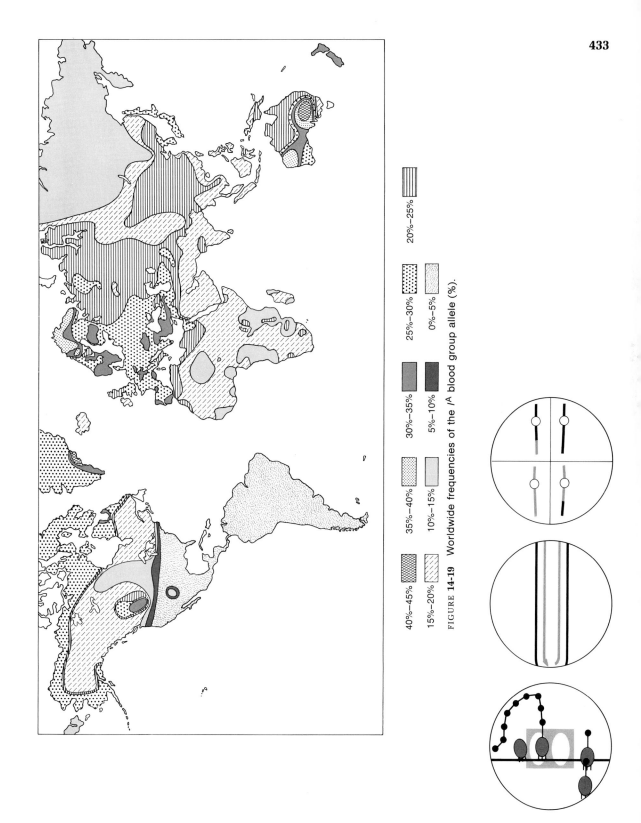

FIGURE **14-19** Worldwide frequencies of the I^A blood group allele (%).

40%–45%   35%–40%   30%–35%

25%–30%   20%–25%

15%–20%   10%–15%   5%–10%

0%–5%

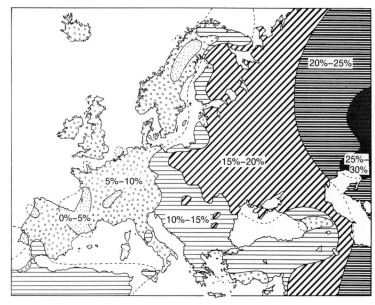

FIGURE **14-20**
Frequency of I^B allele (%) in Europe.

Environmental Mutagens

One of the main areas of public awareness of population genetics is in the subject of environmental mutagenesis, especially in the issue of the supposed hazards of radioactivity and mutagenic materials in foods, pesticides, and so forth. There has been much trumpet-blowing about this issue; all that we want to do is to make a conservative statement. The fact is that it is very difficult to point to actual examples of increases in genetic disease through environmental mutagen exposure. The Atomic Bomb Casualty Commission findings on Hiroshima/Nagasaki did not demonstrate any statistically significant increases in heritable disease. Undoubtedly, there is a possibility that these will turn up in later generations. Yet mammalian populations exposed to radiation for many generations show no decreases in fitness—in fact, for some of the parameters measured, there were slight increases! Why is this? As geneticists, we would have to predict the certain occurrence of mutations, but their effect and persistence in mammalian populations is clearly not so easy to predict. No doubt, our difficulties lie in the extrapolation of results obtained from simple models and organisms to complex organisms like humans. Possible reasons for doubting our simple models are:

1. We have not considered the effects of mutation and selection on polygenic traits. The effects of mutation are potentially less severe.

2. There may be very powerful selective forces acting at stages that we do not understand in full, which could obliterate mutated genomes, or prevent them from entering the sexual gene pool.

3. Perhaps heterozygosity per se is advantageous in some instances. There is some evidence for this in *Drosophila*.

4. Perhaps many loci are under mutational equilibrium: as we have seen, an equal increase of u and v has no effect on q.

Our conclusion: We must be very careful in our use of environmental mutagens *not* because of their *certain* hazards to man, but because we are to a very large degree *ignorant* of their precise effects in our population.

We have only touched on the subject of population genetics with the most basic of analytical tools. We hope, however, that you have gained an understanding of the forces that are capable of driving evolution. The coverage in this chapter has not been complete; much material relevant to evolution is contained in other chapters. The processes of amphidiploidy, duplication, and inversion, for example, are known to have been instrumental in evolution.

Our coverage of evolution has been by and large a coverage of *microevolution,* or within-species evolution. Most biologists believe that the formation of new species (speciation) results from cumulative effects of microevolution occurring over long periods. One exception, of course, is seen in the "instant speciation" in plants through the occurrence of amphidiploids.

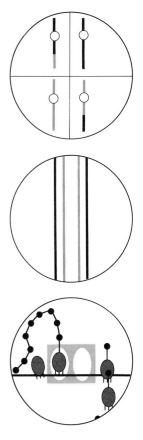

Problems

1. In 1958, it was found in the mining town of Ashibetsu in Hokkaido, Japan, that the number of people with the *M-N* alleles of the *L* gene was:

$L^M L^M$	406
$L^M L^N$	744
$L^N L^N$	332

In the town, there were 741 married couples who had the following distribution of genotypes:

$L^M L^M \times L^M L^M$	58
$L^M L^M \times L^M L^N$	202
$L^M L^N \times L^M L^N$	190
$L^M L^M \times L^N L^N$	88
$L^M L^N \times L^N L^N$	162
$L^N L^N \times L^N L^N$	41

    a. Show whether the population is in HWE with respect to M-N blood type.
    b. Show whether mating is random with respect to M-N blood type.[1]

2. In a survey of Indian tribes in Arizona and New Mexico, it was found that, in most groups, albinos were completely absent or very rare. (There is 1 albino per 20,000 North American caucasians.) However, in three populations, albino frequencies were exceptionally high: 1 per 277 Indians in Arizona, 1 per 140 Jemez Indians in New Mexico, and 1 per 247 Zuni Indians in New Mexico. All three of these populations were culturally, but not linguistically, related. What possible factors might explain the high incidence of albinos in these three tribes?

3. Some women's liberationists suggest that males should be eliminated because there will soon be techniques allowing parthenogenic birth of children. Ignoring the sociological, psychological, and other aspects of such action, discuss from a genetic standpoint, the biological pros and cons of such a unisexual society.

---

[1] Problem 1 is from *Genetics: Questions and Problems* by J. Kuspira and G. W. Walker. McGraw-Hill, 1973.

4. a. Which of the following populations are in HWE?

	AA	Aa	aa
(i)	1.0	0.0	0.0
(ii)	0.0	1.0	0.0
(iii)	0.0	0.0	1.0
(iv)	0.5	0.25	0.25
(v)	0.25	0.25	0.5
(vi)	0.25	0.5	0.25
(vii)	0.33	0.33	0.33
(viii)	0.04	0.32	0.64
(ix)	0.64	0.32	0.04
(x)	0.986049	0.013902	0.000049

b. What are p and q in each population?

c. In population x, it is discovered that the mutation rate from A to a is 5×10^{-6} and that reverse mutation is negligible. What must be the fitness of the aa phenotype?

d. In population vi, the a allele is detrimental, and furthermore the A allele is incompletely dominant so that AA is perfectly fit, Aa has a fitness of 0.8, and aa has a fitness of 0.6. If there is no mutation, what will p and q be in the next generation?

5. AA and Aa individuals are equally fertile. If 0.1% of the population is aa, what selection pressure exists against aa if the mutation rate $A \rightarrow a$ is 10^{-5}?

6. If two cells constitute the sample that made up the germ line of an organism, and a mutation was fixed at an early cleavage division of the zygote, the probability of a gonad that was 50% mutant (q) and 50% wild type (p) would be:[2]

a. p^2
b. q^2
c. $1 - (p^2 + q^2)$
d. $2pq$

7. Given the equilibrium $A \underset{b,}{\overset{m}{\rightleftharpoons}} a$, in which the forward-mutation rate is $m = 10^{-6}$ and the back mutation rate is $b = 10^{-8}$, what will the frequency of gene A be at equilibrium? (Assume that both alleles are neutral.)

8. Gene B is a deleterious autosomal dominant. The frequency of affected individuals is 4.0×10^{-6}. Such individuals have a reproductive capacity about 30% that of normal individuals. Estimate μ, the rate at which b mutates to its deleterious allele B.

---

[2] From *The Challenge of Genetics* by E. H. Simon and J. Grossfield, 1971, Addison-Wesley, Reading, Mass.

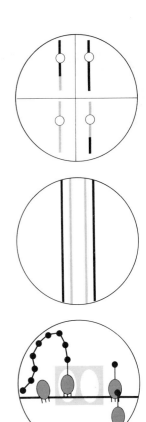

9. Given that color blindness is due to a sex-linked recessive allele and that one male in ten is color-blind,
   a. What proportion of women are color-blind?
   b. By what factor is color blindness more common in men (or, how many color-blind men are there for each color-blind woman)?
   c. In what proportion of marriages would color blindness affect half the children of each sex?
   d. In what proportion of marriages would all children be normal?
   e. In a population that is not in equilibrium, the frequency of the allele for color blindness is 0.2 in women and 0.6 in men. After one generation of random mating, what proportion of the female progeny will be color-blind? what proportion of the male progeny?
   f. What will the allele frequencies be in the male and in the female progeny in part e?[3]

10. In a wild population of beetles of species X, you notice that there is a 3:1 ratio of shiny to dull wing covers. Does this prove that shiny is dominant? (Assume that the two states are caused by the alleles of one gene.) If not, what does it prove? How would you elucidate the situation?

11. a. What fraction of genes does a parent have in common with one of his children?
    b. What fraction do two sibs have in common?
    c. What fraction do two cousins have in common?

12. Of thirty-one children born of father-daughter matings, six died in infancy, twelve were very abnormal and died in childhood, and thirteen were normal. From this information, calculate roughly how many recessive lethal genes we have in our human genomes on the average. For example, if the answer is 1, then a daughter would stand a 50% chance of having it, and the probability of the union producing a lethal combination would be $1/2 \times 1/4 = 1/8$. (So, obviously, 1 is not the answer.) Consider also the possibility of undetected fatalities in utero in such matings. How would they affect your result?

13. In comparing cytochrome c from various organisms, one can calculate a minimal mutation distance necessary to get from one cytochrome c molecule to another by single nucleotide pair changes. Some representative mutation distances are seen in the following table, which shows numbers of nucleotide changes:

	Turtle	Pigeon	Duck	Penguin	Chicken
Turtle	0	10	10	10	10
Pigeon		0	6	6	6
Duck			0	4	4
Penguin				0	2
Chicken					0

From such data it is possible to calculate an "evolutionary tree" for which the length of each line represents mutational distance. Build one. How does it compare with taxonomic relationships based on conventional characters?

---

[3] Courtesy of Clayton Person.

14. The following table shows the number of amino acid differences between various hemoglobins and myoglobin. From the data draw an "evolutionary tree." If you need to postulate any "ancestral forms," specify their exact positions. Also specify where the various "forks" or evolutionary divergences occurred.

Differences in Amino Acid Sequences

Total amino acids	Horse α 141	Human α 141	Horse β 146	Human β 146	Human δ 146	Human γ 146	Whale My 153
Horse α	0	18	84	86	87	87	118
Human α	18	0	87	84	85	89	115
Horse β	84	87	0	25	26	39	119
Human β	86	84	25	0	10	39	117
Human δ	87	85	26	10	0	41	118
Human γ	87	89	39	39	41	0	121
Whale My	118	115	119	117	118	121	0

In humans, α, β, γ, and δ are all coded by *separate* genes. How could this genetic situation have arisen in evolution? (Interpret your evolutionary tree in *genetic* terms.) How could the changes in the total *size* of the protein occur?

15. In an animal population, 20% of the individuals are *AA*, 60% are *Aa*, and 20% are *aa*. What are the allele frequencies? In this population, mating is always with *like phenotype* but is random within phenotype. What genotype and allele frequencies will prevail in the next generation? Such *assortative mating* is common in animal populations. Another type of assortative mating is that which occurs only between *unlike* phenotypes: answer the above question with this restriction imposed. What will the end result be after many generations of mating of both types?

16. In *Drosophila*, a stock isolated from nature had an average of 36 abdominal bristles. By selectively breeding only those flies with more bristles, the mean had been raised to 56 in twenty generations! What would be the source of this genetic flexibility? The 56-bristle stock was very infertile and so selection was relaxed for several generations and the bristle number dropped to about 45. Why did it not drop to 36? When selection was reapplied, 56 bristles were soon attained but this time the stock was *not* sterile: how could this situation arise?

17. You are studying protein polymorphism in a natural population of a certain species of a sexually reproducing haploid organism. You isolate many strains from various parts of the test area and run extracts from each strain on electrophoretic gels. You stain the gels with a reagent specific for enzyme "X," and find that in the population there are a total of, say, five electrophoretic variants of enzyme X. You speculate that these variants represent various alleles of the structural gene for enzyme X.

   a. How would you demonstrate that this is so, both genetically, and biochemically? (You can make crosses, make diploids, run gels, test enzyme activities, test amino acid sequences, etc.) Lay out the steps and conclusions precisely.

   b. Name at least one other possible way of generating the different electrophoretic variants, and say how you would distinguish this possibility from the one mentioned above.

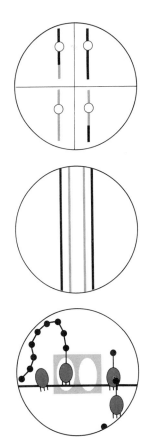

18. Ten percent of the males of a large and randomly mating population are color-blind. A representative group of 1,000 from this population migrates to a South Pacific Island, where there are already 1,000 inhabitants, and where 30% of the males are color-blind. Assuming that Hardy-Weinberg conditions apply throughout (i.e., in the two original populations before emigration, and in the mixed population immediately following immigration), what fraction of males and females are expected to be color-blind in the generation immediately following the arrival of the immigrants?

15 / Conclusion

In the preceding chapters, we have tried to stress the operational techniques that characterize a genetic approach to a problem. In the recent proliferation of new disciplines and interdisciplines, it has often become difficult to distinguish what is a purely genetic approach. It is our view that studying the chromosome damage after radiation (radiation genetics), the DNA content of chromosomes (cytogenetics), or the amino acid sequence of a repressor (molecular genetics) do not illustrate genetic analysis. Although genetic tools such as mutations or special crosses may be used in such studies, the methods of analysis differ from the abstractions of genetic studies.

In a sense, genetics has been consumed by its own success. The spectacular success in molecular biology, utilizing the analytical tools of genetics and biochemistry, has elucidated the basic features of gene replication and function. The fundamental evolutionary unity of all living organisms as embodied in the genetic code of nucleic acids is now universally accepted. The importance of genetic material in development, behavior, and populations is no longer an issue of debate. Furthermore, virtually all areas of biological study now make use of genetic tools in some way or other.

With the fundamental nature of the hereditary material and its transmission from generation to generation and cell to cell already described, it is doubtful that there are many important areas in which genetic analysis alone will be the source of new insights into biological mechanisms. This is especially true for the mechanics of DNA replication, crossing over, chromosome structure, cytoplasmic inheritance, mutation, and eukaryote control systems, which remain to be solved predominantly with molecular techniques. But, as a way of approaching a biological question, and as a major analytic tool, genetics pervades studies in molecular, cell, developmental, behavioral, and population biology. In other words, it is our firm belief that a solid grounding in genetics provides a key to studying virtually all kinds of biological problems.

At this point it is our greatest hope that you are hooked on the subject of genetics. That is not to say that every reader should now be bent on becoming a professional geneticist (Heaven forbid!). Rather, our hope is that you have in your heads a portable genetics that in years ahead you can recall and apply as the need arises. Genetics is simply another way of acquiring knowledge, and as such does not differ from any area of biology. However, the way in which it does differ from most areas in biology is that it lays out a set of concise guidelines or principles for approaching a novel situation with a view to extracting new knowledge. It is this set of guidelines that we called a portable genetics. If you are still uncertain of what they are, riffling through the "messages" should make it clear. Most of them are rather simple, but they represent the essence of the discipline.

If you are now about to take an examination in genetics, the principles that have been treated in this book should be very useful. If you concentrate on remembering them rather than whether chickens have X-Y or W-Z sex determination, or whether XO is called Turner's or Down's syndrome, we think that you will be better off. But education is not only passing exams. When you hear a politician advocating the sterilization of welfare recipients, or when your local hospital spokesman makes a statement about the genetic grounds for abortion, or when you read the word "hybrid" on the packet of seeds you are buying, or when you see your favorite houseplant sectoring or variegating into weird colors, or when the local education authorities are about to impose the teaching of evolutionary theory on the local school system, or when the March of Dimes asks you for a donation for research in birth defects, or when you first hold your child in your arms and see yourself, the portable genetics should click into position and help you to evaluate the situation, or to enrich the experience.

General Questions

The following three sets of questions have been designed for the purpose of review of the entire book.

SET 1

1. A trihybrid $AaBbCc$ is selfed. If all loci are autosomal and unlinked, what proportion of progeny will be phenotypically different from the parent?

   a. 1/8   b. 25%   c. 37/64   d. 7/8   e. 27/256

2. What proportion of the progeny in question 1 will be true breeding?

   a. 1/8   b. 25%   c. 9/64   d. 63/64   e. 0%

3. In rabbits, a $c^{ch}c^h$ ♀ is crossed with a c^hc ♂. What proportion of progeny will look Himalayan?

   a. 1/4   b. 3/4   c. 0   d. 1   e. 1/2

4. Orange in flowers is due to carotenoids. One pure-breeding white plant, blocked in one enzyme stage in carotenoid synthesis, is crossed with a pure-breeding yellow plant blocked in another. If the loci of the genes controlling these enzymes are unlinked, do you expect any orange progeny?

   a. Yes   b. No   c. Can't tell

5. If the answer to question 4 is yes, what proportion of orange do you expect?

   a. 9/16   b. 3/16   c. 1/16   d. 100%   e. My answer to 4 was *No*

6. A double heterozygote is crossed with a double homozygous recessive. To prove that the white block comes before the yellow block, what proportion of progeny would have to be white?

   a. 1/2   b. 9/16   c. 0   d. 3/4   e. Can't tell

7. One pure-breeding white plant was crossed with another pure-breeding white of different origin and all the F_1 were orange. Assuming that there is only one enzyme that will give white when blocked, how many cistrons are there in the corresponding gene?

   a. 2   b. At least 2   c. 2 or less   d. 1   e. Can't tell

8. One pure-breeding white was crossed with another pure-breeding white of different origin and all the F_1 were white. Does this necessarily change your answer to question 7?

   a. Yes   b. No   c. Ambiguous

9. The F_1 in question 8 were selfed and 5,000 progeny obtained, 4,999 of which were white and 1 of which was orange. If this were due to back mutation what would the back-mutation frequency be?

   a. 0.125×10^{-3} b. 0.5×10^{-3} c. 10^{-3} d. 2.0×10^{-4} e. 10^{-4}

10. Could crossing over account for the single orange progeny?

    a. Yes   b. No   c. Meaningless

11. Could a suppressor mutation account for the single orange progeny?

    a. Yes   b. No   c. Meaningless

12. A plant heterozygous for a nonsense suppressor mutation is also heterozygous for the nonsense mutation it suppresses (a UAG mutation in a certain enzyme that normally causes a tall plant, but in mutant form causes a dwarf.) If the plant is selfed, what proportion of progeny will be dwarf? (Assume that the two loci are unlinked.)

    a. 7/16   b. 1/16   c. 9/16   d. 3/16   e. None of these

13. In humans, the recessive diseases phenylketonuria and alkaptonuria are caused by blockages in different enzymes in the same linear metabolic pathway. Do you think it is possible to have both diseases phenotypically?

    a. Yes   b. No   c. Ambiguous

14. In a polygenic system, the effect of each dominant allele is equal and additive. If $A_1a_1 A_2a_2 A_3a_3 A_4a_4$ is selfed, what proportion of progeny will resemble the parent phenotypically? (Assume no linkage.)

    a. 70/256   b. 15/64   c. 3/8   d. 0.0765   e. 1/2

15. A χ^2 test on goals in the Canada/Russia hockey series gives a p value of 0.05%. Would this indicate a significant difference in the teams' abilities?

    a. Yes   b. No   c. Borderline   d. Stupid question

16. If interference is 33.3%, what is the ratio of observed doubles to expected doubles?

    a. 1/2   b. 20.6%   c. 5/6   d. 2/3   e. 1.5

17. Barring gene conversion and mutation, in a *Neurospora* cross $+ + + \times a\,b\,c$ what is the maximum number of different products from a single specific meiosis?

    a. 8   b. 16   c. 4   d. 2/3   e. 1.5

18. If a gene shows 66.6% M_{II} frequency, what proportion of asci will show an $AA\,aa\,AA\,aa$ pattern (do not include its upside-down alternative, $aa\,AA\,aa\,AA$).

    a. 66.6%   b. 6%   c. 33.3%   d. 1/6   e. 1/2

19. From $a/+$ to centromere is 8 map units and from $+/b$ to centromere is 7 map units. The distance from $+/a$ to $+/b$ is therefore

    a. 15 m.u.   b. 1 m.u.   c. 30 m.u.   d. 7.5 m.u.   e. Can't tell

20. What is the maximum recombinant frequency between two loci on the same chromosome *arm?*

    a. 100%   b. 66.6%   c. 25%   d. 50%   e. Meaningless

21. An autotetraploid $AA\,aa$ is selfed. What proportion of the progeny will be a phenotype?

    a. 1/36   b. None   c. 1/4   d. 1/22   e. Need more information

22. A plant of 2n = 20 is crossed with a related one of 2n = 22, and an amphidiploid is produced. How many chromosomes will it have?

    a. 42   b. 21   c. 84   d. 168   e. Can't be done

23. Two normal people have a child who has Turner's syndrome and is color-blind. Where did the gene for color blindness probably originate?

    a. Mother   b. Father   c. Child   d. Can't tell

24. A trisomic *Aaa* has the gene close to its centromere. What proportion of its gametes will be *aa*?

    a. 66.6%   b. 16.7%   c. 1/2   d. None   e. Can't tell

25. Three point mutants *A*, *B*, and *C* are put in heterokaryons with each of a pair of overlapping deletions 1 and 2. Point mutant *A* shows its phenotype with 1 or 2, but *B* shows its phenotype only with 2 and *C* only with 1. What is the order of the point mutants?

    a. *ABC*   b. *BCA*   c. *BAC*   d. *CBA*   e. Can't tell

26. In the cross *a b c d e* × + + + + +, RF values come out as follows: *a–b* is 5 m.u., *a–c* is 5 m.u., *a–d* is 5 m.u., and *a–e* is 5 m.u. What does this suggest?

    a. Deletion   b. Inversion   c. Episome   d. Translocation
    e. Impossible

27. In *Chlamydomonas*, all the progeny of a cross were streptomycin sensitive.* This shows that

    a. *mt*+ was *str*r   b. *mt*+ was *str*s   c. Both parents were *str*r
    d. Both parents were *str*s   e. Several possibilities

28. Which of these populations are in Hardy-Weinberg equilibrium?

	RR	Rr	rr
(i)	1	0	0
(ii)	0	1	0
(iii)	0	0	1
(iv)	1/4	1/2	1/4

    a. (i) only   b. (ii) and (iii)   c. (i), (iii), and (iv)
    d. (ii), (iii), and (iv)   e. (iv) only

29. What recessive allele frequency will be produced at equilibrium by a u/s ratio of 10^{-6}?

    a. Negligible   b. 10^{-12} c. 1/1000   d. 1/10   e. 0.0765

30. An environmental change suddenly makes a recessive allele fully lethal. How many generations will it take to reduce its frequency from 10^{-2} to 10^{-3}?

    a. 900   b. 10   c. 1   d. 30   e. 3,000

31. In *E. coli*, gene *P* enters before *Q*. A recombination experiment showed 72 P^+Q^+ and 3 P^-Q^+. How many recombination units separate the *P* and *Q* loci?

    a. 69   b. 25   c. 75   d. 5   e. 4

32. In a bacterium, the gene order is either *jkl* or *jlk*. Two crosses are made: + + *l* (D) × *jk* + (R), and the same cross except that D and R are switched. In both crosses + + + colonies were selected for, but no difference in frequency was obtained. The gene order is

    a. *jkl*   b. *jlk*   c. *klj*   d. Not enough information

33. A double infection of two T4 *rIIA* mutants will show plaques on *E. coli* strain

    a. B only   b. K only   c. B or K   d. Neither B nor K

---

*Streptomycin resistance/sensitivity is uniparentally inherited in *Chlamydomonas;* only the genotype of the *mt*+ parent is transmitted to products of meiosis.

34. In an *E. coli* phage, five point mutants $(a_1 a_2 a_3 a_4 a_5)$ of a gene are tested for complementation in pair-by-pair combination. The results are

	a_1	a_2	a_3	a_4	a_5
a_1	−	+	+	+	+
a_2		−	−	+	+
a_3			−	+	+
a_4				−	+
a_5					−

What is the minimum number of cistrons in this gene?
   a. 1   b. 2   c. 3   d. 4   e. 5

35. How many mutons in a codon?
   a. 1   b. 2   c. 3   d. 4   e. Variable

36. How many recons in a codon?
   a. 1   b. 2   c. 3   d. 4   e. Variable

37. What level of protein structure is affected mainly by the DNA in a gene?
   a. Primary   b. Secondary   c. Tertiary   d. Quaternary

38. Is it possible to cross two mutants altered in the same codon and get progeny with a wild-type amino acid at that point?
   a. Yes   b. No   c. Has never been done, but possible

39. If 20% of the base pairs of a DNA molecule are TA or AT, what will the G + C content of RNA transcribed from it be?
   a. 0.2   b. 0.5   c. 0.8   d. 0.4   e. 1/4

40. Thymine-requiring *E. coli* cells are put into a medium containing tritiated thymine for one generation only. After two additional generations, what proportion of cells will be labelled?
   a. 75%   b. 1/2   c. 1/8   d. None e. 1/4

41. How many amino acids will be incorporated into protein form when this piece of RNA is read L to R? AUGAUGAAUAAUUUAGUUUAAAAUAGG
   a. None   b. 9   c. 6   d. 27   e. 8

42. Is it possible to revert nonsense codons with hydroxylamine?
   a. Yes   b. No   c. Not enough information

43. Indicate whether, in this partial diploid, β-galactosidase production is

$$\frac{i^c \, o^+ \, z^+}{i^+ \, o^c \, z^-}$$

   a. Repressed   b. Inducible   c. Constitutive

44. How many cistrons are there in an operon?
   a. 1   b. 2   c. 3   d. 4   e. Variable

45. Acridine-induced mutations can be reverted best by

    a. UV   b. X rays   c. Acridine   d. EMS   e. Hydroxylamine

46. Snake venom diesterase (SVD) breaks the phosphodiester bond at the 3′ position. If AP* − PP (* = labelled) is used in a DNA synthesizing system and the resulting DNA is broken up with SVD, the label will come out on

    a. A   b. T   c. G   d. C   e. Several nucleotides

47. Many genes are known to be pleiotropic. Does this invalidate the one-gene-one-enzyme hypothesis?

    a. Yes   b. No   c. Meaningless

48. A 5:3 pattern of a pair of alleles in an ascus could be due to conversion or mutation and the two cannot be distinguished a priori.

    a. True   b. False   c. Possible with further observations
    d. Impossible, even with further observations

49. Given the operon *o A B C*, indicate whether the following changes in *B* will have an effect in either *A* or *C*.

    a. Transversion of one base pair
    b. Transition of one base pair
    c. Deletion of nine base pairs
    d. Inversion of four base pairs
    e. None of the above

50. Which of the following statements is false?

    a. Two deletions will never complement.
    b. Two mutations affecting separate polypeptide strands will almost always complement.
    c. Two mutations affecting the same polypeptide strand will sometimes complement.
    d. If two mutations affecting the same polypeptide strand do complement, it is likely that the final gene product is a multimeric protein.
    e. Several of the above.

SET 2

PLEASE NOTE: *This set of questions has been designed to test your ability to relate what has been covered in many parts of the book to one genetic system. Although all the questions concern one system, they are largely autonomous and the inability to answer one part will not seriously affect your answers to other parts. Read carefully and think before you write.*

In a certain haploid eukaryotic organism, a mutation experiment produced a batch of mutants that were unable to make their own biotin (*bio⁻* mutants); consequently their growth medium had to be supplemented with biotin. It was known that the synthesis of biotin in the cell involves three precursors, which we can call prebiotins 1, 2, and 3. An attempt was made to supplement with these compounds to see if they could substitute for biotin. From their reactions to these tests, the mutants could be grouped into four sets (a plus indicates growth; a minus, no growth):

	Supplement			
Set	Prebiotin 1	Prebiotin 2	Prebiotin 3	Biotin
A	−	−	+	+
B	−	−	−	+
C	+	−	+	+
D	+	+	+	+

1. Using this table, decide in what sequence the precursors of biotin occur in its synthetic pathway.

2. Ascribe each mutant set to a blockage of one of the four enzymic steps defined by these data.

Mutants within any set showed very low recombinant frequencies when intercrossed (e.g., *bio-A₁* × *bio-A₂*), and thus behaved like alleles of a gene. However, crosses between mutants of different sets showed higher recombinant frequencies typical of intergenic crosses:

Cross	Percentage of *bio⁺* recombinants found on minimal medium
Any A × any B	5
Any A × any C	25
Any A × any D	25
Any B × any C	25
Any B × any D	25
Any C × any D	2

3. Is there any evidence of linkage between any of the genes represented by the four sets? Illustrate your answer with linkage maps labelled in map units. (Call the genes *bio-A*, *bio-B*, *bio-C*, and *bio-D*).

Luckily, in this organism the four products of meiosis remain attached as a tetrad. A cross between a *bio-D* and a *bio-C* mutant (*bio-D bio-C⁺* × *bio-D⁺ bio-C*) produced the following (nonlinear) tetrad.

> Meiotic product 1: requires biotin
> Meiotic product 2: requires biotin
> Meiotic product 3: requires biotin
> Meiotic product 4: wild type

4. Is this ascus PD, NPD, or T?

5. The meiotic products 1, 2, and 3 were crossed with *each* of the parents of this cross. Only 3 showed no *bio⁺* recombinants with either parent. What is the genotype of 3?

For several reasons, the experimenters concentrated on the gene *bio-A* and a great deal was learned of its genetics. X-ray treatment produced four partly overlapping deletion mutations that together span the entire *bio-A* gene. The deletions were crossed with each other and the crosses were scored for the appearance of *bio-A⁺* recombinants.

The data obtained were as shown in the following matrix, in which a plus indicates that $bio\text{-}A^+$ recombinants were obtained.

```
            1   2   3   4
      1  |  -   +   -   +
      2  |      -   -   -
      3  |          -   +
      4  |              -
```

6. Draw a deletion map of the *bio-A* gene using this matrix.

There were ten point mutants of the *bio-A* gene available; they were crossed with each of the four deletion mutants, and the crosses were scored for the appearance of $bio\text{-}A^+$ recombinants. The data follow, in which a plus again indicates that $bio\text{-}A^+$ recombinants were obtained.

Deletion mutants	Point mutants									
	i	ii	iii	iv	v	vi	vii	viii	ix	x
1	−	+	+	−	+	−	−	+	−	−
2	+	−	−	+	−	+	+	−	+	+
3	−	−	−	−	+	−	−	+	−	−
4	+	+	+	+	−	+	+	−	+	+

7. Assign each of these point mutants to one of the areas defined by your deletion map. (*Note:* If you *couldn't* do part 6, make up a deletion map and fit these data into it, but be sure to state clearly that this is what you are doing.)

The next experiment was designed to specifically order *some* of the mutant sites in relation to each other and to the two genes flanking the *bio-A* gene, one on each side of it and closely linked to it. The flanking genes were for nicotinic acid requirement ($nic/+$) and pantothenic acid requirement ($pan/+$). The procedure was to intercross some of the *bio-A* alleles and plate meiotic products onto minimal medium *plus* nicotinic and pantothenic acids. Colonies of $bio\text{-}A^+$ recombinants were scored for the presence of the flanking markers by subsequent tests.

Cross			Flanking markers of $bio\text{-}A^+$ recombinants	
nic	*bio-A*$_{ii}$	+		
			nic	*pan*
+	*bio-A*$_{iii}$	*pan*		
nic	*bio-A*$_{viii}$	+		
			nic	*pan*
+	*bio-A*$_{v}$	*pan*		
nic	*bio-A*$_{v}$	+		
			+	+
+	*bio-A*$_{iii}$	*pan*		

Use this information and the deletion analysis and

8. State the order of the alleles (mutant sites) ii, iii, v, and viii in relation to each other.

9. State the order of these alleles in relation to each other *and* in relation to the block of alleles i, iv, vi, vii, ix, and x (which you don't have enough information to order) *and* in relation to the flanking genes; for example, your answer might be *nic*/+-iii-v-block-ii-viii-*pan*/+ (which would be incorrect).

A complementation analysis was performed using the same ten *bio-A* alleles. The following results were obtained, in which a plus indicates strong complementation and a minus, no complementation at all.

	i	ii	iii	iv	v	vi	vii	viii	ix	x
i	−	+	+	−	+	−	−	+	−	+
ii	+	−	−	+	−	+	+	−	+	−

The rest of this matrix is unnecessary. From this information alone

10. How many cistrons are there in the *bio-A* gene?

11. Which alleles are in which cistron?

12. If these data allow you to expand your answer to part 9, expand it.

13. How many types of polypeptide chains would you predict might constitute enzyme A?

Being so easy to isolate, enzyme A, coded by *bio-A*, has been extensively used in analyzing amino acid sequences. Most work has concentrated on *one* of the types of polypeptide chains that constitute the enzyme: the so-called β chain. The most common species of our organism (the one just analyzed) is "known" to be evolutionarily the most primitive (base species), and several other species are known, occupying rather specialized outlying habitats. The β chain in all these species has been analyzed and the relevant sequences are shown in the following chart. The related species are identical in sequence with the base species (shown first in its entirety) except at the positions indicated, where the substituted amino acid is written in.

											Sites										
Species	1	2	3	4	5	6	7	8	9	10	11	12	13	14	15	16	17	18	19	20	21
Base	Gly -	Ile -	Val -	Glu -	Gln -	Cys -	Cys -	Ala -	Ser -	Val -	Cys -	Ser -	Leu -	Tyr -	Gln -	Trp -	Glu -	Asn -	Tyr -	Cys -	Thr -
1		Met			Pro		Trp				Tyr										
2			Ala					Thr													
3		Met			Pro																
4			Ala																		
5		Met																			
6			Ala					Thr			Arg										

14. Draw the simplest "family tree" that describes the evolution of these species.

15. One of the point mutants of the base species (which was chemically induced) was found to be identical with the base species except that sites 16 through 21 were all different; they were Leu-Gly-Lys-Leu-Leu-Tyr. What name is given to the type of mutational event that led to this kind of altered polypeptide?

16. Another chemically induced nucleotide substitution mutant was found to have amino acids 16 through 21 missing completely. What is the name given to this kind of mutational event?

17. Knowing that the mRNA triplet for the tryptophan at position 16 was UGG, what kind of nucleotide pair substitution in the DNA could have been responsible for the mutation event? (Indicate which DNA strand is transcribed.)

18. The mutant in part 15 was vii, one of the original ten *bio-A* point mutants. Which of the original ten were mutants affecting the β polypeptide chain?

Humans have an identical biotin synthesizing pathway, and mutations corresponding to *bio-A*, *B*, *C*, and *D* and their enzymes have all been identified. People who cannot make their own biotin owing to homozygosity of one or more of these mutations are said to have "bionemia." Although this disease is not lethal if untreated, people with bionemia need their diets supplemented with biotin shots. The genes in humans are all autosomal and show independent assortment.

19. Are the alleles for bionemia dominant or recessive?

20. If two individuals who are *both* heterozygous at the *bio-B* and the *bio-D* loci (i.e., both are of the genotype $+/bio\text{-}B$, $+/bio\text{-}D$) marry, what proportion of their children will suffer from bionemia?

21. If this couple has sixteen children, what is the probability that exactly half of them will need biotin shots? (Set up only.)

On a remote island, the only mutant alleles known are at the *bio-A* locus; *all* of them are called *bio* and normal alleles are called $+$ by simple-minded population geneticists. Because the natural vegetation is rich in biotin, the *bio* alleles have not been selected against at all and a 1940 survey showed that the frequency of *bio* was 0.6 (i.e., q = 0.6). If the population mates at random with respect to this gene,

22. What proportion of the matings can produce no homozygous *bio/bio* offspring?

23. If the forward-mutation rate from $+$ to *bio* is 10^{-5}, what reverse-mutation rate is needed to maintain the *bio* frequency at 0.6?

24. If the population is 10 million, how many $+$ alleles change to *bio* alleles per generation?

25. How many *bio* change to $+$?

26. A new form of virus swept the island in 1950. For some unknown reason, only *bio* homozygotes were susceptible and *all* were killed. What was the frequency of *bio* in the remaining population?

27. A study on electrophoretic variants of the *bio* gene product has revealed eight isozymes. How is this compatible with the existence of only two alleles, *bio⁺* and *bio?*

SET 3

These crossword puzzles first appeared in the *Journal of Biological Education.* They were devised by L. Mullenger and D. M. Hawcroft and are reproduced here by permission of the authors.

Most of the words in the puzzles have been used in this book; the few that have not can be pieced together using the other clues.

A Genetic Crossword

A Genetic Crossword is from the *Journal of Biological Education* (1974) 8(4):240-241.

Across

1. Variation in the number of chromosome sets (8)
5. Those genetically controlled characters which express themselves in both homozygous and heterozygous conditions (8)
9. The total genetic complement of a cell (6)
11. The various expressions of 9 across (10)
12. An organism with more than two sets of identical chromosomes which may therefore freely homologise (13)
15. A quantitative expression of Mendelian segregations (5)

17. A rare, recessive genetic disease manifested in the inability to produce homogentisic acid oxidase which leads to the incomplete breakdown of phenylalanine and the production of black urine (12)
18. The smallest element in a gene capable of independent mutation (4)
21. The compartmentalisation of the chromosomes into this is a distinguishing feature of eukaryotes (7)
22. with 27 across. A genetic test to determine the zygosity of an individual, made by crossing with its parents (4-5)
24. Large baglike enclosing structure (3)
25. Stationary female gamete often supplied with food reserves for developing zygotes (3)
27. See 22 across
29. A type of fleshy fruit produced by a considerable swelling of the receptacle of the flower. A great deal of genetic breeding has been involved in producing present day varieties (4)
30. An internal reversal of part of a chromosome which may prevent homologous pairing and influence phenotypic expression (9)
33. One of 28 down (abbrev.) (3)
34. Perhaps the present day ultimate species produced by natural selection (3)
35. Chromosomal crossing over and interchange of genetic material make these apparent (9)
37. Allelic sites on the chromosome (4)
38. Accepted notation for that part of the human chromosome complement which is typically male (2)
39. The result of reassociation of genes (11)
41. Microtubules responsible for chromosome segregation (8)
42. The pairing of homologous chromosomes (8)

Down

1. The improvement of the human race by social selection rather than evolutionary selection of characters (8)
2. A floral characteristic, intended to promote outbreeding and shown by about 50% of flowers of many *Primula* species (3)
3. A statistical term indicating the number of factors that are free to vary independently (abbrev.) (2)
4. The ability to taste this substance depends on the presence of a dominant gene and is present in about 70% of the American white population (19)

6. This type of breeding preserves heterozygosity (3)
7. Different homologous chromosomes of a diploid do this when separating into gametes and this results in a large part of the variability seen in the offspring (6)
8. A polyploid; normally sterile (8)
10. A phenotypic mutant on linkage gp IV of *Neurospora crassa* giving rise to colonial growth (3)
11. The study of human genetics often begins with this individual (7)
12. A red stain commonly used in cytogenetics during microscopical examination of chromosomes (12)
13. This gives rise to progeny ratios which would not be predicted on the Mendelian principle of complete assortment of genes during meiosis. It provides a very useful way of mapping gene distribution on chromosomes (7)
14. Fluctuations in gene frequency in a population during succeeding generations (5)
16. An abbreviation sometimes used to refer to a homozygote for a dominant gene (2)
19. A double-stranded polynucleotide (abbrev.) (3)
20. A descriptive term for the cross between heterozygous and homozygous recessive individuals used to determine the degree of linkage between loci (4)
22. The name given to chromatin containing bodies situated just inside the nuclear envelope. They are usually restricted to female cells and are sufficiently obvious to be used in sex determination (4)
23. Two mutations on the same chromosome are said to be in this configuration (3)
26. The replacement of sexual reproduction with asexual mechanisms in which there is no fusion of gametes (8)
28. Of the same parents (8)
31. Chromatin may exist in this geometrical arrangement during cell interphase (3)
32. In classical genetics thought to be the particulate indivisible independent determinant of a character (4)
35. A population; the individuals of which are genetically identical (5)
36. The quotients produced when the sums of various series of quantities are divided by the numbers of quantities in the series (5)
37. A geometric figure often shown during pairing of only partially homologous chromosomes (4)
40. Indicates the proportion of cells in mitosis in a population (abbrev.) (2)

A Molecular Biology Crossword

A Molecular Biology Crossword is from the *Journal of Biological Education* (1973) 7(3):56-57.

Across

1. A fundamental property of DNA giving rise to the stability of the molecule and its potential for replication and transcription (13, 4, 7)

7. In this type of *E. coli* mutant amino acid starvation leads to the inhibition of RNA synthesis as well as blocking protein synthesis. A further mutation at a single locus (RC^rel) gives rise to a "relaxed" mutant and RNA synthesis continues normally (9)

11. Amino acid linkers, the synthesis of which requires factors G, Tu and Ts, GTP and the enzyme peptidyl transferase (7, 5)

12. A blood factor governed by a single pair of genes, dominant in 85% of white males. Under certain conditions it may give rise to a foetal condition known as erythroblastosis (6)

15. A term often applied to newly formed proteins or nucleic acids (7)

16. A bacterial strain prototrophic with respect to tryptophan (3, 11)

18. Yanofsky demonstrated that polypeptides and polynucleotides exhibited this (11)

19. A system of cytoplasmic striations radiating from the centriole and composed of microtubules (5)

20. Scientist associated with early work on codon assignment (5)

21. Proteins that are produced by regulator genes and control operonic activity through combination with the appropriate operator gene (10)

24. A division leading to the production of four monoploids from one diploid (7)

25. One of the first tumour viruses isolated; gives rise to Avian sarcomas (4)

26. The result of the fusion of two haploid gametes (1, 6)

27. The degradation of damaged DNA by endo- and exonucleases and its replacement by incorporation of new nucleotides (6)

29. The product of a genetic process either inter- or intrachromosomal (11)

32. An individual having an altered genetic constitution as a result of the uptake of DNA fragments from solution (12)

33. The original gene pool (5)

34. Pyridine phosphonucleotide involved in the degradation of pyrimidine nucleotides although its normal role as a coenzyme is in anabolic metabolism (abbrev.) (4)

37. The separation of a newly synthesized polypeptide chain from the terminal tRNA is a response to this (10, 5)

38. with (50) across. Archaic term used to describe protozoan asexual multiplication (6, 7)

39. A protein factor responsible for the termination of RNA polymerase synthetic activity (3)
40. A genetic element having the property of leading an autonomous existence in the cytoplasm or alternatively of becoming integrated into the genophore (7)
42. The junction of two chromosomes which often leads to an interchange of material (7)
43. A temporal or spatial gene distribution as determined by experiment (3)
45. The average used in computing doubling or generation times (4)
48. Having injected their DNA, phages leave this outside the host (5)
50. See 38 across.
52. An RNA-containing, helical, rod-shaped, plant virus the nucleic acid and protein capsid of which can be completely separated and reassembled with the retention of infectivity (abbrev.) (3)
53. The "life-support" system for a virus (4, 4)
55. DNA can act as this because of the nature of its specific complementary base pairing (8)
56. The shifting of segments of chromosome either inter- or intrachromosomally. The term is also applied to the relative movement of the ribosome and mRNA (13)
59. The smallest living unit of biological structure and function (4)
60. Chargaff's method for determining the frequency with which nucleotides are adjacent to each other in a polynucleotide (7, 9, 8) British spelling.

Down

1. Polynucleotides are said to be this when they may be subdivided operationally into functional units by the complementation test (9)
2. Those genes used in the detection of recombinants (7)
3. Oögamete (3)
4. Transcription demonstrates this, unlike replication, as only one strand of the DNA is copied (9)
5. That stage of viral growth during which no infective particles may be detected in the host cell (7, 6)
6. A suppressible mutation leading to polypeptide chain termination; applied to the codon UAG (5)
8. Nonphosphorylated 34 across required as a cofactor for the activity of *E. coli* polynucleotide ligase. (abbrev.) (3)
9. Being in the state of forming 57 down although the exact mechanism is still in dispute (11)

10. An extracellular effector molecule which intracellularly gives rise to some protein synthesis (9)
13. The DNA strand which is transcribed (5)
14. The type of recombinant formed by genetic recombination through crossing over (10)
17. The formation of a polyribonucleotide on a polydeoxyribonucleotide template by the enzyme RNA polymerase—occasionally reversed (13)
21. The specific inhibition of enzyme synthesis (10)
22. A polynucleotide, usually single stranded although demonstrating specific base pairing. Very low in thymine content (11, 4)
23. The mode of DNA replication as demonstrated by Meselson and Stahl (16)
26. This is required to activate amino acids before their attachment to their specific transfer RNA molecule (abbrev.) (3)
27. The smallest genetic unit capable of undergoing a recombinational event (Benzer) (5)
28. That locus where RNA polymerase commences transcription (8)
30. The nucleoside found in RNA replacing thymidine (7)
31. The renaturation of DNA after melting. This property is used to demonstrate that DNA and mRNA are complementary (9)
35. Genes from different linkage groups in the act of associating together (13)
36. Term used to describe two identical chromosomes. Also used to describe structures which are similar in different organisms because of a common ancestral origin (10)
41. Transcriptions (10)
42. A substance distributed throughout the interphase nucleus which later gives rise to visible chromosomes (9)
44. The end product of the Central Dogma of molecular biology (7)
46. The site of formation of ribosomal RNA in eukaryotes (9)
47. The stage of nuclear division involving the movement of chromosomes from the equator to the poles (8)
49. Scientist who proposed the relationship between genes and enzyme production (5)
51. Scientist who demonstrated that transformation of the *Pneumococcus* was the result of the transfer of genetic material (5)
54. Membranous structures around which chloroplast DNA is dispersed (5)
57. The polynucleotide forming the double helix (abbrev.) (3)
58. This may act in an informational, structural or amino acid carrying role (abbrev.) (3)

Other Reading

It is customary in introductory texts to include long lists of reference material. We intend to break with this custom because in our experience these bibliographies rarely attain the use that would merit their inclusion. Furthermore, the sophistication of modern library indexing and information retrieval systems (many computerized) enable the curious student to generate bibliographies in his own specific area of interest with remarkable ease. Nevertheless, some sources are so useful that they demand attention, and we have included a sampling of these.

General

Virtually all *Scientific American* articles on genetics are useful and interesting to the introductory student. Several specific ones are given below, but a complete catalog of these articles can be obtained from W. H. Freeman and Company, 660 Market Street, San Francisco 94104.

We have decided that probably the most useful book to have in addition to this one is *Towards an Understanding of the Mechanism of Heredity,* 3d ed., by H. L. K. Whitehouse (1973, Edward Arnold Limited, London), which contains an excellent detailed discussion of classical and molecular genetics, but not of population genetics.

Other more detailed texts than ours that we recommend are *Genetics* by U. Goodenough and R. P. Levine (1974, Holt, Rinehart and Winston, Inc., New York) and *Genetics* by M. W. Strickberger (1968, The Macmillan Company, New York).

Specific

CHAPTER 1

Carlson, E. A. 1966. *The Gene: A Critical History*. Saunders.
Olby, R. C. 1966. *Origins of Mendelism*. Constable and Co., London.
Sturtevant, A. H. 1965. *A History of Genetics*. Harper and Row.

CHAPTER 2

McLeish, J., and B. Snoad. 1958. *Looking at Chromosomes*. Macmillan.
Peters, J. A., ed. 1959. *Classic Papers in Genetics*. Prentice-Hall.

Bodmer, W. F., and L. L. Cavalli-Sforza. 1970. Intelligence and Race. *Sci. Amer.* 223(4):19–29. Offprint 1199.
Hutt, F. B. 1964. *Animal Genetics.* Ronald Press.
Lerner, I. M., and W. J. Libby. 1976. *Heredity, Evolution, and Society,* 2d ed. W. H. Freeman and Company.
Peters, J. A., ed. 1959. *Classic Papers in Genetics.* Prentice-Hall.

CHAPTER 4

Peters, J. A., ed. 1959. *Classic Papers in Genetics.* Prentice-Hall.
Sturtevant, A. H., and G. W. Beadle. 1962. *An Introduction to Genetics.* Dover.

CHAPTER 5

Fincham, J. R. S., and P. R. Day. 1971. *Fungal Genetics,* 3d ed. Blackwell, London.
Kemp, R. 1970. *Cell Division and Heredity.* Edward Arnold, London.
Ruddle, F. H., and R. S. Kucherlapati. 1974. Hybrid Cells and Human Genes. *Sci. Amer.* 231(1):36–44. Offprint 1300.

CHAPTER 6

Hayes, W. 1968. *The Genetics of Bacteria and Their Viruses,* 2d ed. Wiley.
Stent, G. S. 1971. *Molecular Genetics.* W. H. Freeman and Company.

CHAPTER 7

Dupraw, E. J. 1968. *Cell and Molecular Biology.* Academic Press.
Dupraw, E. J. 1970. *DNA and Chromosomes.* Holt, Rinehart and Winston.
Stern, C. 1973. *Principles of Human Genetics,* 3d ed. W. H. Freeman and Company.
Swanson, C. P., T. Mertz, and W. J. Young. 1967. *Cytogenetics.* Prentice-Hall.

CHAPTER 8

Lawrence, C. W. 1971. *Cellular Radiobiology.* Edward Arnold, London.
Peters, J. A., ed. 1959. *Classic Papers in Genetics.* Prentice-Hall.
Stent, G. S. 1971. *Molecular Genetics.* W. H. Freeman and Company.

CHAPTER 9

Benzer, S. 1962. The Fine Structure of the Gene. *Sci. Amer.* 206(1):70–84. Offprint 120.
Stahl, F. W. 1969. *The Mechanics of Inheritance,* 2d ed. Prentice-Hall.
Watson, J. D. 1970. *The Molecular Biology of the Gene,* 2d ed. Benjamin.

CHAPTER 10

Cohen, S. N. 1975. The Manipulation of Genes. *Sci. Amer.* 233(1):24–33. Offprint 1324.
Kornberg, A. 1974. *DNA Synthesis.* W. H. Freeman and Company.
Watson, J. D. 1970. *The Molecular Biology of the Gene,* 2d ed. Benjamin.

CHAPTER 11

Crick, F. H. C. 1962. The Genetic Code. *Sci. Amer.* 207(4):66–74. Offprint 123.
Nirenberg, M. W. 1963. The Genetic Code: II. *Sci. Amer.* 208(3):80–94. Offprint 153.
Crick, F. H. C. 1966. The Genetic Code: III. *Sci. Amer.* 215(4):55–62. Offprint 1052.
Stent, G. S. 1971. *Molecular Genetics.* W. H. Freeman and Company.
Watson, J. D. 1970. *The Molecular Biology of the Gene,* 2d ed. Benjamin.

CHAPTER 12

Drake, J. W. 1970. *The Molecular Basis of Mutation.* Holden-Day.
Fincham, J. R. S., and P. R. Day. 1971. *Fungal Genetics,* 3d ed. Blackwell, London.
Whitehouse, H. L. K. 1973. *Towards an Understanding of the Mechanism of Heredity,* 3d ed. Edward Arnold, London.

CHAPTER 13

Benzer, S. 1973. The Genetic Dissection of Behavior. *Sci. Amer.* 229(6):24–37. Offprint 1285.
Britten, R. J., and D. E. Kohne. 1970. Repeated Segments of DNA. *Sci. Amer.* 220(4):24–31. Offprint 1173.
Markert, C. L., and H. Urpsprung. 1971. *Developmental Genetics.* Prentice-Hall.
Sager, R. 1972. *Cytoplasmic Genes and Organelles.* Academic Press.
Wood, W. B., and R. S. Edgar. 1967. Building a Bacterial Virus. *Sci. Amer.* 217(1):60–74. Offprint 1079.

CHAPTER 14

Briggs, D., and S. M. Walters. 1969. *Plant Variation and Evolution.* McGraw-Hill.
Cavalli-Sforza, L. L., and W. F. Bodmer. 1971. *The Genetics of Human Populations.* W. H. Freeman and Company.
Crow, J. F., and M. Kimura. 1970. *An Introduction to Population Genetics Theory.* Harper and Row.
Falconer, D. S. 1970. *Introduction to Quantitative Genetics.* Ronald Press.
Ford, E. B. 1971. *Ecological Genetics,* 3d ed. Chapman and Hall, London.
Lewontin, R. C. 1974. *The Genetic Basis of Evolutionary Change.* Columbia University Press.

CHAPTER 15

Etzioni, A. 1973. *Genetic Fix.* Macmillan.
Hilton, B., D. Callahan, M. Harris, P. Condliffe, and B. Berkley, eds. 1973. *Ethical Issues in Human Genetics.* Plenum.
Mertens, T. R., ed. 1975. *Human Genetics: Readings on the Implications of Genetic Engineering.* Wiley.
Ramsey, P. 1970. *Fabricated Man.* Yale University Press.

Figure Credits

FIGURE 1-1 From *Botany* by J. B. Hill, H. W. Popp, and A. R. Grove, Jr. McGraw-Hill Book Company, 1967.

FIGURES 2-1, 2-2 Photographs by C. J. Marchant and A. M. Adamovich.

FIGURE 2-12 From C. Stern, W. R. Centerwall, and S. S. Sarkar, *The American Journal of Human Genetics* 16(1964):467. By the permission of Grune & Stratton, Inc.

FIGURE 3-2 From Claussen, *Zeitschrift für induktive Abstammungs- und Vererbungslehre* 76 (1939).

FIGURE 3-15 From *Heredity, Evolution, and Society,* 2d ed., by I. M. Lerner and W. J. Libby. W. H. Freeman and Company. Copyright © 1976.

FIGURES 3-16, 3-17 From "Intelligence and Race" by W. F. Bodmer and L. L. Cavalli-Sforza. Copyright © 1970 by Scientific American, Inc. All rights reserved.

FIGURE 3-19 From *Heredity, Evolution, and Society,* 2d ed., by I. M. Lerner and W. J. Libby. W. H. Freeman and Company. Copyright © 1976.

FIGURE 5-2 From *Introduction to Biostatistics* by R. R. Sokal and F. J. Rohlf. W. H. Freeman and Company. Copyright © 1973.

FIGURE 5-24 From "Hybrid Cells and Human Genes" by F. H. Ruddle and R. S. Kucherlapati. Copyright © 1974 by Scientific American, Inc. All rights reserved.

FIGURE 5-25 Photograph by Fred Dill.

FIGURE 5-26 From "Hybrid Cells and Human Genes" by F. H. Ruddle and R. S. Kucherlapati. Copyright © 1974 by Scientific American, Inc. All rights reserved.

FIGURE 6-2 From *Molecular Biology of Bacterial Viruses* by G. S. Stent. W. H. Freeman and Company. Copyright © 1963.

FIGURES 6-4, 6-5 From E. L. Wollman, F. Jacob, and W. Hayes, *Cold Spring Harbor Symposia on Quantitative Biology* 21(1950):141. Electron micrograph by T. F. Anderson.

FIGURE 6-11 From *Molecular Genetics* by G. S. Stent. W. H. Freeman and Company. Copyright © 1971.

FIGURE 6-12 From "The Genetics of a Bacterial Virus" by R. S. Edgar and R. H. Epstein. Copyright © 1965 by Scientific American, Inc. All rights reserved.

FIGURES 6-13, 6-14 From *Molecular Biology of Bacterial Viruses* by G. S. Stent. W. H. Freeman and Company. Copyright © 1963.

FIGURE 6-16 From A. Lwoff, *Bacteriological Reviews* 17(1953):269.

FIGURE 6-17 Electron micrograph by Edouard Kellenberger.

FIGURE 7-3 Photograph by Tom Kaufman.

FIGURE 7-11 From J. Lejeune, J. Lafourcade, H. Berger, and R. Turpin, *Comptes Rendus. L'Academie des Sciences, Paris* 258(1964).

FIGURE 7-32 From *Down's Anomaly* by L. S. Penrose and G. F. Smith. Little, Brown and Company, 1966.

FIGURES 8-2, 8-3, 8-4, 8-11, 10-4 From *Molecular Genetics* by G. S. Stent. W. H. Freeman and Company. Copyright © 1971.

FIGURE 10-8 From "Gene Structure and Protein Structure" by C. Yanofsky. Copyright © 1967 by Scientific American, Inc. All rights reserved.

FIGURE 10-26 From E. B. Gyurasitis and R. G. Wake, *Journal of Molecular Biology* 73(1973):55.

FIGURES 10-28, 10-29 From *DNA Synthesis* by A. Kornberg. W. H. Freeman and Company. Copyright © 1974.

FIGURE 10-30 From Bird and Caro, "Origin and Sequence of Chromosome Replication in *Escherichia coli,*" *Journal of Molecular Biology* 70(1972):557.

FIGURE 10-31 From *DNA Synthesis* by A. Kornberg. W. H. Freeman and Company. Copyright © 1974.

FIGURE 11-4 Electron micrograph by O. L. Miller, Jr., and Barbara A. Hamkalo, Biology Division, Oak Ridge National Laboratory.

460

Figure Credits

Index